Ecology

To G. E. Hutchinson, who gave us so much ecology

ECOLOGY

Stanley I. Dodson
Timothy F. H. Allen
Stephen R. Carpenter
Anthony R. Ives
Robert L. Jeanne
James F. Kitchell
Nancy E. Langston
Monica G. Turner

University of Wisconsin–Madison

New York   Oxford
OXFORD UNIVERSITY PRESS
1998

Oxford University Press

Oxford • New York
Athens • Auckland • Bangkok • Bogota • Bombay • Buenos Aires
Calcutta • Cape Town • Dar es Salaam • Delhi • Florence • Hong Kong
Istanbul • Karachi • Kuala Lumpur • Madras • Madrid • Melbourne
Mexico City • Nairobi • Paris • Singapore • Taipei • Tokyo • Toronto

and associated companies in
Berlin • Ibadan

Copyright © 1998 by Oxford University Press, Inc.

Published by Oxford University Press, Inc.,
198 Madison Avenue, New York, New York, 10016

Library of Congress Cataloging-in-Publication Data
Ecology / Stanley I. Dodson ... [et al.]
    p. cm.
  Includes bibliographical references and index.
  ISBN 0-19-512079-5 (cloth : alk. paper)
  1. Ecology. I. Dodson, Stanley I.
QH541.E31925 1998   97-39029
577–dc21.   CIP

9 8 7 6 5 4 3 2 1

Printed in the United States of America
on acid-free paper

CONTENTS

v

An ecologist
on the shores of the Wisconsin River at Merrimac.

PREFACE

This book focuses on the ideas and techniques characteristic of different approaches to the study of ecology. Ecology is made up of a number of distinct disciplines; each of our authors, an expert in a specific kind of ecology, speaks for the distinctness of that specialization. We discuss the relationship of people and nature, and six kinds of ecology: ecosystem, physiological, behavioral, population, and community. For each kind of ecology, we explore how it is unique, the appropriate theory and technology, the kinds of questions asked, and the successes and possibilities for the future. This discussion is set in a context of the human relationship to nature, with an emphasis on practical applications of ecology.

This text is unique in that it presents ecology as being composed of a number of different ways of knowing, seeing, or asking questions. We focus on the kinds of questions ecologists ask about their world, rather than on accumulated ecological knowledge.

This text is for students who have taken an introductory biology course and are ready for an overview of general ecology, a "user's guide" to ecology that will help readers understand the diversity of the field and navigate the sometimes tricky currents among the subdisciplines. Each chapter is a gateway, not an advanced course in itself.

Our goal is to communicate our excitement about ecology and to provide readers with a basis for continued learning in the area of ecology. The book can be easily read in one semester, allowing students to grasp the distinctions among the various kinds of ecology, and to use these distinctions, for example, to analyze scientific papers, to know what different kinds of ecology can do and what they are inappropriate for, and to express how different kinds of ecology can be used to answer ecological questions.

ACKNOWLEDGMENTS

Two people planted the seeds for this book: Ginny Dodson, who said it was enough for beginning ecology students to be able to distinguish among the different kinds of ecology, and Jill Persick, who kept reminding Dodson of the possibility of an ecology text that would make a difference for students. Thanks are also due to our teachers and colleagues who prepared us to write this text, and to the students who asked questions about ecology and who experienced the early stages of this approach to ecology. University of Wisconsin–Madison students attending the Fall, 1997, General Ecology course were especially helpful during the revision and polishing phases of this effort.

We thank those who read our proposal and early drafts: Tim Wootton, Charles Southwick, David Lodge, Susan Harrison, David Marsh, John Alcock, Bill Lewis, and especially Bob Paine. Many reviewers provided valuable suggestions, including: Joy Bergelson, Mary Bremigan, David Bunnell, Theo Colborn, Larry Crowder, Andy Dobson, Sarah Dodson, James Garvey, Frank Golley, John Havel, Martha Groom, Peter Kareiva, Mark Kershner, Stuart Ludsin, James Rice, Roy Stein, David Takacs, Gary Vinyard, Don Waller, David Wallin, John Wiens, Donald Worster, and Rusty Wright. Nadine Ives and Val Smith provided particularly careful readings of the entire book.

Kandis Elliot provided exceptional assistance in the production of the book. Responsible for the whole production, she created the book's design and artwork. She interfaced and conceptualized art with each of the authors, provided instant feedback during the review process, and, with James Jaeger, did the typesetting and copy editing. We are profoundly grateful to Kandis for her assistance!

We had outstanding support, encouragement, and assistance from the Oxford University Press staff, including Elyse Dubin, Manager of Editorial, Design, and

Production, who made sure our technical details meshed with press requirements, and especially Bob Rogers, Senior Editor, who was quick to see our vision and willing to accommodate our desire to have eight authors and to allow Kandis Elliot to handle the production.

Chapter One, Introduction, benefited from readings and conversations with Matt Brewer and Tom O'Keefe, Hugh Iltis, Bill Cronan, Sarah Dodson, Nancy Langston, and Tim Allen. Stanley Dodson is grateful to the Zoology Department of the UW–Madison for giving him the opportunity to teach general ecology.

Chapter Two, People and Nature, benefited immensely from careful readings by Donald Worster, David Takacs, and Dave Perry. The National Humanities Center provided a wonderful place to write this chapter, and a generous fellowship from the Center funded the work.

Chapter Three, Landscape Ecology, benefited from thoughtful reviews and comments provided by Jeffrey Cardille, Andy Dobson, Sarah E. Dodson, Stanley Dodson, Frank B. Golley, Nadine E. Ives, Peter Kareiva, Rebecca A. Reed, Val Smith, Roy Stein, Gary L. Vinyard, and David O. Wallin. Monica Turner's understanding of landscape ecology has been shaped by discussions with many people over the years, but she especially acknowledges the influence of her ongoing research collaboration with Robert H. Gardner, Robert V. O'Neill, and William H. Romme; and collaborative teaching of landscape ecology with Hazel R. Delcourt and David J. Mladenoff. Preparation of this chapter was supported in part by funding from the National Science Foundation.

Chapter Four, Ecosystem Ecology: Stephen Carpenter especially acknowledges R. A. Carpenter, S. G. Carpenter, J. N. Houser, M. L. Pace, C. Scheele, and the students of the UW–Madison Ecosystems Concepts course, Spring 1997.

Chapter Five, Physiological Ecology, benefited from reviews by Jim Rice and Theo Colborn. Background for this chapter derives from a variety of research projects sponsored over time by several agencies. Foremost among these are the National Science Foundation and the University of Wisconsin Sea Grant Institute. James Kitchell genuinely appreciates the chance to share these pages with a diverse and interesting group of ecologists. This chapter was prepared during a sabbatical leave that was primarily supported by the University of Wisconsin Graduate School. The chapter was written while the author was in residence and partially supported through the National Center for Ecological Analysis and Synthesis in Santa Barbara, California.

Chapter Six, Behavioral Ecology, benefited from reviews by Sarah Day, Karen London, and Louise Jeanne for reading earlier drafts. Darryl Gwynne, Jane Brockmann, and Paul Sherman provided especially useful comments. Lukas Keller, Darryl Gwynne and David Queller provided the slides used in Plates 6.1, 6.3, and Figure 6.12, respectively. Jane Brockmann and David Queller provided especially helpful discussions. Steve Krauth and the Insect Research Collection, Department of Entomology, UW–Madison, loaned specimens of the great golden digger wasp and the rhinoceros beetle for illustrations.

Chapter Seven, Population Ecology: Many people generously provided ideas and inspiration for this chapter—Nicola Anthony, Joy Bergelson, Johannes Foufopoulos, Peter Grant, Martha Groom, Kevin Gross, John Havel, John Losey, Stuart McNeill, Dennis Murray, Barry Noon, Anders Olsen, Kim Rauwald, Don Waller, Tim Wootton, and especially Nadine Ives and Jen Klug. For Figure 7.1, Elizabeth Breuhl provided data from her student project in Zoology/Botany 152, Introductory Biology. Joy Bergelson, Linda Graham, Lukas Keller, and Tim Wootton provided great photographs. Over the last several years, students of Zoology/Botany 260, Introductory Ecology, at the UW–Madison have taught Tony Ives how to teach, and he is very grateful for their diligence. Much of the author's perspective on ecology arose from research projects funded by the National Science Foundation and the U.S. Department of Agriculture. Finally, Mary Ann Fitzgerald provided patience.

Chapter Eight, Community Ecology: Tim Allen especially thanks Tom Brandner, Linda Puth, Tanya Havlicek, Jim Yount, Nikki Law, Chris Pires, and other members of the Tuesday noon seminar, affectionately known as the sandbox. For useful conversations, the author thanks Tom Hoekstra, Bruce Milne, Bob O'Neill, and Susan Will-Wolf. He especially acknowledges the support of Valerie Ahl and Josephine Allen; and heartfelt thanks of course to John Harper and Peter Greig-Smith, for superb training in population and community ecology 35 years ago.

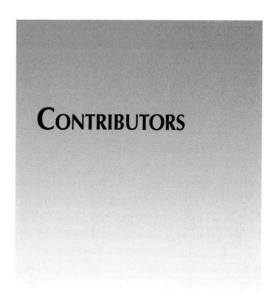

CONTRIBUTORS

STANLEY I. DODSON received his Ph.D. in Zoology from the University of Washington at Seattle, a student of W. T. Edmondson. He has taught courses in ecology and limnology at the Rocky Mountain Biological Laboratory and at the Flathead Lake Biological Station in Montana. He is currently Professor of Zoology at the University of Wisconsin–Madison, and formerly served as chair of the UW–Department of Zoology. He presently serves on the editorial board of *Hydrobiologia* and *Ciencia Ergo Sum*. With David G. Frey, he co-authored "Cladocera and Other Branchiopoda" for *Ecology and Classification of North American Freshwater Invertebrates* (Thorp & Covich, eds.). His research focuses on predator-prey interactions, chemical communication, and aquatic toxicology. He specializes in aquatic community ecology and the biology of aquatic invertebrates, especially zooplankton.

NANCY E. LANGSTON received her Ph.D. in Environmental Studies at the University of Washington, and was an Eddy Postdoctoral Fellow in Ornithology at the Burke Museum, University of Washington. She is Assistant Professor in Forest Ecology and Management and the Institute for Environmental Studies at the University of Wisconsin–Madison. Her book on the environmental history of western forests, *Forest Dreams, Forest Nightmares: The Paradox of Old Growth in the Inland West,* was the 1997 winner of the Forest History Society Book Award. Her research has encompassed the evolution of gender differences in African birds, life history strategies in albatrosses, and the environmental history of the forest health crisis. Her current research examines human effects on landscape change.

MONICA G. TURNER received her Ph.D. in Ecology in 1985 from The University of Georgia in Athens, and is Associate Professor of Terrestrial Ecology at the University of

Wisconsin–Madison. She serves as a member of the editorial boards of *Bioscience, Ecological Applications, Landscape Ecology, Conservation Ecology,* and *Climate Research* and co-edits the journal, *Ecosystems.* She has recently published *Landscape Heterogeneity and Disturbance* (Springer-Verlag). Her research focuses on landscape ecology, in particular the causes and consequences of broad-scale disturbances and land-use changes.

STEPHEN R. CARPENTER received his Ph.D. in Botany/Oceanography and Limnology from the University of Wisconsin–Madison. He is now the Halverson Professor of Limnology at the Center for Limnology, Department of Zoology, UW–Madison. He is a Pew Fellow in conservation and the environment, and received the Per Brinck award in limnology from Lund University, Lund, Sweden, and currently serves as co-editor for the journal, *Ecosystems.* Among many other publications, he edited the book. *Complex Interactions and Lake Communities.* His research interests include analysis of ecosystems, ecological modeling, ecological economics, and limnology, especially of plankton, fishes and macrophytes.

JAMES F. KITCHELL received his Ph.D. in Biology in 1970 from the University of Colorado at Boulder, and is now the Arthur D. Hasler Professor of Limnology at the UW–Madison. He presently serves as coordinator of The Living Resources Program of the University of Wisconsin's Sea Grant Program. He also serves on the advisory committee of the Fisheries Centre, University of British Columbia and is on the dean's advisory committee of the College of Fisheries and Marine Sciences, University of Alaska. With Stephen Carpenter, he co-edited *The Trophic Cascade in Lakes* (Cambridge University Press). His research and teaching interests focus on food web dynamics in an ecosystem context, and the bioenergetics of fishes.

ROBERT L. JEANNE received his Ph.D. in Biology in 1971 at Harvard University under the tutelage of Edward O. Wilson. A Guggenheim Fellow and a National Science Foundation predoctoral Fellow, he is currently Professor of Entomology and Zoology at the University of Wisconsin–Madison. He serves as associate editor for *Insectes Sociaux* and was the organizer and first president for the Western Hemisphere Section of the International Union for the Study of Social Insects. His publications include over one hundred articles and reviews. His research focuses on the behavioral ecology of social wasps, with special attention to communication, nesting behavior, and the organization of colony labor.

ANTHONY R. IVES received his Ph.D. in Biology in 1988 from Princeton University, and currently is Associate Professor in Zoology at the University of Wisconsin–Madison. He was a Life Science Research Foundation Postdoctoral Fellow sponsored by Glaxo, Inc. at the University of Washington, Seattle and a Lilly Teaching Fellow at the University of Wisconsin-Madison. He serves on the editorial board of *American Naturalist.* He is a theoretical ecologist whose many research investigations include a model of snowshoe hares, predators, and parasites; metapopulation dynamics and pest control in agricultural systems; and measuring aggregation of parasites at different host population levels.

TIMOTHY F. H. ALLEN received his Ph.D. in Botany in 1968 at the University College North Wales, and currently is Professor of Botany at the University of Wisconsin-Madison. His work on prairie data, forest simulation output and plankton communities gave him a synthetic view of the scale issue, and led to his first book with Thomas Starr, *Hierarchy: Perspectives for Ecological Complexity*. With the theory group at Oak Ridge National Laboratory he sharpened the theory of scaling, hierarchy theory, producing a second book with R. D. O'Neill, *A Hierarchical Concept of Ecosystems*. Continuing to broaden the scope of hierarchy, he published *Toward a Unified Ecology* with Thomas Hoekstra, and has continued to work on this subject. He has served on the National Science Foundation panel for Ecology and has organized many symposia for the Ecological Society of America. On the Madison campus he has received numerous distinguished teaching awards.

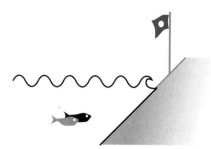

Chapter 1

WHAT IS ECOLOGY?

Looking at Nature from Different Perspectives

Stanley I. Dodson

- ■ **What Is Ecology?**
- ■ **Ecology as a Group of Subdisciplines**
- ■ **How This Book Is Organized**
- ■ **Themes Common to Ecological Subdisciplines**
- ■ **Summary**

…Like all real treasures of the mind, perception can be split into infinitely small fractions without losing its quality. The weeds in a city lot convey the same lesson as the redwoods; the farmer may see in his cow-pasture what may not be vouchsafed to the scientist adventuring in the South Seas. Perception, in short, cannot be purchased with either learned degrees or dollars; it grows at home as well as abroad…

—*Aldo Leopold*

WHAT IS ECOLOGY?

Ecology today is a complex field. The eight authors of this book each address the question "What is Ecology?" However, it is useful to start with a short definition that sketches the broad limits of the field: **Ecology** is the study of the relationships, distribution, and abundance of organisms, or groups of organisms, in an environment.

Although ecology is a major interest and concern of modern society, the science did not even exist two hundred years ago. The word ecology was originally coined in 1866 in *General Morphology*, a book on biology and philosophy by Professor Ernst Haeckel. Haeckel based his new word on the classical Greek words *oikos*, meaning "a household or homestead," and *logos*, meaning "study."

In his definition, Haeckel wanted to publicize the concept of the "economy of nature" discussed by Charles Darwin in his 1859 book *Origin of Species*. By "economy of nature," Darwin meant that all of nature appears to be an orderly, well-regulated system of interactions among plants and animals, and with their environment. Darwin argued that the appearance of organization was the result of a natural process of evolution based on a struggle for existence by each individual organism. Haeckel liked this idea, and used it in his definition of the science he called *oecologia*. Haeckel defined ecology as:

> *The economy of nature—the investigation of the total relations of the animal both to its inorganic and its organic environment; including, above all, its friendly and inimical relations with those animals and plants with which it comes directly or indirectly into contact—in a word, ecology is the study of all those complex interrelations referred to by Darwin as the conditions of the struggle for existence.*

ECOLOGY AS A GROUP OF SUBDISCIPLINES
Unified Ecology?

Haeckel's definition represented all of ecology in 1866. However, as other biologists applied Haeckel's basic concept to their own interests, the science of ecology quickly began to grow and diversify. The fact that different scientists have very different philosophies of life is reflected in their work. Some biologists prefer to ask mathematical questions, while others prefer an engineering or chemical perspective. Similarly, some people like to do their investigations in a library armchair; while others prefer to be outside walking through woods, looking at plants and animals.

Despite these differences, ecologists retain the dream that ecology is, or could be, a single unified field of study. The idea is that a unified field would contribute to our understanding, as if ecology were a single giant clock and we are learning how it works. One significant reason for striving toward a unified ecology is that all ecologists, regardless of their area of specialization, would know the basic principles that would allow them to answer any ecological question. Specialization would be merely an outgrowth of the unity of general ecology. For example, an individual would specialize on a part of the clock, but still understand the general principles of how the entire clock works.

A problem with this view of ecology as a single whole is, although ecologists agree there are basic ecological principles, they might not agree on what those principles are. Today, a unified ecology is more of a goal than a reality. However, this dream is only one way of looking at modern ecology. Another way of looking at ecology is to take the view that it is made up of *a number of different subdisciplines*. In this model, ecology is made of several distinct concept-based subdisciplines, each with its own set of concepts (or world view). These subdisciplines are different ways (or perspectives or views) of looking at the same thing, ecology. Because the perspectives are different, they produce different questions, require different techniques, and result in different conclusions about the relationships, distribution, and abundance of organisms, or groups of organisms, in an environment.

Viewing ecology from different perspectives is an excellent way to get a handle on the diversity and complexity of modern ecology. Modern ecology is characterized by having more facts than any one person can possibly know. How do we make sense of this mountain of information? The questions and concepts characteristic of different subdisciplines provide a structure for getting to know ecology—an approach similar

Ernst Haeckel

to the way modern ecologists actually work. These days, every ecologist is a specialist, with their favorite and specialized questions, data sets, and even special journals for publishing their research results.

Once a person has learned to see ecology from different perspectives, the next step is to remember to look at how the parts are related to make a whole. After several perspectives are mastered, it is then possible to begin putting different ideas from different perspectives together in new and imaginative ways. Throughout this book, we will stress the distinctiveness of different kinds of ecology, but also give examples of how those different kinds overlap and are linked together to ask interesting questions.

Before we can run with unified ecology, we need to walk through some of the different subdisciplines of ecology. What are these subdisciplines? No two ecologists would give exactly the same answer to this question. We will present subdisciplines that are generally recognized and that lead to understanding ecology.

Labels Ecologists Give Themselves

Ask any five ecologists what kind of ecologist each is, and you will probably get five different answers. The kinds of labels ecologists give themselves say a lot about the definition of ecology. Ecology can be sliced up into specializations in at least four ways (Box 1.1): by ways of looking at the world (perspective), by the kind of organisms looked at, by habitat, or by application.

These different labels can be used alone (as in "Terrestrial Ecologist") or in combination. I, for example, could call myself an "Experimental Zooplankton Community Ecologist." This means I prefer to use experiments to study the relationships among dif-

---

BOX 1.1. EXAMPLES OF DIFFERENT WAYS OF APPROACHING THE STUDY OF ECOLOGY

Kinds of ecology as defined by **concept** or **perspective**	Kinds of ecology as defined by **organism**
Landscape	Plant
Ecosystem	Animal
Physiological	Microbe
Population	Zooplankton
Behavioral	Human
Community	Deer
	Tree
Other kinds of ecology	Many other organisms

Kinds of ecology as defined by **habitat**	Kinds of Ecology as defined by **application**
Terrestrial	Theoretical
Lakes and streams (limnology)	Conservation
Marine (oceanography)	Agricultural
Arctic	Public Policy
Rain forest	Academic
Benthic thermal vents	Management
Urban	Restoration
Many other locations	Many other applications

ferent kinds of small animals that swim in the open water of lakes. The *concepts* are those of community ecology, the *location* is freshwater lakes, *organisms* are zooplankton, and the *applications* are academic.

This book is organized according to definitions of ecology based on concept or perspective (upper left column in Box 1.1). The following chapters are focused on concepts, that is, on how one thinks about ecology. The standard approach is to emphasize organism-, location- or application-based definitions of ecology. Why did we choose concepts for the organizational structure? Concepts are the best way to organize ecological knowledge. Organizing ecology by organism, location or application leads to complexity and interferes with clear communication. For example, just because two ecologists are studying the same kind of organism does not mean they are speaking the same language. Two zooplankton ecologists will have trouble discussing their work if one is a community ecologist and one is a physiological ecologist. On the other hand, ecologists with the same way of looking at the world are much less likely to be talking at cross purposes, because they will have common basic concepts, examples, techniques and questions, regardless of the organisms they study.

A second consequence of choosing organisms, locations or applications as a way of giving structure to ecology is that these categories each contain a nearly infinite number of parts. If ecology is organized around organisms, for example, then what limits the number of categories of ecology? A stupendous amount of knowledge has accumulated for any given group of organisms. This information is fascinating but overwhelming to everyone, from undergraduates to ecological experts. For decades, publication volume has doubled about every ten years, with new scientific journals appearing like mushrooms after a summer rain. It is no longer possible to make sense out of ecology by knowing all there is to know. Instead, the best approach is to know what are interesting questions and how to find the information we need to answer these questions.

Being clear about concepts is one way to make sure we are speaking the same language, and that we have a way to organize the mass of ecological data. For this reason, we are organizing our text using the concept-based kinds of ecology listed in the upper left-hand column of Box 1.1.

Ecological Concepts: The Kinds of Ecology Defined

Box 1.2 provides short definitions for the six ecological subdisciplines emphasized in this book. These are the basic perspectives that will be explored in the rest of the chapters. Although the definitions are different, they are all consistent with the general definition of ecology, which is the study of the relationships, distribution, and abundance of organisms, or groups of organisms, in an environment.

While there are certainly more than six ecological perspectives, the six chosen are practiced by large numbers of ecologists and contain enough of the basic concepts of ecology to be a gateway to understanding other areas of specialization.

BOX 1.2. SHORT DEFINITIONS OF ECOLOGY, PEOPLE AND NATURE, AND SIX ECOLOGICAL PERSPECTIVES

Ecology	The study of the relationships, distribution, and abundance of organisms, or groups of organisms, in an environment.
People and Nature	All environments change, and people have been important forces shaping these changes for only a tiny fraction of Earth's history. In the last ten thousand years, however, human effects on ecosystems have often overwhelmed the ability of these ecosystems to respond to change. Historical ecology examines these environmental changes by focusing on questions about how humans have affected the environment, how our cultural attitudes affect how we do ecology, and the history of how humans have attempted to manage the environment.
Landscape	The landscape can be thought of as being made up of different patches, characterized by different organisms and environments. Landscape ecology examines the interaction between this pattern of patches and ecological process—that is, the biological causes and consequences of a patchy environment.
Ecosystem	Ecosystem ecology is the study of the interactions of organisms with the transport and flow of energy and matter. Ecosystem size and shape depends on the specific questions being asked about energy flow or chemical cycling. The "system" part of an ecosystem is a description of how energy or matter moves among organisms and parts of the environment.
Physiological	Physiological ecology is the study of how individual organisms interact with their environment to carry out the biochemical processes and express the behavioral adaptations that accomplish homeostasis and survival. Homeostasis involves the maintenance of time, matter and energy budgets that allow for growth and reproduction by the individual.
Behavioral	The goal of behavioral ecology is to understand how a plant or animal's behavior is adapted to its environment. That is, behavior is understood to be the result of an evolutionary process.
Population	A population is a collection of individuals from the same species that occupy some defined area. Population ecology focuses on how and why populations change in size and location over time.
Community	Community ecologists examine the patterns and interactions seen in groups, or aggregations, of different species. The distributions of species are influenced both by biological interactions (such as predation and competition) and by environmental factors (such as temperature, water and nutrient availability).

Whatever perspective is used to view it, ecology is often assumed to be something that exists "out there." While it is true that things, such as birds and bees and trees and mountains do exist, ecology exists only in our language. Ecology is an interpretation of our perceptions of organisms and the environment. As with any interpretation, ecology depends completely on the history and culture of the people making the interpretation.

HOW THIS BOOK IS ORGANIZED AROUND SUBDISCIPLINES OF ECOLOGY

The chapters in this textbook are organized to emphasize similarities among the different subdisciplines. To the extent that the subdisciplines are distinct, each chapter can stand on its own. However, while ecology is made up of a series of distinct concept-based perspectives, these subdisciplines bear striking similarities.

This introduction is designed to present the basic philosophy of our approach to ecology, by defining and illustrating the subdisciplines, giving examples of the kinds of questions ecologists ask, and by presenting themes common to the subdisciplines: scale, evolution, statistics, modeling, and applications. Chapter Two explores connections among ecology, human culture and history. Chapter Three focuses on landscape ecology, which analyzes maps or other two- or three-dimensional representations of nature to understand the interaction between organisms and their environment. This is a good place to start, because landscapes are familiar to us all. Ecosystem ecology, Chapter Four, is next because patterns of energy flow and chemical cycles are often related to the environmental environmental patterns of landscape ecology. Physiological ecology, Chapter Five, discussed after ecosystems, resembles ecosystem ecology, but the emphasis is on individuals rather than systems of organisms and the environment. Chapters Six and Seven, on behavioral and population ecology respectively, are different ways of looking at the ecology of groups of organisms. Finally, community ecology, Chapter Eight, is about the structure and function of multi-species systems.

Different Kinds of Ecology Are Different Ways of Looking at Organisms in an Environment

Plate 1.1 pictures a rural Wisconsin landscape. We see a stream, fields, some forest on hills, and a farm. Different kinds of ecologists ask different questions about this scene. The following are examples of typical questions from different subdisciplines.

- *Landscape ecologist:* How does the two-dimensional pattern of forest, field, and farm buildings affect the ability of deer to move from one forest patch to another?
- *Ecosystem ecologist:* In this watershed, how much phosphorus is stored in the soil of the forest and fields, how much is applied to the fields each year, and how much moves annually into the stream?
- *Physiological ecologist:* Is the local climate optimal for the genetic strain of the corn growing in the fields?
- *Behavioral ecologist:* How does the size, condition and age of male redwing blackbirds affect their ability to defend breeding territories along the stream bank, and how in turn does this impact their breeding success?
- *Population ecologist:* What factors control the size of the trout population in the stream?
- *Community ecologist:* How many species of native plants and insects live in the woodlot, and are there enough pollinators to maintain the plant diversity?

The above series of questions are asked at fairly broad size scale. This is natural, given the size of the scene pictured, but questions could also be asked at quite different scales. For example, a community ecologist might ask questions about species diversity over a wide range of scales, from species diversity of the community of soil bacteria (centimeter scale) to the species diversity of forest trees (tens of meters scale).

Graphical Representations of Different Perspectives of Ecology

Different subdisciplines of ecology also have characteristic and distinct graphical representations. Up to this point, our explanation of what we mean by different kinds of ecology has been verbal. Let's look at some typical graphical representations (Figure 1.1 A–L) associated with the different kinds of ecology defined in Box 1.2.

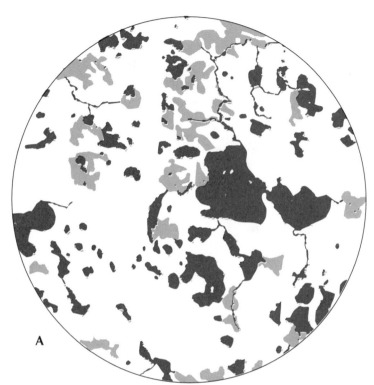

A

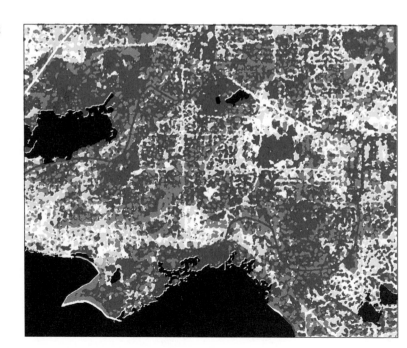

B

Figure 1.1, A & B. Landscape Ecology uses maps and pictures of surfaces more than any other branch of ecology. Landscape images typically include several different subdivisions of the environment. **A.** A map of a 20-km diameter circle in northern Wisconsin (Trout Lake area), showing the distribution and connectedness of lakes and streams (color) and wetlands (gray) within a terrestrial environment (white). The map is recreated from a digitized USGS topological map. Thanks to Sarah Gergel. **B.** Recreated from a satellite image (taken from a Landsat Thematic Mapper satellite) of south-central Wisconsin (Madison area) taken in June, 1992. Black areas represent lakes and wetlands, with Lake Mendota at lower left. Vegetation is represented by color. Patterns of light gray and white delicately show the city radiating from the hub of the State Capitol at upper left-center. Thanks to David Bolgrien.

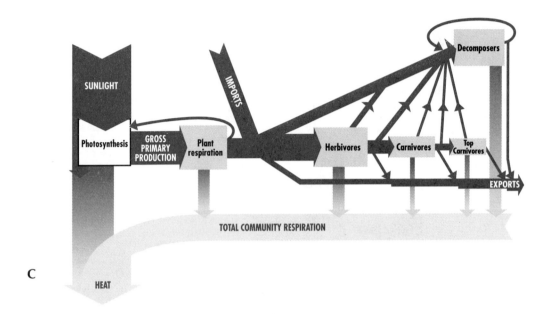

C

Figure 1.1. **C & D. Ecosystem Ecology** images often look like wiring or plumbing diagrams. The graphical model made up of boxes and arrows, now widely used in business plans, programming flow sheets, and many kinds of science, was first used by Charles Elton (1927), an early animal ecologist. **C** is a classic model, developed by H. T. Odum in 1957 to represent the flow of energy from the sun to photosynthetic organisms, to herbivores and predators. This model was intended to be applicable to streams, lakes, and coral reefs. It was developed in a paper that focused on the energetics of a stream flowing out of Silver Springs, Florida. In this figure boxes indicate "trophic levels" (see Chapter 4 for further explanation). **D,** a model of the "phosphorus cycle," is made up of boxes representing phosphorus reservoirs, with arrows indicating flow of phosphorus between reservoirs. The model is based on observations taken at the University of Notre Dame Environmental Research Center in the Upper Peninsula of Michigan. The numbers in the boxes represent the amount (kg) of phosphorus in each reservoir for the entire lake. Thanks to Steve Carpenter.

D

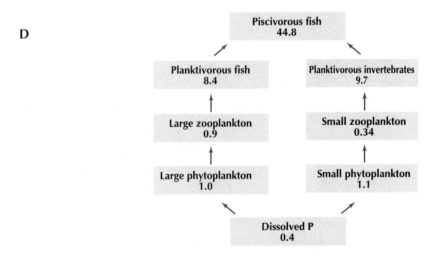

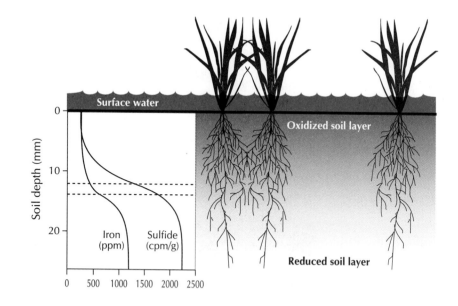

E

Figure 1.1. **E & F. Physiological Ecology** images typically show an individual organism as part of a larger physical or chemical system, or how an organism responds physiologically to an environmental variables. **E** shows a pictorial model illustrating how an organism, in this case a marine grass *(Spartina),* faces environmental constraints such as nutrient limitation and lack of oxygen in the mud in which it grows (redrawn from Mitsch and Gosselink 1986). **F**, a diagram based on a largemouth bass, is a representation of the "bio-energetic" model developed by James Kitchell. The model uses metabolic processes (inputs and outputs of energy) to predict survival, production, and distribution of organisms.

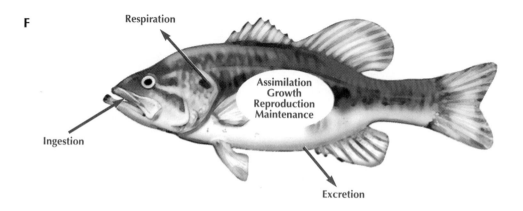

F

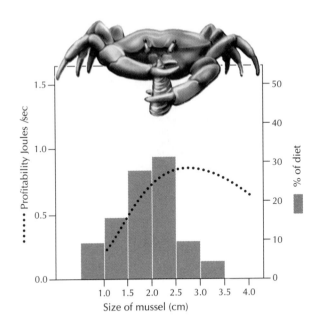

G

Figure 1.1. G & H. Behavioral Ecology images often illustrate a specific behavior, or summarize how changes in behavior affect ecological variables such as feeding rate or reproductive success. **G** graphs the optimization of prey capture behavior in a crab. Redrawn from Krebs and Davies 1993. The assumption is that the behavior is optimized as a result of a long process of natural selection. **H** graphs the effect of helping on breeding success in black-backed jackals. Redrawn from Moehlman 1979.

H

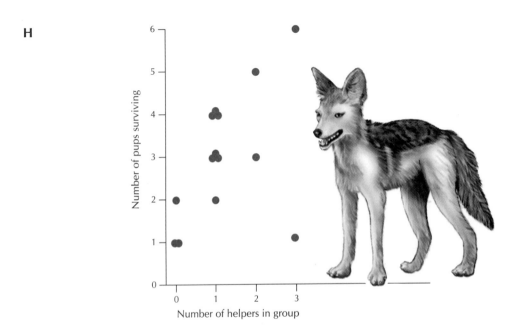

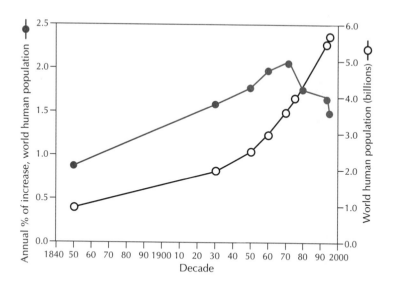

I

Figure 1.1. I & J. Population Ecology often uses graphs that show how a population changed in size over a period of time. **I** is a graph of the global human population between about 1850 and 1995. The data for this graph are given in Erlich et al. 1973, and *World Population Data Sheets* from the Population Reference Bureau. **J** shows the size of a population of Canadian lynx and snowshoe hares for the period 1845–1935. Redrawn from MacLulich 1937.

J

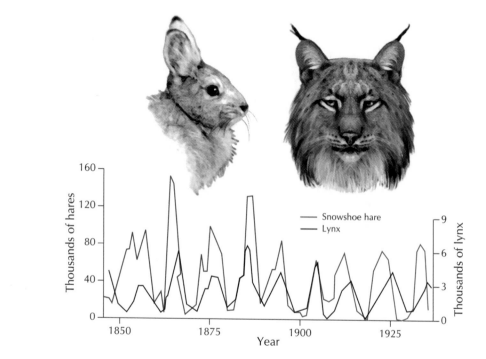

K

Figure 1.1. K & L. Community Ecology images usually express either the complexity of a group of species or relationships among them. **K** shows community ecologist Timothy Allen in 1980 admiring a group of species in the intertidal zone of the coast of northern Wales. **L** is a pictorial model indicating (upper graph) the history of growth of a beech forest starting with bare soil, and (lower graph) the distribution of trees and understory herbs after the mature forest is left to itself. Redrawn from Watt 1947.

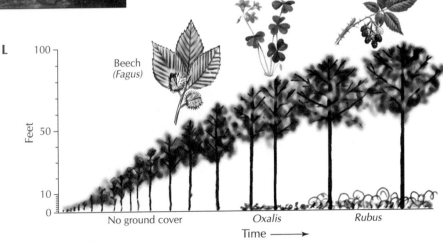

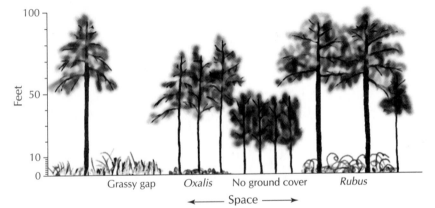

BOX 1.3. EXAMPLES OF TOOLS OR TECHNIQUES CHARACTERISTIC OF DIFFERENT KINDS OF ECOLOGY

Landscape	Satellites, photos, maps, and computers are essential, especially for geographic information systems (GIS).
Ecosystem	Calorimeter pressure bomb, quantitative chemical analysis.
Physiological	Respirometer, treadmill, infrared gas analyzer (IRGA), stable isotope chemistry, light sensors, thermocouples.
Population	Sampling traps, computer, greenhouse.
Behavioral	Video equipment, event recorder, binoculars, radiotags, geographic position satellites, computer, DNA fingerprinting.
Community	Quadrat sampling, species identification book, enclosures.

Techniques Specific to Different Kinds of Ecology

Different kinds of ecology require expertise in using and understanding different technologies. These different technologies require different tools (Box 1.3).

It is possible to do excellent ecology using very simple tools. However, development of new materials and technologies affects all branches of ecology. It is difficult to overemphasize the breakthroughs that have occurred in ecology with the advent of micro electronics, computers, video technology, and new materials. In his retirement speech at an annual meeting of the American Society of Limnology and Oceanography, the renowned American limnologist Bob Pennak said that the greatest thing to happen to **limnology** (the study of lakes) during his career was the invention of plastics, which permitted the development of a wide range of new sampling gear.

The computer revolution made landscape ecology possible and greatly assisted all other aspects of ecology. Computers have made possible the acquisition and analysis of huge amounts of data. Statistical analyses are now relatively quick and simple instead of an onerous chore. Complex computer models, once impossible because of computational requirements, now routinely simulate ecological systems, testing hypotheses in virtual reality.

Experiments, models, and observations are all used to gain ecological knowledge. However, the mix of these techniques differs greatly among the subdisciplines. All three can be applied to any kind of ecology, but typically there are subdiscipline-based biases. Observation is used widely in all subdisciplines. Community ecologists tend to do manipulative experiments at broad scales, population ecologists favor fine-scale experiments, and ecosystem and landscape ecologists tend to test hypotheses using simulation models rather than experimental manipulations of nature.

Linkages Among Different Kinds of Ecology

Although the previous discussion has focused on the distinctions among different kinds of ecology, it is important to also understand the connections. The different kinds of ecology are related like a clump of flowers (Figure 1.2). Ecology can be divided into several subdisciplines, as a plant might be divided into several stems, and like the petals of different flowers, the parts continue to grow and overlap.

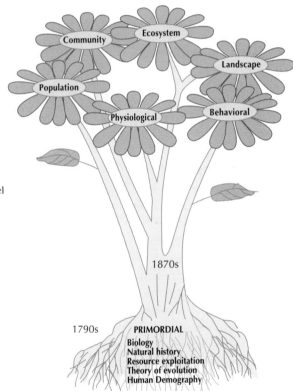

Figure 1.2. A graphical model of ecology, showing that the different subdisciplines have developed over the last few centuries, and are to some degree overlapping.

Like overlapping daisy petals, ecology subdisciplines are not totally isolated. It is common for an ecological study to include more than one kind of ecology as different parts of the study, or for complex questions to be asked about areas of intersection of one or more kinds of ecology. Different ecological "schools" have twisted together different stems. For example, community ecologist Robert MacArthur and his students melded population, community and behavioral ecologies. Similarly, Raymond Lindeman of the University of Minnesota combined aspects of ecosystem and community ecologies (you can see a description of his work in the paper by Cook, 1977). Box 1.4 gives some examples of overlapping ecological subdisciplines.

BOX 1.4. EXAMPLES OF QUESTIONS THAT USE CONCEPTS FROM DIFFERENT KINDS OF ECOLOGY

Do annual fluctuations in population size of plants and animals living together affect the species composition of the community over time?	Population and Community
Does the number of plant species in a meadow affect the rate of annual production (because of total photosynthesis) of the meadow?	Ecosystem and Community
Do social insect species (such as ants, bees and wasps) have a competitive edge over closely related solitary insect species (such as solitary bees)?	Behavioral and Community
Are small patches of suitable habitat associated with large fluctuations in population size?	Landscape and Population
Is there a global latitudinal gradient in species diversity?	Physiological and Community

THEMES COMMON TO ECOLOGICAL SUBDISCIPLINES
Scale and the Kinds of Ecology

Scale is the unit on your measuring stick or clock. When ecologists study small organisms or systems, they use a small (fine) scale of measurement, and when studying large organisms or large areas of habitat, they use a large (course) scale. **Resolution** is another way of expressing scale. Resolution is the smallest meaningful measurement that can be made of space or time. For example, if in a digitized satellite image a single pixel represents a square 30 m on a side (as in Figure 1.1B), then the smallest possible scale or resolution is 30 m. **Size** is the number of units measured. Size or extent refers to an entire length, area or time period. For example, at the cm scale, resolution is 1 cm and the size of a human is about 160 cm.

This book takes the position that the different kinds of ecology are not defined by scale. In other words, the definitions of the kinds of ecology are scale-independent, and each kind of ecology is currently studied at a wide range of scales. For example, ecosystem ecology is studied from millimeters to global scales (Box 1.5).

The Role of Evolution in Ecology

Nothing in biology makes sense except in light of evolution.
—*Theodosius Dobzhansky*

Evolution is change over time. In biology, we are usually interested in changes in characteristics such as form, physiology, or behavior that can be passed from one generation

BOX 1.5. AN EXAMPLE OF SCALE INDEPENDENCE OF ECOLOGICAL CONCEPTS, IN THIS CASE FOR ECOSYSTEM ECOLOGY

Example of an Ecosystem Studies at Different Scales	Scale
Storage and movement of the plant nutrient phosphate at the mud-water interface of lakes (see Figure 5.3)	Millimeters
Carbon storage and movement in the stomachs of a cow, or the nitrogen cycle in the insect-eating pitcher plant	Centimeters
Phosphorus cycle between soil and plants (such as trees and shrubs) in a forest	Meters
Nitrogen cycle among lakes, streams, soil, and organisms in a watershed	Kilometers
Carbon cycle for the Earth	Biosphere or Global

to another. Evolutionary change is the natural result of 1) characteristics being inherited, 2) variation in characteristics among offspring, and 3) production of too many offspring, so that some live to reproduce and some do not (this non-random process of mortality is also called "**natural selection**"). Organisms that survive to reproduce are said to have **adaptive** characteristics. An adapted organism fits into its environment, in the sense of being able to survive and produce offspring. **Fitness**, a concept related to adaptation, is defined as the number of viable and fertile offspring produced by a specific genotype: more fit organisms have more viable and fertile offspring than less fit organisms of the same species.

Evolution is an underlying concept found in all branches of ecology; a concept that is one of the common principles or themes that tends to unite the different kinds of ecology. However, while the concept of evolution is strongly integrated into population, behavioral, and community ecology, other kinds of ecology put less stress on evolution (Box 1.6). Organismal ecology makes a moderate use of evolutionary theory, in that many physiological traits are discussed as being "adaptations" (inherited traits associated with survivorship or fecundity in a specific environment). In other areas of ecology, such as ecosystem and landscape, the use of the concept of evolution is peripheral to other major themes.

The Role of Models in Ecology

Modeling is a common approach used in all kinds of ecology. An ecological **model** is a simplified representation of a system whose purpose is to understand the relationships, distribution, or abundance of organisms, or groups of organisms, or their interactions with the physical or chemical environment. The model is typically a system that includes objects and the relationships or interactions among those objects. Models come in many forms, including physical structures, pictures, graphs, analogies, narratives, equations, and computer programs.

> **BOX 1.6. REPRESENTATIVE QUESTIONS USING THE CONCEPT OF EVOLUTION IN DIFFERENT KINDS OF ECOLOGY**
>
> | | |
> |---|---|
> | Ecosystem | Are rates of energy transfer from plants to animals the result of natural selection? |
> | Population | Is the coevolution of prey and predator species likely to produce population dynamics with large boom-and-bust cycles? |
> | Physiological | How is the organism adapted to live in a specific environment with a limited range of temperature, water, and food? |
> | Behavioral | Are behaviors inherited and how do specific behaviors influence individual survival and reproduction? |
> | Community | Is the huge number of plant and animal species found in tropical rain forests due to low extinction rates, high speciation rates, or some other factor? |
> | Landscape | How does the size and spatial distribution of islands affect the evolution of bird species? |

Some models are simply used to summarize what we know about causes and effects, and to identify gaps in our knowledge. Other models are used to predict future relationships, distributions, and/or abundances, given the current state of our knowledge.

One common model, first used by Elton (1927) is composed of boxes and arrows (for example, see Figures 1.1C, 1.1K, and 1.3). The boxes contain things, such as number of individuals, a list of species, or amount of a chemical or energy. The arrows indicate either 1) how the members of one box influence those of another box (for example, by consuming other species, or by giving birth), or 2) the flow of energy or chemicals from one box to another.

Models vary from being realistic to abstract. Realistic models tend to be detail-specific and therefore complex, with many parts and interactions. These models simulate the real world, and are often called simulation models. To make predictions, a simulation model starts with a specific box-and-arrow model (or its mathematical equivalent, a group of equations) and then asks what will be the content of the boxes in the next time interval, given the contents of the boxes and the interactions. Simulation models are typically so complex they necessitate the use of a computer to make predictions. An example of a complex, detail-specific model is the computer simulation model created by a group of population ecologists at the University of Wisconsin (Luecke et al. 1992). The goal of this model (Figure 1.3) is to predict the amount (grams per liter) of the algae-eating *Daphnia* (a small crustacean also called the water flea) in the open water of lakes. The model predicts the daily amount of *Daphnia* based on how fast the animals grow depending on food supply, and how fast they are eaten by predators. This is an ecologically important model, because *Daphnia* is an important food item for fish that often has a large effect on the amount and kinds of algae living in a lake.

Abstract models are based on simplified analytical (algebraic) equations that are highly abstract. These equations attempt to capture the essence of complex interactions in a few variables, which summarize rather than simulate the real world. Before computers were widely available, ecologists used analytic models because of the relative ease of computation compared to simulation models. Abstract models are still favored by some population, community, and ecosystem ecologists. An example of a highly abstract model is the exponential population growth rate model, in which the average percent increase in a given year can be used to predict population growth in future years (see Figure 1.1I).

The Role of Statistics in Ecology

Statistical tests are an important tool used by most ecologists. In the following chapters, each assertion of relationship or mechanism is very likely based on statistics or the results of statistical tests. Because statistical tests are so widely used, it is valuable to say a few words about statistics here.

The word *statistics* is commonly used in two different ways. Descriptive statistics are used to convey information, data or facts; inferential statistics are used to test hypotheses.

Descriptive statistics

Ecological data values are always variable. No matter how careful you are in making measurements, there will be differences among measurements made on replicate samples. Replicate samples are chosen to be separate things but otherwise as similar as possible. It is not always obvious how to make replicate samples (Hulbert 1984). Some of the variability among replicate samples is due to actual differences among individual things or organisms and some is due to measurement error. For example, if you weigh 20 male 2-year-old bluegill sunfish, all from the same lake, they will not all be the same weight.

Variable data are often summarized or represented by a measure of central tendency (average or median) and a measure of variability (variance or range):

- *Data:* can be qualitative (as in black or white), ranks (as from 1 to 10), or quantitative (as in measurements or counts). Descriptive statistics are appropriate for quantitative data.
- *Average:* add up all the values and divide by the number of values. The average of 2, 3, 4, 6 and 10 is 5.
- *Median:* find the middle value. The median of the five numbers given above is 4.
- *Variance:* find the difference between each value and the average. Square this difference, add up the squares, and divide by one less than the number of values. For the string of numbers given above, the variance is 10.
- *Range:* the difference between the smallest and largest values. The range of the example is 8.

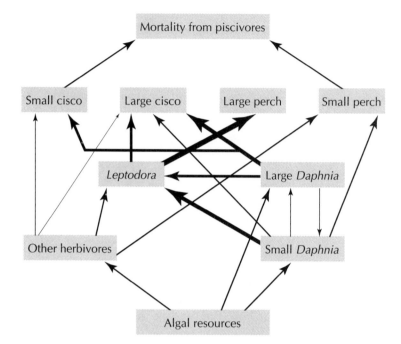

Figure 1.3. A box-and-arrow simulation model of inter-actions among algae, inverte-brates and fish living in Lake Mendota, Wisconsin. The arrows indicate the amount of food going from one box to another, and the thickness of the arrows indicates the amount of food. Redrawn from Luecke et al. 1992.

Information can often be interpreted in a number of ways, depending on who is using the information. Human ingenuity, not the information itself, can lead to differ-ent interpretations, resulting in the often expressed opinion that "statistics can be twisted to prove any point." Or, as Darwin pointed out, data themselves have no meaning except when they are used to argue for or against some idea.

Inferential statistics

The second usage of "statistics" refers to formal statistical tests, which have been designed by mathematicians to be as independent of personal preferences as possible. A statisti-cal test tests the significance of observed relationships of measurements or association. Examples of questions that can be tested are:

- Are the fish in this lake longer than the fish in that lake?
- Do chickadee eggs in this forest weigh more than in that forest?
- In the winter, do chickadees tend to occur in the same tree with crows?
- Is there a relationship between sugar content of fruit and growth rate of fruit-eating birds?
- Do small nature preserves have fewer species than large preserves?

Each of these questions is another way of stating a **hypothesis**, which is an "educated guess" about a relationship. For example:

The fish in this lake are longer than the fish in that lake.

The system of statistics that most ecologists subscribe to is based on the **null hypothesis**. Although this is not the only approach (for more information, see the books by Pickett and by Hilborn and Mangel listed in the "Recommended Reading" section at the end of this chapter). The null hypotheses states that any apparent relationship is due only to chance resulting from the sampling technique. An example of a null hypothesis is:

> The fish in this lake are on average just as long as the fish in that lake,
> and any apparent difference is due only to the variation of fish length
> and the number of fish collected.

The statistical test always asks, "Given the null hypothesis, knowledge of the variability of the individual measurements, and the sample size, what is the probability that we could have observed by chance the relationship we did observe? To continue the fish example, the test would ask:

> Given our null hypothesis of no real difference in fish lengths between
> the two lakes, the amount of variation in fish size, and the number of fish
> we measured, what is the probability that we could have collected data
> showing that the fish are on average longer in this lake than in that lake?

If the observed relationship has a low probability, then it is acceptable to reject the null hypothesis (that is, decide it is not true). Ecologists typically take a probability (called a *p-value*) of 0.05 (1 in 20) as the boundary between probable and improbable. Rejecting a null hypothesis can be taken as support for the hypothesis about the relationship. However, statistical tests only allow the scientist to reject the null hypothesis or fail to reject the null hypothesis, not to directly test any other hypothesis about a relationship.

Neither a course in statistics nor experimental design are necessary for an introduction to ecology. However, anyone who plans to learn more about ecology will quickly find it desirable to become familiar with these important tools. Hypotheses and statistics are discussed in many different contexts in the following chapters.

Ecological Application

Applications of different kinds of ecology are as varied as the subdisciplines. Each has its own characteristic questions to ask of nature, and messages for human society. General ecological questions include "How does this work?" and "How can this be managed?" For example, while conservationists often look to ecology for direction, the recommendation returned will vary depending on what kind of ecologist is asked. A population ecologist might stress the importance of genetic diversity, while community ecologists stress biological diversity and the importance of saving rare species. Landscape ecologists are concerned with the design of habitats to preserve species or groups of species. Ecosystem ecologists ask how much of the global energy and nutrients humans co-opt, and study the effect of increasing fertilization (**eutrophication**) of streams and lakes. Organismal ecologists are concerned about global warming and increasing ozone as limits of tolerance are approached or exceeded. Behavioral ecologists find that animal behavior depends on the environment, and lament that the appropriate habitat is in many cases disappearing. Animal behavior in human-impacted habitat can be inappropriate and potentially lethal, as when birds or turtles become entangled in or eat plastic bags.

The following chapters include discussions of ecological applications. Recognizing the distinctions among different kinds of ecology contributes to clearer communication between ecologists and the conservation advocates and policy makers who look to ecologists for advice.

SUMMARY

An ecologist who understands the distinctions among the different kinds of ecology is a person empowered to do ecology anywhere, in any location or context. There is urban ecology as well as ecology in wilderness areas; study of ecology is possible both at the university and in the back yard or woodlot.

The purpose of this book is to focus on "Ways of Looking at the World." The relationship of humans to the rest of nature is investigated in Chapter Two. Chapters Three through Eight deal with six different kinds of ecology. The kinds of ecology discussed in this text are only six of the many possible ways in which people look at the world to ask questions about the distribution and abundance of organisms, but they represent dominant themes in the practice of ecology.

Our strategy is to describe the different ways of looking at the world (giving definitions, techniques, questions, and applications), stressing the different concepts and using concept-specific examples or case histories. The student is encouraged to become familiar with the six kinds of ecology, use the local library or electronic database to explore the fascinating plethora of ecological knowledge, and use local surroundings to apply these ways of looking at the world, especially the ones that are most appealing to you.

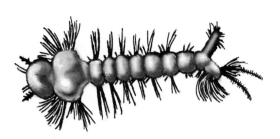

RECOMMENDED READING

TIMOTHY F. H. ALLEN AND THOMAS W. HOEKSTRA. *Toward a Unified Ecology*. Columbia University Press, New York. 1992.
> *The organization and content of Allan and Hoekstra's text inspired the organization of the text you are now reading.*

CHARLES ELTON. *Animal Ecology*. Sidgwick & Jackson, London. 1927.
> *This excellent classical text reads quickly and contains many modern ecological concepts in seed form.*

NORMAN HAIRSTON. *Experiments in Ecology*. Cambridge University Press. 1997.
> *An important book that complements many of the things addressed in this chapter.*

RAY HILBORN AND MARC MANGEL. *The Ecological Detective: Confronting Models with Data*. Princeton, New Jersey. 1997.
> *An excellent introduction to quantitative techniques in ecological modeling.*

STEWARD T. PICKETT, JUREK KOLASA AND CLIVE G. JONES. *Ecological Understanding*. Academic Press, San Diego, California. 1994.
> *This book provides further background on the current state of ecology.*

ROBERT C. STAUFFER. "Haeckel, Darwin, and Ecology." *Quarterly Review of Biology* 32:138-144. 1957.
> *This short article presents the early history of ecology.*

EUGEN WARMING. *Oecology of Plants*. Oxford University Press, London. 1927.
> *This book gives a good idea of the state of plant ecology in the early part of this century.*

Chapter 2

PEOPLE AND NATURE

Understanding the Changing Interactions Between People and Ecological Systems

Nancy E. Langston

A new day has begun on the crane marsh. A sense of time lies thick and heavy on such a place. Yearly since the ice age it has awakened each spring to the clangor of cranes. The peat layers that comprise the bog are laid down in the basin of an ancient lake. The cranes stand, as it were, upon the sodden pages of their own history.
—Aldo Leopold

DEFINITIONS

Environmental history examines the history of human interactions with the non-human world. All environments change, and people have been important forces shaping these changes for only a tiny fraction of Earth's history. In the last ten thousand years, however, human effects on ecological systems have often over-whelmed the ability of these systems to respond to change. This chapter will explore the study of environmental change, asking what the interactions between people and nature mean for the study of ecology.

QUESTIONS ASKED BY ECOLOGICAL HISTORIANS

Why consider humans in an ecology textbook? Isn't ecology supposed to be about natural systems, not about people? The first answer is simple: the world is dramatically affected by humans. Understanding how ecological systems function requires understanding the history of those systems. Changes caused by humans have been—and will continue to be—an important part of that history.

The second major reason to pay attention to humans is that the relationship between science and culture is more complex than most people imagine, and understanding the links between the two helps bring into focus the controversies in modern ecology. Ecology is not only a set of facts, but also a set of questions that people ask about the world. Ecologists have particular worldviews that lead them to ask particular kinds of questions, and therefore know the world in certain ways. Understanding these questions, and the different ways people have answered them, requires knowledge of the relationships between cultural factors and scientific hypotheses. Although good scientists test hypotheses without thought of political gain, the ways scientists formulate their hypotheses reflect their own cultural perspectives, as well as the natural world the scientists study.

Finally, scientists need ecological historians to help answer a question which is not purely scientific, but nevertheless motivates many ecologists in their work: why is the Earth in a state of global environmental crisis, and what should be done about it? For example, ecologists can trace numerous ways that extensive deforestation affects forests, but ecological theory alone cannot explain why people cut the trees (Plate 2.1). The most important causes of ecological degradation lie in human culture, and so understanding the links between culture, economics, and ecology can help devise better strategies for conserving and restoring nature.

Environments Shaped by People

How does a forest, a grassland, a wetland, or an estuary come to work the way it does? Exploring these questions means paying attention not just to current interactions between individuals, species, or energy flows, but also to the history of that system. Any ecological system, no matter how you define its boundaries, is a product of all the events, processes, and **disturbances** that led to its current state.

Ecologists long assumed that they could study the laws of nature apart from human history. Human-influenced systems seemed to be aberrations, because human history seemed too brief in evolutionary time to worry about. Most ecologists believed that over time, natural processes would eventually erase the effects of different initial stages in the life of an ecosystem (Christensen 1989). But recent research shows that past environmental conditions play a continuing role in most ecological systems: you cannot erase or ignore history, and people are one of many sources of historical disturbances that shape environments.

The modern world—no matter how much like a wilderness it may appear—has been measurably altered by humans. Human activities affect the entire globe from the depths

Figure 2.1. The ecology of Australia's bush has been shaped both by ecological processes and by thousands of years of management and intensive cultural interaction by aboriginal peoples.

of the ocean to the highest levels of the atmosphere. Many places on Earth that we view as pristine—the Wyoming wilderness, the Australian bush, the Brazilian rain forest—are actually the results of long interactions between people and place.

The Amazon rain forest, a place North Americans often envision as an Edenic paradise threatened by slash-and-burn peasant farmers, has actually had an extensive history of management by people who used slash-and-burn farming to shape those landscapes (Simmons 1996; Hecht and Cockburn 1990). Australia's bush (Figure 2.1), seen by European conquerors as a desolate wilderness, also had a complex history of aboriginal management and intensive cultural interaction (Goudie 1994; Simmons 1996). The great treeless moors that cover much of upland Scotland and England (Plate 2.2) are not purely natural, but were shaped by the burning activities of prehistoric hunter-gatherers (Goudie 1994; Simmons 1996). As the environmental historian Donald Worster eloquently argued,

> *Scientists must acknowledge, as many have begun to do, that the nature they describe in their textbooks often seems unreal and contrived to the historian. Typically, it lacks any connection to human history and all its contingencies, accidents, cycles, ideas and social forces. Too often science seems oblivious to the fact that human beings have been interacting with nature over a very long period of time, at least over two million years—some would say four million years—and that what we mean by nature is, to some extent, a product of history (Worster 1996).*

Although people have influenced the entire globe, no ecosystem is entirely an artifact of humans. For all the efforts that people have made to understand, manage, and ultimately control nature, a world of ecological processes and complex interrelationships flourishes outside of our control. To say that places have been "managed" by peo-

Figure 2.2. The forces driving environmental change include natural processes, but cultural processes (such as ethics, religion, and science), and political and economic processes (including political, technological, and market change) also shape environmental change. Each of these sets of processes in turn affects the other sets, so understanding the links between them is also critical.

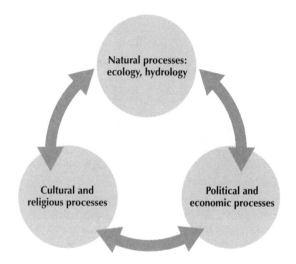

ple means that humans have been only one among many influences on them. Human management, even in the most intensely cultivated tree farm or garden, never completely replaces ecological interactions and constraints.

Environmental change comes about not just because people cut down trees, plow prairies, or burn fossil fuels, but because they do these things in a world where nature, culture, science, and markets tangle in complex ways. The reasons for environmental change fall into three interwoven categories (Figure 2.2):

- *Cultural:* How did cultural ideals affect the ways different groups of people changed the land? What kinds of visions of the relationship between humans and nature did people bring to the land? Whose vision of the land determined how the land was shaped? In particular, what scientific visions of the forest shaped people's work? In which political and cultural contexts did these scientific theories develop?
- *Political:* Over the course of several centuries, many ecological systems in America and across the globe have been transformed into collections of resources exported out of the region to feed the demands of distant markets. Economics, industrialization, the development of global markets have all had profound effects on the world's ecosystems, and the people who inhabit those ecosystems.
- *Ecological:* What were the biological and physical factors that shaped the landscape? Plant communities, animals, disturbance processes such as fires, floods, insect epidemics, soil processes, nutrient cycles, erosion, and the movement of water are major players in ecological history.

Example 1: Nature and culture on Easter Island

Although many people like to think that all pre-industrial peoples lived in harmony with nature, they were perfectly capable of transforming their environments—often to the point where those environments could not continue to support their cultures. Even in some of the places most remote from industrial civilizations, humans have had profound effects on the places they inhabited. The rise and fall of civilization on Easter Island illustrates some of the ways pre-industrial peoples could alter their environment, and the ways these environmental changes could in turn affect culture (Bush 1997; Ponting 1992).

More than 3,500 km from Chile and 2,200 km from the nearest inhabited land, Easter Island is one of the most isolated islands on earth (Figure 2.3). About 400 AD, Polynesians in enormous dugout canoes found their way to the island, a forested landscape with palm trees, no mammals, and abundant birds. To provide food while at sea, the settlers carried with them a species of rat. When they reached the island, the rats jumped ship and spread throughout Easter Island.

Figure 2.3. Easter Island is extremely isolated from other land, and this isolation has affected its environmental and human change.

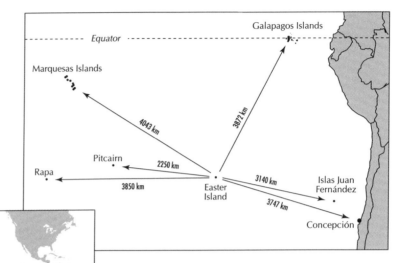

Figure 2.4. The enormous statues on Easter Island have long fascinated people who wondered what they meant and how they could have been built.

The people who had landed on Easter Island existed in complete isolation for over a thousand years, evolving a set of cultural practices distinct from their Polynesian ancestors. They revered their ancestors and erected great statues several meters high carved from volcanic rock (Figure 2.4). The population was probably divided into farmers, stonemasons to build the religious statues, and fishers, who provided much of the protein sources.

Archaeologists and paleoecologists (Bahn and Flenley 1992; Flenley and King 1984) recently reconstructed the forest and human history of Easter Island (Figure 2.5), using archaeological records from island caves and pollen records from lake mud. Both sources document the rise and fall of an ecosystem and civilization. When the Polynesians arrived, the island supported a species of palm that provided wood large enough for the construction of seaworthy dugout canoes. These canoes enabled islanders to fish for sharks and large fish and to visit small, uninhabited islands over 400 kilometers away. These islands were home to dense bird colonies, which provided an important pro-

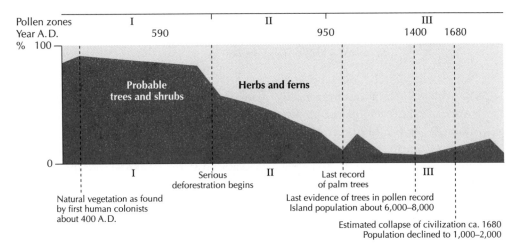

Figure 2.5. The pollen history of Easter Island illustrates how trees began to decline after people arrived on the island. Redrawn from Flenley and King 1984.

tein source to supplement the Easter Islanders' fish catches. The islanders rapidly logged the palms for shipbuilding, and new palms failed to replace them, probably because the escaped rats rapidly multiplied and devoured so many palm nuts that the forest could not regenerate. The pollen records show that for more than 30,000 years before humans arrived, the islands were forested. When the Polynesians arrived, forests began a steady decline, until by the late 1600s, when Europeans arrived, the islands were almost entirely treeless.

Soil erosion followed deforestation, fertility in the fields declined, and farming deteriorated. When the last palms were cut, islanders could no longer build the canoes to get fish from the sea or eggs from the bird colonies. Famine resulted, leading to warfare, cannibalism, a population crash, and finally the end of the Easter Island civilization.

Nothing, unfortunately, is unique about this story. Similar cycles of over-exploitation of resources, deforestation, erosion, famine, warfare, and societal collapse are evident from the archaeological record of many islands. Humans have a long history of transforming their environments, and those transformations may make it impossible for humans, along with many other species, to persist.

Example 2: The Eastern North American forest

While Easter Islanders destroyed their own home and culture through resource over-exploitation, the story is not always so grim. In eastern America, forest exploitation was followed not by ecological and societal collapse, but by recovery.

When English settlers arrived at the Plymouth colony in New England in 1620, they thought they were stepping into what one observer called,

*...a hideous & desolate wilderness, full of wild beasts & wild
men...The whole country, full of woods & thickets,
represented a wild & savage hue.*

—*William Bradford, 1620*

The settlers thought they were gazing upon a desolate wilderness unaltered by people, but actually the forests of eastern America had a long history of human transformations. Although the northern forest had never been clearcut or extensively logged, Native American practices had radically affected its development, as William Cronon (1983) and William Denevan (1992) have shown. Native Americans moved into the northern forest soon after the retreat of the glaciers 11,000 years ago. Soon after their arrival, huge Pleistocene-era mammals such as mammoths, mastodons, armadillos, ground sloths, giant beavers, dire wolves, and saber-toothed tigers went extinct (Figure 2.6). Scientists debate whether Native American hunting, climate change, disease, or a combination of all three destroyed the animals. Whether or not Native Americans were responsible for the demise of the huge mammals, they soon altered the forest, clearing plots of land for shifting agriculture, and burning forests in patches to keep them open, parklike, and full of berries. These burns created excellent habitat for deer and other wildlife, and likely increased deer populations (Cronon 1983). The tribes were mobile, allowing farm plots to regenerate back into forest and giving soils time to recover from cultivation.

Figure 2.6. Soon after the arrival of people to North America, most of the huge Pleistocene-era mammals such as the mastodons illustrated here went extinct.

Europeans introduced an economy which rested not on hunting and shifting cultivation, but on the extraction of four primary resources: timber, furs, fish, and agricultural products. Settlers exported timber to Europe and cleared large plots of forest land for settled agriculture (Cronon 1983; Merchant 1993; McKibben 1996; Whitney 1994). Although Native American farm plots had quickly recycled back into forest cover, the demands of colonial property ownership required that farm plots remain fixed on the landscape. The effects on soil fertility and erosion were often dramatic. Forest cover across the Northeast dropped from 70 percent to 25 percent or less (Figure 2.7). As the trees ran out, loggers moved from New England to New York, Pennsylvania, the Great Lakes, the South, and eventually to the forests of the West. As soil fertility declined, agriculture left the region as well.

What can we learn from this contrast between Native Americans and Euro-Americans in the eastern forest? Both groups affected the forests they lived within, but Native Americans extracted resources without depleting the resource base their cultures depended upon. Europeans had effects that were much more dramatic in the short term, for they, like the Easter Islanders, extracted resources too quickly to sustain their own economies. Yet the outcome was very different than on Easter Island. While Easter Island's forests have never

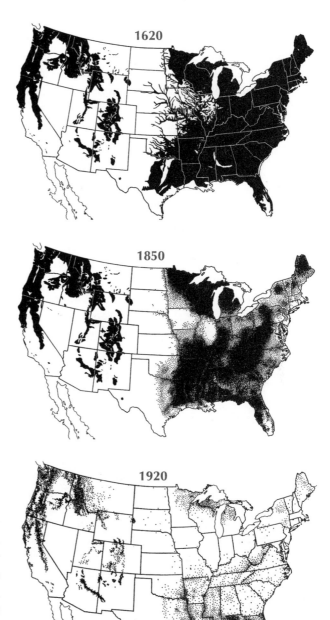

Figure 2.7. The extent of unlogged forest across North America declined with the spread of Euro-Americans, beginning with the logging of the New England forests. Each dot represents 25,000 acres. Adapted from Greeley 1925.

returned, the landscapes of New England are once again forested; for example, trees now cover nearly 90 percent of Vermont. These recovered forests are not the same as their predecessors: the dominant tree species have shifted, and stands of trees are often smaller and denser, leading to changes in soil conditions, temperature, and water availability (McKibben 1996). Nonetheless, many wild animals that were nearly extirpated from the region in the nineteenth century—including wolves, bear, cougar, and beaver— now reinhabit landscapes where people thought they had vanished for good. As forests come back, they create a new set of ecological relationships, leading to new landscapes and new choices and responsibilities for human management.

Example 3: The Mediterranean

As an example of the ways that cultural, economic, and ecological forces interact to shape the history (and future) of an ecosystem, consider the Mediterranean (McNeill 1992; Hughes 1993). Scholars have argued that one of the three worst environmental disasters in human history happened during the era of the Ancient Greeks in the Mediterranean— at the same time their civilization was constructing the foundations of western rationality and science (Worster 1979; the other two disasters were soil erosion in China around 3,000 B.C. and the Dust Bowl on the American Great Plains in the 1930s).

Figure 2.8. The lovely views of bare limestone ridges, such as this scene in the Greek uplands, are in part a product of deforestation caused by centuries of farming, goat grazing, shipbuilding, and logging for mine smelters. From a photo by Johannes Foufopoulos.

When you look at the mountains of the Mediterranean today, (Figure 2.8) you see beautiful landscapes, with bare limestone ridges between which lie picturesque villages. Both the villages and the limestone ridges are lovely, but both are dying. The villages are now empty shells where few but the very old live; the others have gone off to cities to find jobs. Over thousands of years, people have struggled to make a living in these hills, grazing goats, logging trees, planting wheat, fighting battles, hunting wildlife, writing poetry, founding empires, creating much of the philosophy that forms the basis of western culture. In the process, people stripped the hillsides of their trees (hence the lovely views), their soil, and their ability to support much life, human or otherwise. People built empires from the resources extracted from these ecosystems, but in extracting without limit, the empires eventually destroyed themselves from within. To understand why, one needs to understand two interconnected factors usually seen as separate: ecology and economics.

Just after the end of the last ice age, 12,000 years ago, pollen records show that many Mediterranean areas, low elevation and high, supported forests of oaks and pines that have long since dwindled. Focusing on Greece, the geology can be roughly divided into three main zones. The *plains* are low-lying basins filled with silts washed off the hills by erosion—a zone that is readily farmed. At higher elevations are *hills* of volcanic lavas; still higher are *mountains* formed of hard limestone and thin, easily erodible soils (Rackham 1990).

Humans have inhabited the Mediterranean basin for at least 500,000 years (McNeill 1992), but populations in the region remained relatively low for most of that time. For about 490,000 years of inhabitation, people were largely on the move, preferring the lowlands to the mountains for gathering resources. They hunted forest-dwelling animals, mostly deer, and burned woodlands as part of their hunt. These fires probably had strong effects on forest development, keeping forests open, free of dense undergrowth, and dominated by species tolerant of fire and intolerant of shade. Fires set by hunters initiated the long process of human-induced erosion, but because people had no livestock grazing in the forests, vegetation recovery after fire was probably rapid (McNeill 1992).

During the Neolithic period (11,000 years before present), the climate became increasingly arid, and many groups of people in the region began to shift from hunting and gathering to agriculture. Early farmers cleared many of the lowland forests for fields, and by 3,000 years ago, the plains of Northern Greece had lost most of their forests. Although the first Greek farmers did remove large swathes of forest, their farming practices seem to have been relatively ecologically stable, for archaeologists have evidence that farmers developed conservation techniques to sustain the soil and nutrients in their fields (Rackham 1990). Farming on steep slopes is hard on the farmer as well as the soil, and most farmers tried to leave hill slopes in forest, or else plant orchards and vineyards (which do not need to be plowed each year, thus reducing labor and erosion). When farmers did have to plant on slopes, they tried to plow on contours and terrace their fields, minimizing soil loss. By spreading lime on the soil, adding animal manure, and planting legumes which fix nitrogen into the soil from the air, farmers may have maintained nutrient cycles in their fields (Rackham 1990).

But farming soon led to an increase in human population, and in search of more land to put into production, people began moving up the hillsides, clearing forests on steeper hills and hard rock soils—places that are slower to recover from disturbance and soil loss. Farmers also introduced livestock, particularly goats, into their forests and farms. Although forests can recover from many disturbances, repeated grazing by goats seems to have overpowered the forests' ability to regenerate on cleared land. Livestock grazing, while providing manure for returning organic matter to farm fields, hastened ecological deterioration on overgrazed pastures and scrublands. Manure was available only part of the year, since during the hot summers, herders brought goats and sheep up to the hill lands. Political boundaries restricted herders to limited pastures, resulting in more overgrazing. Shepherds created more grazing land by burning forests and cutting rings of bark off trees to kill them (girdling). Between 500 B.C. and A.D. 25, many Greek writers such as Herodotus and Aristotle commented on the rapid replacement of forests with pastures and fields, while Plato warned about severe soil erosion and the associated loss of precious water sources (Simmons 1994).

Cultural changes sparked even more rapid ecological changes. The Greek empire that ruled much of the Mediterranean during this era focused on the sea, using their power on the ocean to colonize distant lands. Constructing ships to support this empire required spectacular quantities of timber from a land already losing its forests. Loggers climbed ever higher up into the mountains to find good trees for building ships. Entire forests became ocean fleets. As ship-building technologies developed, the cultures shifted from locally-producing farmers to an empire based on a seagoing civilization and extensive trade. Growing markets for agricultural goods were coupled with increasing specialization of agricultural products. Farmers in different regions focused on either wheat or olives and vines, depending on their local ecology, and this in turn fostered accelerating economic and population growth (McNeill 1992). As mining developed, deforestation accelerated, for the technology of metal smelting demanded ever larger quantities of timber to keep the fires burning. Mining, logging, and farming reduced wildlife habitat, and by 200 BC, the lion and the leopard no longer inhabited Greece; wolves and jackals survived only in the mountains. So much soil eroded that silt began to fill in river deltas and harbors along the Mediterranean Sea, ruining ports that the empire depended upon and creating low-lying swamps, which were perfect habitat for the malarial mosquitoes that plagued city dwellers (Simmons 1994).

How does economics fit into this story? Although as the Easter Island example illustrates, global markets are not a necessary condition for environmental degradation, extensive trade often hastens ecological problems. When resources get siphoned out of a region, maintaining the fertility of the region becomes difficult. Local peasants, farmers, and herders are continually forced to extract more and more simply to survive, and they lose the ability to respond to signs of ecological deterioration. For example, in Mediterranean Spain during the Middle Ages, cooperatives run by large families managed sheep grazing (McNeill, 1992). Grazing made a few people very rich; most people, however, became impoverished. The king supported grazing monopolies, because much of the money went to the crown to support his empire. These grazing monopolies gained a great deal of political power, and the king allowed them to cut any trees they wanted to increase forage for sheep. Instead of keeping sheep in the same low-elevation pastures year round, Spanish shepherds moved them up into the mountains for the summer, where grass and water were more abundant, a practice called **transhumance.** Transhumance was good for the local pastures, for it rested the grasslands enough to allow some recovery from grazing. But transhumance could have dramatic effects on other ecosystems. When shepherds moved their vast herds of sheep hundreds of miles each spring to summer grazing areas, those sheep ate everything along the way (Figure 2.9), infuriating the peasants who lived along the route. Peasants planted trees to try and reduce the erosion from these migrations, but sheepherders cut those trees to replace grasslands that were deteriorating from overgrazing. When peasants protested that their lands were overrun by a plague of sheep, the King had the peasants killed or tossed into jail. Peasants found it increasingly difficult to farm sustainably, since they had to meet the demands of outsiders.

Sustainable use of land requires that the people using the land be able to respond to clues that tell them when ecological conditions are changing. A good shepherd

Figure 2.9. Large bands of sheep, such as this group resting in an Oregon meadow in 1913, could dramatically damage local vegetation. From a photo probably by M. N. Unser, 1913, USDA Forest Service.

learns to read the land, learning that when a certain species of grass gets too short in a field, it is time to move the sheep on, otherwise little forage will return next year. A good farmer also learns to read ecological signs, adding a little more manure or allowing a field to lie fallow when plant growth slows. This sensitivity to changing ecological conditions gives one the ability to respond to the limit of a specific place—a process now called **adaptive management** in resource management circles. Adaptive management becomes impossible when one is forced to meet the demands of outsiders who are not under those ecological constraints, as happened in Mediterranean Spain.

Across much of the Mediterranean, forests did not return after they were cut and grazed; scrubby vegetation took their place. Why did much of the Mediterranean fail to reforest, when many forested landscapes such as the New England forest have recovered well from deforestation? Mediterranean landscapes were less resilient to human activity than New England landscapes for reasons related to geological activity, ecology, climate, and soil. Ecological constraints are often much more severe in semiarid environments such as the Mediterranean than in humid regions such as New England. Greece is also a tectonically active country: the mountains are still rising, and the limestone landscape of the uplands is also quite dynamic; water percolates through the rocks, eroding the mountains from within, leading to landslides, sink holes, caves, and vanishing springs. Erosion is an inherent property of these landscapes, with or without people (Rackham 1990).

Erosion is not always a bad thing—past erosion from the mountains created the fertile soils of the plains—but human activity can accelerate erosion to the degree that plants cannot easily reestablish. People can inadvertently establish **positive feedback cycles,** where an event creates conditions that favor the recurrence of that event, thus leading to rapidly accelerating changes. For example, in the dry climate of the Mediterranean, when erosion, grazing, farming, or drought led to a decline in vegetation, that decline created conditions that made future declines in vegetation more frequent and more intense. Tree roots on steep slopes in Greece had held onto the soil; when plows and goats destroyed those roots, new trees had difficulty regenerating, so fewer roots existed to hold onto soil, and more soil eroded off, leading to fewer roots, more erosion, and so forth, until much of the soil was in the sea instead of on the hills. As the soil declined, so too did the quality of fodder for animals. Since animals were in poor condition, people had to graze more and more of them to survive, further damaging what soil was left. The resulting positive feedback cycles seemed to have pushed many Mediterranean ecosystems over a threshold where forests could not recover for thousands of years.

The science of ecology has its roots in these Mediterranean ecosystems, for it was here that western science has its philosophical origins. You are studying this book because of a civilization that once flourished in an fragile, dry landscape. When the ecosystems supporting it were depleted, the civilization collapsed, but the habits of thought that flourished there have remained fundamental to western science. Ironically, the culture of science has long been an important factor contributing to environmental degradation, but science also offers some of the best tools for reversing that degradation.

Example 4: Beaver and models of nature

Global markets for goods, such as those that developed in the Mediterranean, concentrate demand on a small, distant producing area, often overwhelming the ability of that area to extract resources. This can have cascading effects not just on the resource extracted, but on the entire ecosystem, as the removal of beaver from the Rocky Mountain West illustrates. In the nineteenth century, the European fashion of beaver hats transformed ecosystems halfway across the globe. When Europeans arrived in North America, between 60 and 400 million beavers ranged over about 6 million square miles of varied habitats on the boundary between water and land, from Arctic tundra down to the deserts of Northern Mexico (Naiman et al. 1986, Naiman 1988). Beavers still occupied nearly every body of water when fur trappers began to work the mountain waters in the early nineteenth century. Within just thirty years the beaver had nearly vanished.

To understand the ecological effects of removing beaver from the landscape, consider first what beavers did to the ecosystem. They cut down trees and built dams—as many as 15 to 25 dams per mile of creek in prime habitat—and those dams had wide-ranging effects on the landscape (Plate 2.3). Dams slowed the water flow, and by creating wetlands, they buffered floods and helped prolong the late summer flow of streams—both critical factors in allowing dry forests to persist. The dams retained tremendous amounts of sediment and organic matter in the stream channel (a single dam could gather 6500 cubic meters of sediment behind it), and this too had critical effects on the streams as well as the surrounding forests. The great heaps of sediment provided a massive reservoir of carbon—20 times the carbon in free-flowing stream sections—and this buffered nutrient flows, because the sediment piles released carbon more slowly than surrounding areas. Flooding the soil increased the amount of nitrogen accessible to plants, so beaver activity enhanced nitrogen availability across the landscape. Because many mountain forests are nitrogen-limited, beavers may have indirectly increased forest productivity (Naiman 1988).

Beavers also shaped the direction of forest history. When they toppled trees, they were not indiscriminate clear cutters. They favored certain trees and shrubs and ignored others, so eventually the less-preferred trees came to dominate the forest along the streams. If beavers cut down enough of the trees they liked, they would end up with a forest they could not live in. Then they would have to abandon that stream reach for a while, until their favored trees came back. Should beaver density grow high enough, it might virtually eliminate preferred hardwood species, thus creating patches of bright light in the riparian areas where shrubs could come in thickly. The effect of all this cutting and moving about was an increase in diversity along the streams (Naiman et al. 1986; Naiman 1988).

Beavers preferred smaller streams because spring floods would knock out their dams in larger streams. Yet their effects were not limited to the small streams they dammed. Beaver-cut wood from upstream dams swept into bigger streams, adding large woody debris that formed an integral part of salmon habitat. These logs in turn accumulated debris and sediment, which acted as sources of carbon and nitrogen for the larger streams. Beaver alterations made watersheds more resistant to disturbance, as well as more resilient—quicker to recover from disturbance (Naiman et al. 1986, Naiman, 1988). Beaver modifications rippled across the landscape, and could last for the centuries it might take an abandoned beaver pond to change back to forest.

Removing beaver changed all this—but not in clear or simple ways, since even when beaver vanished, their ponds persisted. **Succession** in abandoned beaver ponds was rarely straightforward. Instead of moving in an orderly fashion from pond to marsh to meadow to forest, regrowth occurred in complex jumps and pauses. Some marshes, bogs, and forested wetlands remained in surprisingly stable condition for a century; others quickly reverted to forest. Beaver ponds in various stages of creation and decay formed a shifting mosaic of diverse patterns across the landscape.

No one at the time of the fur trappers realized that beaver's presence had been instrumental in shaping the landscape. Nearly all the streams that American ecologists have ever studied were streams that trappers had stripped of their prime shapers. The models of nature that stream ecologists have used to formulate their ideas about how "normal" streams function are therefore not normal at all—they grow out of impoverished waters.

What Is the Relation Between Science and Culture?

As the beaver example illustrates, understanding the current ecological functioning of a system requires first examining the human history of that system. To understand ecology, one also has to understand the history of the science itself. Ecology is not merely a collection of details gathered by brave scientists who go off to the Arctic and watch wolves, or who live in the Costa Rican rain forest and study bats. Ecology, like all sciences, reflects a set of views about the world, a set of choices about what is worth observing and measuring. Trying to disassociate ecology from the people asking the questions leads to a skewed view of the science. As the historian of science Peter Bowler put it, science is a process that

> …*mediates between the scientists' creative thinking—stimulated by a host of cultural factors—and their efforts to observe and interpret the external world.*

To understand this, think of a person looking out at the world through a pair of eyeglasses or contact lenses. Everyone views nature through lenses, whether their vision needs correction or not. These lenses are made not of glass or plastic, but instead of cultural beliefs, perceptions, ideas, desires, dreams, ethics, religious ideals, and scientific paradigms. These **cultural lenses** profoundly shape the practice of science and the ways science is used to transform the Earth. Figure 2.10 shows a scientist embedded first within a cultural world and then within a larger natural world. Her perceptions of that natural world are refracted through a set of cultural lenses. Saying that "nature is culturally constructed" does not deny the independence and importance of the larger world. But the scientist cannot know that "real" world directly; what she knows is her perception of it.

Scientific theories are models of nature; they are ways of making sense of the world by reducing complexity to a subset of measurable factors. People create those theories, and those people live within a particular culture and society. For example, during the 1930s and 1940s, a group of ecologists from Chicago led by Warder Allee challenged the contemporary paradigms in community ecology that assumed **competition** structured relationships between organisms. Allee and his fellow ecologists argued that cooperation was more prevalent and important than other ecologists recognized. Historian of science Gregg Mitman (1992) has demonstrated that the Chicago School conceived first of an idealized human society—one where people cooperated rather than competed—

Figure 2.10. A scientist's perceptions of the natural world she is studying are refracted through the lenses of her culture.

and then framed their theories about the ecological world within the context of their ideals for the human world. As this example suggests, ecologists see the world through a set of complex cultural and historical lenses, and those lenses frame the patterns they perceive as significant.

Animal behavior

To make this concept clearer, we turn briefly to animal behavior, looking at a few of the changing ideas about primate social systems. After World War I, when evolution was generally accepted among biologists, many animal behaviorists argued that studies of the primates should throw light on human behavior. Suggesting that some aspects of human behavior might have been shaped by evolution brought biologists into sharp conflict with anthropologists and sociologists, most of whom believed that cultural and social evolution was entirely unconstrained by biology. When primate biologists argued that the behavior of apes was dominated by males within a family group, this only seemed to strengthen the stereotype of male dominance that was the norm for European and American culture at the time. Critics accused biologists of trying to justify cultural inequities with an appeal to nature, arguing that biologists were simply projecting their own prejudices onto the behavior of the animals (Bowler 1993, Haraway 1989).

Early observers studying primates focused on dominance behavior. They developed a set of ideas about dominance structure and hierarchy, based on their observations of fierce alpha males who fought down the other males, ostensibly using physical aggression to keep the group in order. In the 1920s and 1930s, men such as Solly Zuckerman and Robert Yerkes argued that a family group wherein conflict over resources was resolved through aggressive displays (Figure 2.11), and a single powerful male was dominant, was the natural model for primates. Their work was done largely on primates in captivity, and these studies argued that traditional values of male dominance were part of our evolutionary heritage (Bowler 1993).

Studies of aggression and dominance grew out of the work of early ethologists such as Konrad Lorenz, who won a Nobel Prize for his work on animal behavior. Lorenz saw dominance hierarchies as an image of the ideal state, with a dominant male leader (the alpha male) keeping lower-ranking ani-

Figure 2.11. Chimpanzee threatening with a stick. The ways that scientists have interpreted the role these aggressive displays play in primate societies have shifted as more women have come into primatology.

mals in line, and the interests of individuals subsumed to interests of the group or the state. To a certain degree, early ethologists were projecting their own ideals of culture onto the animals when they focused on aggressive interactions between males: Lorenz came out of a long European cultural tradition that valued dominance and hierarchy in human society, and therefore tended to focus on it in the rest of nature. Male primates, particularly the baboons on which much of the early work was done, can certainly be aggressive, so those observations were not incorrect. Yet people interpret their observations, and culture influences scientists' interpretations. With primates, the hypothesis that aggressive interactions formed the basis of primate society was a hypothesis deeply shaped by cultural assumptions.

Scientists tended to ignore what female primates were doing, and so overlooked the importance of female coalitions and mother-child relationships within the groups. The focus changed when more women came into the field and began observing females, looking at the roles of communication, coalition building and sex in structuring social interactions (Altmann 1980). For example, Jane Goodall and Dian Fossey were the first women to do formal research on primates in the wild, Goodall focusing on chimpanzees and Fossey on gorillas. In her early work, Goodall argued that chimps were peaceful, loving, tool-using vegetarians. Fossey likewise rejected the idea of gorillas as savage killers, claiming that they were gentle creatures. Yet later research by Goodall showed that chimps engaged in hunting, group violence, and cannibalism (Goodall 1986). Goodall was devastated by these results, for they overturned her hopes that gentle, loving cooperation was the "natural" primate social structure, and that brutal warfare was only a recent human innovation that was not part of true primate nature. If violence, warfare, and cannibalism were natural to chimps, what did that say about human society?

Primate studies clearly offer a number of models of what might be "natural" in humans. Some primatologists argue that male dominance is built into all primates; others argue that interactions between females and children structure primate society, and that male dominance means little except to other, mid-ranking males. The ways scientists see the world are shaped by their expectations and their cultural history. These cultural lenses do not mean early ethologists were doing bad science. Lorenz transformed and energized the study of animal behavior with his work on aggression; the early work on primates in captivity provided the groundwork for later field work. But people do science. The ways scientists have looked at primates reflects something about primates, but also a great deal about people, and people's relationship to nature and science.

Ecology, like all sciences, is affected by the cultures and ideologies of the people who study it. The complexity of the world is difficult for humans to imagine—there are millions of behaviors an ethologist could observe, measure, respond to at any moment. Science, like all human activity, is a way of making sense of that bewildering complexity by choosing what to notice and what to filter out and ignore. No one can process every single thing happening. Problems only arise when scientists deny that they have cultural biases, and that what they see may not be all there is to see.

How Have Ecologists Viewed Nature?
History of the Balance of Nature Concept

Over millennia, human views of nature have shifted in ways that reflect changing cultural assumptions about the human role on Earth, and those views in turn change the ways people transform the Earth. To explore this dialectic between scientific ideas, culture, and ecology, we will examine changes in the concept of the balance of nature.

Figure 2.12 shows a stand of old-growth forest (forest that has never been logged). Along with many large, mature trees, in such a forest one might see a few deer, tracks left by a wolf pack, some young trees, *understory vegetation*—the plants beneath the trees—and a small opening. Such a picture suggests many questions:

- Will a forest still be there in 100 years?
- Will a fire or insect epidemic kill the trees?
- When the old trees die, will they be replaced with the same species, with other tree species, or with no trees at all?
- Will the opening get bigger or fill in with trees?
- Will deer multiply and destroy the understory vegetation?
- Will a wolf pack eat all the deer and then starve to death when their food is gone?

Figure 2.12. Old growth forest, looking out onto a sunny opening. From a photo by Nancy Langston.

Over thousands of years, people have looked at forests like these and asked similar questions. In the last several decades, however, what scientists believe to be the answers have changed dramatically, for a revolution has occurred in the worldview of ecologists. Instead of seeing a world in balance, most ecologists see a world of flux and uncertainty.

Until recently, most ecologists would have looked at this old-growth forest and believed that it was in stable **equilibrium** with its environment, meaning that dead trees were replaced with the equivalent biomass of young trees. A young forest, they believed, developed over time to attain a constant and predictable endstate, the **climax community,** which in some ways resembled an organism. Just as you would expect a puppy to develop into a dog instead of a chicken or a cucumber, ecologists predicted that a young forest would eventually develop into a specific climax forest community. If a disturbance such as defoliating insects entered the forest, they might interrupt the forest's development toward climax, but scientists believed "…this was a transitory state, a foreign intrusion into an otherwise balanced and unchanging community that in the healthy ecosystem was quickly 'disposed of' in the same sense a healthy organism disposes of disease" (Perry 1995). In other words, ecologists believed in the **balance of nature**.

What exactly the balance of nature means to people has changed over the centuries, as the following sections will describe, but underlying most beliefs about nature was an assumption that ecosystems were essentially static. Any change would occur in a predictable manner and would lead to a constant endpoint. This was a **deterministic model:** ecologists believed that the direction of change was determined by laws that people could understand (and manipulate, if they were applied-forest ecologists).

These ideas made their way into popular culture, and most people felt "that nature undisturbed displays miraculous order and balance" (Pollan 1992). They hoped that if people only left nature alone, it would tend toward a "healthy and abiding state of equilibrium." In other words, people typically assumed that the forest was ruled by natural law, capable of preserving its balance if only humans could avoid disturbing it. Most people would have assumed that to protect the forest in Figure 2.12, one simply needed to draw a line around it and keep other people out.

Although most non-scientists typically still believe in balance within nature, over the last several decades a revolution has occurred in ecologists' views. Instead of seeing a world of equilibrium and stability when they look at Figure 2.12, the average ecologist now sees a forest marked by flux, instability, and unpredictability. What has changed is not the forest, but the ideas within the scientist's head. In this view of the world, there are no fixed laws tending toward greater balance and stability in nature; chance events are extremely important, so the direction of changes in the young forest would be impossible or difficult to predict. Would trees, shrubs, or grass come into the opening? Hard to say. Would the forest still be there in 100 years? It all depends. Would predators keep herbivores in balance with the habitat? Probably not. These ecologists believe "that the natural world is far more dynamic, changeable, and entangled with human history than popular beliefs about 'the balance of nature' imply" (Pollan 1992). Protecting the forest in Figure 2.12 would mean, to such an ecologist, less emphasis on protecting the structure that is already there, and much more emphasis on the processes both inside and outside the boundaries.

Who is right? Is nature essentially orderly, or is it chaotic and random? Are the living and non-living members of a community linked together into a harmonious, interdependent system? Or are all the inhabitants of a place simply there by chance, blindly striving for the best each one can do on its own? Is that forest essentially a static place, or a place of random, unpredictable changes? There are no single right answers to these questions (and indeed, most current ecologists would argue that the truth probably lies somewhere in between). People have wondered about the balance of nature for thousands of years, and for thousands of years people have searched for order and pattern in nature (Bowler 1993, Egerton 1973, Egerton 1977, McIntosh 1985, Tobey 1981, Takacs 1996, Worster 1994). The next several sections of this chapter will address the ways ideas about the balance of nature have developed in the last two thousand years. At the end of the chapter, two still unresolved questions will be discussed:

- Is the new vision of dynamic nature any closer to the truth about the real world, or is it instead yet another set of models that reflect current cultural assumptions?
- How can these changing models help solve current environmental problems?

Early Theories About Balance in Nature
Greek view of balance

Ancient Greeks certainly recognized the fact of change in the world—in human affairs, in populations of animals and plants, even in climate. But they believed that these changes cycled around a stable point of equilibrium. Seasons came and went and came again, in much of the world. When the winter rains came, spring would eventually return. Food might become scarce for a time, but eventually food returned. The natural world of the Greeks, overall, seemed balanced, even though stability was rarely true for the human world.

Nonetheless, the Greeks occasionally perceived that stability was threatened. A plague of locusts might descend upon the crops and devour the grain. Eventually, however, species numbers usually returned to normal. These observations led to a set of questions: Why did population numbers so often seem stable? And why, when explosions and collapses did arise, did those so often lead, not to complete disaster, but back to stability?

One place people looked for answers was in the regulation of animal numbers (Figure 2.13; see Chapter Seven for much more on **population regulation**). The Ancient Greeks asked a simple question: how can diversity be maintained in a world where some species eat others? The Greek philosophers believed that nature worked not by magic or arbitrary powers, but by laws that human reason could hope to under-

Figure 2.13. For millennia, people have wondered what regulates animal numbers. Populations of animals sometimes explode and overwhelm their habitats.

stand. Yet this did not preclude a faith in divine order: they assumed that divine providence was the source of the laws of balance. For example, the Greek philosopher Herodotus believed that natural laws comprehensible to human reason kept predators from driving prey populations to extinction. He argued that differences in reproductive rates maintained numbers of predators well below the available numbers of prey, so that predators would not entirely wipe out prey numbers (Egerton 1973). His explanations of the mechanisms that keep predator reproductive rates low and therefore nature in balance were based on hearsay, rather than on observation of the natural world. For example, he argued that lions (which roamed much of Europe and Mediterranean before human pressure reduced their range) were kept in check by limited fecundity. Herodotus suggested that the mechanism limiting lion reproduction was this: since lions had such sharp claws, each lioness could only have one cub in her life—the cub must rip out its mother's womb while being born. While his mechanism was not religious, Herodotus believed that difference in reproductive rates were ordained to serve a divine purpose—in other words, the gods created predators with low reproductive rates so that balance in the universe would be maintained. Natural history proceeded according to a divine plan. Human history might seem random and uncontrolled, but not nature.

Christian views

Although the rise of Christianity had profound effects on western cultures, a faith in the balance of nature remained strong. This idea was so fundamental that it became what the historian of science Frank Egerton (1973) called a "background assumption." Rather than trying to analyze or test hypotheses about the balance of nature, scientists simply assumed it must exist. Balance was a belief central to western faith in the human place in a divinely created world, for a lack of balance seemed to suggest a lack of divine order. If God were all-powerful, as most Europeans believed, then how could He allow the patterns of nature to be random and unpredictable? How could He allow a species He had created to go extinct? A belief in divine order led to a belief in natural order.

The seventeenth-century rationalists of the Scientific Revolution believed, even more strongly than had the Ancient Greeks, that laws governed nature, and that those laws were accessible to human reason. In thinking about the balance between predators and prey, they returned to the Ancient Greeks and, in criticizing their ideas, developed hypotheses of their own. For example, the English scientist Thomas Browne (1646) decided that Herodotus was wrong: lion cubs surely would never rip out their mothers' wombs. Instead, what kept predators and vermin in check must be hibernation: God must order all the noxious animals out of sight for the winter to give humans a break from their mischief. In 1662, John Gaunt came up with a new take on the regulation of predator numbers: wolves, lions and foxes were excessively fond of sex, and so much promiscuity limited their fertility.

These arguments may seem absurd now, but they reveal an important point about how strongly scientific ideas were shaped by cultural beliefs. People saw stability because they expected to see stability. Most people were aware of agricultural pests that wiped out entire crops, plagues of animals, and possible extinction of species. While these facts would have been difficult to reconcile with ideas of the balance of nature, the assumptions of balance were so fundamental to society that for centuries no one appeared troubled by the contradictions. Plagues, pests and famine might seem to indicate a world out of balance, but surely that was only in human terms: God had made the world go out of balance for his own divine purposes, perhaps to punish sinners.

What creates balance? Relationship between species

Most popular ecological thought now assumes that what keeps nature balanced is not divine order, but instead relationships between species: the "web of life." Deer numbers won't explode in a pristine system, people assume, because wolves keep their populations in check, leaving just the right number of deer that the habitat could support. Although few modern ecologists would agree with the details, relationships between species is still a topic that absorbs many many ecologists (see Chapter Eight). Before the eighteenth century, however, biologists had shown relatively little interest in the ways different species interacted. In the eighteenth century, with the Swedish scientist Carl Linnaeus (1707–1778), the focus changed from species in isolation to interactions between species, and the role that might play in keeping populations in check and nature predictable and balanced.

Before Linnaeus, most plant biologists had focused on individual species, rather than on plant communities and the complex interrelationships between plants and animals. Scientists asked few questions about why things lived in the places they did, why one tree was here, another there. Trees were there because that was the place God put them; not because of biological relationships—not because one plant altered its habitat, making it possible for another plant to survive. Most scientists thought of nature as unchanging, at least on any time scale that might matter to humans. Seeing the land as a static entity, rather than a place whose history was shaped by complex biological relationships, allowed people to simply pull out certain pieces—beaver, for example— without worrying too much about the indirect effects on the rest of the community. If the Earth was a collection of separate entities, then people had two alternatives: they could either admire those pieces, or extract them and use them. But few scientists expressed the thought that people were there to participate in an ecological community held together by a web of biotic relationships (Langston 1995).

Linnaeus, like his predecessors who focused on individual species, believed that each species fit into a precise place in nature's order, and that classification could help people perceive the underlying pattern in God's design. Linnaeus was determined to reduce the overwhelming complexity of natural history to a semblance of order, and this motivated his enormously influential system of classification for plant and animal kingdoms. Unlike his predecessors, Linnaeus believed it critical to understand relations between species, relations he believed had been designed by the Creator specifically to create order and stability—what he called the **economy of nature** (Egerton 1973). Centuries later, modern ecology dropped the assumption of divine design, but held onto Linnaeus' hypothesis that complex relationships between species created stability.

Linnaeus did not ask how such relationships developed, because like most of his fellow scientists, he assumed that God must have created them through divine wisdom. When, like many before him, Linnaeus asked what kept some populations of animals in check, the hypothesis that competition for resources in short supply might limit populations occurred to him. But he never examined this hypothesis closely, because the very idea of competition seemed to violate divine wisdom and harmony (Egerton 1973).

Linnaeus' interest in relationships between species was shared by the influential British naturalist Gilbert White (1720–1793)—one of the first natural historians to turn his passion for studying nature into an effort to preserve a natural order that seemed threatened by human industry. His 1789 work *The Natural History of Selborne* set out to understand the ways in which plants and animals interact with each other and their environment, creating a complex web of relationships and ordered beauty (Worster 1994). White, like Linnaeus, believed in a balance of nature endowed by God in which predators did not wantonly exterminate prey, but instead killed just the right amount to maintain numbers. Unlike Linnaeus, White realized that people could wreck that order. White realized that the nature he knew was drastically modified—and, he felt, badly damaged—by thousands of years of human activity. White longed for a time in the rural past when people lived in harmony with that nature, instead of destroying the balance of nature. His work was in many ways a precursor to modern environmental and ecological concerns, for White was among the first European biological thinkers to express a sense of the vulnerability of nature.

Succession and the balance of nature

One of critical developments in nineteenth-century science, during the period ecology was developing as a distinct science, was the awareness that nature has a history that shapes its present structure (Bowler 1993, Worster 1994). To us, it may seem obvious that the world changes through time, but for centuries few people believed this. The emergence of an historical view of nature was the source of enormous controversy, for it challenged the traditional view of a divinely created universe. Worster (1994) has argued that a growing sense of the Earth's vulnerability emerged from the realization that extinctions had happened in earth's history. A sense of history allowed people to see that extinction was indeed possible, forcing them to confront the possibility that humans might change the world in ways that could also destroy themselves.

As nineteenth-century scientists began to understand that extinctions had happened in the history of the Earth, other plant biologists began to focus on the roles of change and history on smaller time scales; i.e., within the development of an individual forest or meadow. Biologists began to focus not just on plants in isolation, as plant ecologists had for centuries, but also on the relationships between plant species, and on the ways those relationships affected the history of each in a given place. Noting which species tended to grow near each other, they asked how one kind of plant might affect the presence, absence, or growth of another plant.

In other words, biologists thought about the ways plants fit into communities. They saw the forest not just as a collection of individual trees, but also as changing patterns of trees in groups. This difference was critical. Once people started thinking about plant associations, they soon realized that they needed to consider the ways in which these plant associations formed—the history of communities as they grouped and regrouped. When people moved from seeing the forest as a collection of separate objects to a complex community, they were much more likely to imagine roles for change and history, as well as interconnections and indirect ecological effects.

This growing focus on the history of plant communities led to a critical concept in ecology: the theory of **succession,** which hypothesizes an orderly sequence of changes in plant communities leading to a stable climax community (Clements 1916, Christensen 1989, Finegan 1984, Perry 1995, Shugart 1984, Tobey 1981, Worster 1994). According to succession theory, changes in plant communities should follow orderly laws, and competition for space, light, and water would determine the patterns of those changes. After a disturbance such as fire destroys a forest, succession theory predicts that grasses would first invade the site, followed by shrubs that would crowd out the grasses. Soon certain tree species would displace the shrubs, and under the shade of the first trees, other tree species more tolerant of shade would come in and eventually eliminate the original species (see Chapter Three). Competition—for light, for water, for space—would determine the patterns of forest changes (Langston 1995).

Henry Cowles, working on the sand dunes of Lake Michigan (1899), formalized the idea of dynamic vegetational succession. After a major disturbance, such as a fire that burns down all the trees on a site, the first species to colonize are those that best exploit the conditions of the disturbed site—lots of sun but little water or nutrients. Soon, however, these **pioneer species** change the environment in ways that make their own continued survival difficult—they create so much shade that their seedlings cannot survive. They moreover cool the soil surface, contribute organic matter to the soil, and increase soil moisture, all of which favor the invasion of the community by more shade-tolerant species.

The idea of plant succession was not new; in his 60 A. D. *Natural History,* Pliny the Elder had described something quite similar. In the nineteenth century, Henry Thoreau, in *Natural History of Massachusetts* and *Succession of Forest Trees,* argued that shade tolerance was critical in determining successional changes. He suggested that oaks succeeded pines because young oaks could grow in the shade of pines, whereas pines could not grow in the understory unless the forest was thinned by burning or logging. With the development of succession theory, foresters gradually came to believe that what existed in a particular place was not only a matter of predetermined, abiotic factors, but also a product of biological history. A given plant existed in a particular place because of the other plants that had once been there.

Would succession continue forever, or was there some end to all that change? Frederic Clements (1916), one of the most influential ecologists of the twentieth century, proposed that succession led to a climax community—a stable community in which the vegetation was in equilibrium with the climate. When new species no longer changed the patterns of light intensity and soil moisture, succession stopped. Plants in the climax community could grow as well under their own parents' shade as those parents had grown under the species they replaced. Eventually all communities would arrive at the climax community determined by their regional macroclimate, and this community would have the potential to remain essentially unchanged forever. Different plant communities might begin with different species, but in a given climactic region they would all end at the same climax; i.e., climate and not biotic interactions determined the final community. Individual characteristics of the species on the site, local environment, soil, interconnections between plants and animals, disturbances such as fire and grazing, the plant and animal species available to colonize an area after disturbance, and finally chance—all these mattered little compared to the effects of climate. As Clements argued, the chance accidents of history mattered for only a short time, until finally the end point of succession was reached.

In Clements' theoretical framework, disturbance was a rare, external event, not an intrinsic property of the community. Succession, Clements insisted, was an orderly process. If ecologists could not predict the exact community that would come in after disturbance, they just did not know enough yet about the situation. Clements' ecological framework suggested an inevitability about the development of a community. It became an often-rigid orthodoxy that treated disturbances such as wind, fire, insects, and diseases as external influences that applied ecologists such as foresters could, and should, eliminate.

Clements' focus on holistic, interdependent communities grew partly out of his interest in the social philosopher Herbert Spencer, who argued that human society was similar to a biological organism. Each specialized trade in society was like an organ in the body such as the liver, providing functions that helped the whole but were also dependent on it. A liver could not exist for long outside the whole body, nor could the body exist without its liver. Likewise, the whole human community was more than the sum of its parts. Clements argued that natural communities could also be envisioned in these terms.

Clements was not the first to propose the concept of a biological community. Serious attempts to understand communities began early in marine biology and limnology. In 1877, Karl Mobius had originally suggested the biotic community concept in his study of oyster beds. Victor Hensen examined how plankton in the ocean functions, applying physiological methods to understand the annual cycles of plankton blooms in the ocean. His findings led him to argue for the importance of communities in marine systems. The limnologist Stephen Forbes, in his address on "The Lake as a Microcosm" (1887) explicitly emphasized that all species within a lake were linked into a functioning community that balanced predators and prey and formed something like an organism.

In America, the biologist Victor Shelford began to apply Clementsian concepts of community and succession to animals (Shelford 1913). He became interested in predator-prey relationships, food chains, and population fluctuations, asking many of the same questions about the regulation of animal numbers that had fascinated the Ancient Greeks. In Britain, the animal ecologist Charles Elton was also intrigued by these questions. Using the historical records of Hudson's Bay Company, Elton traced the fluctuations of fur-bearing animal populations (see Figure 1.1J). The cyclic relationship between lynx and hares seemed to suggest that predators controlled prey populations. However, Elton's continued research on fluctuations in animal populations eventually led him to repudiate the balance of nature concept:

> The 'balance of nature' does not exist, and perhaps never had
> existed...Each variation in the numbers of one species causes direct
> and indirect repercussions on the numbers of the others, and since
> many of the latter are themselves independently varying in numbers,
> the resultant confusion is remarkable (Elton 1930).

Wildlife ecology and the balance of nature: Predators and prey on the Kaibab Plateau

Generations of American students have learned about the balance of nature, the regulation of animal numbers, and the relationship between predators and prey from an ecological disaster reported to have taken place on the Kaibab plateau north of the

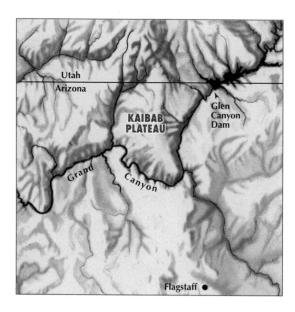

Figure 2.14. The Kaibab Plateau lies north of the Grand Canyon.

Grand Canyon (Figure 2.14; Young 1998). On this plateau, President Theodore Roosevelt established the Grand Canyon National Game Preserve in 1906, hoping to protect mule deer from destruction by hunting. Across North America in the late nineteenth century, unregulated commercial hunting had devastated wildlife populations, and many Americans were convinced that protection was necessary if wildlife was to remain part of the North American continent. Conservationists felt that deer needed to be protected from commercial hunters, sports hunters, Native Americans, and especially predators such as wolves.

Beginning in 1906, hunting was stopped and predators were killed across the plateau. In response, deer populations began to climb. Their populations increased from about 4,000 in 1906, to between 20,000 and 100,000 by 1924, according to various estimates (Young 1998). The native vegetation could not support such large herds, and deer began to die of starvation and disease. In 1924, the Secretary of Agriculture brought together a committee of experts to investigate the situation; members disagreed on policies that would reverse the decline, but all agreed that they needed to develop a plan for "scientific management" of the Kaibab deer population (Young 1998).

Wildlife biologists subsequently focused decades of intense research on these fluctuations in deer and predator numbers. In his 1932 dissertation, Victor Shelford's student D. Irwin Rasmussen graphed the history of population swings on the Kaibab, a figure that has been reprinted in generations of ecology and wildlife textbooks (Box 2.1). The graph showed that as predator numbers collapsed, deer numbers soared: a relationship that seemed to support the hypothesis that deer populations were regulated by predation. Although Rasmussen did not publish his work until 1941, in 1933 Aldo

Box 2.1. PORTRAIT OF POPULATION CHANGE

D. Irvin Rasmussen's original 1941 graphs for "Estimated numbers of deer and removals; predator removal; livestock numbers and feral horse removal, Kaibab Plateau, Arizona." Data for I were collected by "forest supervisors' estimates," "men visiting the Plateau," and organized winter counts. The deer death loss of 1924–26 (line a–b) is based on "a report by United States Forest Service Ranger Benjamin

Swapp, who was in charge of the area where the deer died." The dashed line represents an estimated population trend from 4,000 deer in 1906 to near 100,000 in 1924, an accumulative annual increase of 19.58%. Graph II "gives record of predator removal"; III "shows number of livestock permitted on the area and the number of feral or wild horses removed by hunting". Redrawn verbatim from the original.

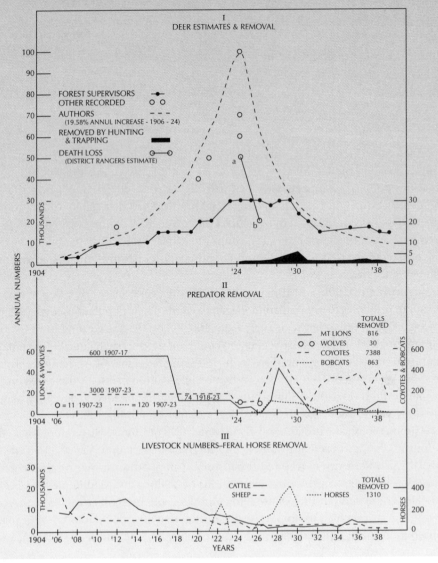

Leopold used Rasmussen's graph to argue for the importance of predators in *Game Management*—the first wildlife biology textbook. As Rasmussen and Leopold well knew, many factors other than predation—such as climate, food, and competition from domestic livestock—also affect the regulation of animal numbers. But in succeeding generations of biology and wildlife textbooks, the message of the Kaibab deer became increasingly simplified. From the Kaibab story, students were taught that natural processes of population regulation such as predation kept nature in balance, and when people removed those natural regulations, disaster resulted.

Eventually, the ecologist Graeme Caughley called a halt to the simplifications in the Kaibab story. In 1970, he published an article in *Ecology* that pointed out how textbooks had distorted Rasmussen's original graph. Caughley called attention to forces other than predators that might have affected the Kaibab deer herd. Moreover, he stressed that even in the absence of human interference, natural forces such as predation may often fail to keep population numbers within limits that the habitat can support. Nature does not necessarily provide perfect, orderly regulation of any population of organisms.

The Kaibab deer story has also been held up as an example of the ways wildlife managers have sometimes manipulated ecological data to persuade the public to accept changes in their management policies (see Dunlap 1988). According to Dunlap, in the early 1920s the experts had easily reached a consensus that the deer herd needed reducing, but political pressures made it impossible for administrators of the preserve to act on the experts' recommendation. The implication is that ecologists performed the basic research demonstrating the importance of predation, but the wildlife biologists then applied that research poorly, failing to control deer herds because of political disputes between different federal agencies over control of the wildlife resource. Yet, as Chris Young (1997) argues, this is too simple an interpretation of the relationship between wildlife biology, ecology, politics and culture. The delays in reducing the deer herd came about as much because of valid scientific uncertainty on the part of the experts, as from political pressures. No clear division between pure and applied science exists, and much valuable basic knowledge about predators and prey has come from the careful work of wildlife biologists who are trying to manage game. As Young notes,

> *…Controversy is as much a part of science as observation or evidence. Scientists do not develop theoretical principles about nature in one setting then apply them unproblematically in another. A process of negotiation (some would call it trial and error) takes place in attempting to make sense of the natural world* (Young pers. com.).

Recent challenges: Is nature in balance?

Soon after Clements published his works on succession, a few scientists challenged his theories of a stable climax. British ecologists were dubious from early on, believing that the contingencies of place and history had more effect on forests than implied by the climax theory of succession. In Britain, the landscape clearly bears the marks of human intervention and history, and vast, stable climax communities were more difficult to envision than in the American landscape. The British ecologist Henry Gleason argued in 1926 that superficially similar plant associations in truth differed substantially in species composition from place to place—thus pointing to an important role of the specific site. Given enough knowledge about the particular site, ecologists still thought they had enough information to know exactly where history would take a forest. A few years later, Arthur Tansley proposed the concept of the **ecosystem,** a complex system consisting of "...the whole complex of physical factors forming what we call the environment" (Tansley, 1935). Few ecologists, however, believed that the chance events of history were very important, until ecologist Robert Whittaker (1953) argued that there was no absolute climax vegetation for any area. In the 1950s, biologists who studied ecological history using pollen also began seriously challenging Clements' work, for they showed that the vegetation of North America has been in continual flux for at least the past 40,000 years (Christensen 1989). Although these ideas about ecosystems, changing climaxes, and the possible confusions of history were enormously important within academic ecology, they had little influence on how most applied ecologists thought about managing nature, as the Kaibab example shows.

After World War II, a perspective on community change developed which saw landscapes not as climax communities, but as patches recovering from disturbances (Pickett and White 1985). Ecologist A. S. Watt in 1947 presented to the British Ecological Society his "Pattern and Process in the Plant Community," a talk many have argued was one of the most influential in ecology. This work led to the **patch dynamics** perspective, which views a community not as a stable, fixed assemblage of species, but as a mosaic of patches differing in successional stages (Loucks and Wu 1996). Within intertidal communities, waves help create a mosaic of shifting patches (Levin and Paine 1971, 1974). Forest ecologists have similarly argued that disturbance processes within many forests create a dynamic system of patches of different ages and composition (Botkin et al. 1972, Shugart 1984). In grasslands, animal ecologists examined the ways bison graze, rodents burrow, fires burn, and plants and animals compete and cooperate; all of these processes shape a dynamic landscape (Loucks et al. 1985; Wu and Levin 1994).

To envision this new model of nature, imagine a landscape as a multi-colored patchwork quilt (the patch perspective), compared to a solid green wool blanket (the traditional view). One can also understand this change in representations of the natural world by considering a similar shift in representations of the political world. Figure 2.15 shows three maps of Africa, one from 1913, one from 1939, and one from 1983. The earlier map is dominated by lands ruled by the British and French empires (parallel to the climax community). But the second map looks different; the new mosaic reflects the ways political boundaries rapidly changed after the post-colonial revolutions and disturbances of the century. The third map shows yet another distribution of patches, as the bound-

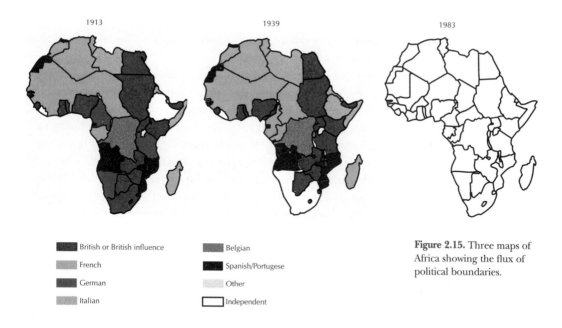

1913   1939   1983

British or British influence

French

German

Italian

Belgian

Spanish/Portugese

Other

Independent

Figure 2.15. Three maps of Africa showing the flux of political boundaries.

aries shifted again after wars in the late twentieth century. These political maps offer one clue to the question: Why do the new ecological models of disturbance and flux seem more plausible to many people than the former equilibrial models? Since World War II, political and cultural changes have been rapid and often bewildering. Disturbance and uncertainty now seem more natural than harmony and equilibrium to many people, and this cultural climate may have helped the new ecological models seem more plausible than the former equilibrial models.

How do ecosystems persist?

Ecologists still debate the question that puzzled the Ancient Greeks: how do ecological systems persist in the presence of forces that seem potentially destabilizing? As we've noted, ecologists now envision the landscape as a set of patches fluctuating in time and space, rather than a set of static climax communities. Few ecologists would argue that those dynamic patches are completely random collections of species, able to change into any other assemblage of species at any time. Many forests persist as forests in spite of thousands of years of disturbance, even though the patches within that forest may be changing. Other forested sites may, after a disturbance, be supplanted by scrub and grass. Most ecologists would now argue that neither of those forests was balanced or random, but instead that they had different levels of **resilience to disturbance**—defined as the ability to undergo change and then return to a similar, but not exact, system configuration (Perry 1995). Resilient forests, such as the New England forests discussed early in the chapter, may experience catastrophic events such as logging or enormous wildfires, but eventually trees return. But on a site that has been degraded past its ability to respond to disturbance, such as the Mediterranean forests, trees may not be able to recolonize and the forest disappears.

Resilience does not mean an absence of change; instead, it refers to what happens after the change. Ecosystems usually gain resilience not just because of the properties of individual organisms, but because of intricate ecological relationships (Perry 1988). These relationships are dynamic rather than stable, fluctuating over time as climates and ecosystems change. Actions that upset those relationships, however, may bring about unpredictable, undesirable, and irreversible changes. For example, even when fires repeatedly volatilized nitrogen from the soil, much of the American West still supported forests, perhaps because such nitrogen-fixing plants as alder and ceanothus came in after fires, returning nitrogen to the soil. Later, when foresters excluded alder and ceanothus from clearcuts because they had no commercial value, they introduced a new disturbance that reduced the system's ability to respond to other disturbances. If people alter a system beyond a threshold by breaking up ecological relationships, that system may lose resiliency and return to a new (and from a human perspective often much less desirable) community after a major disturbance (Perry 1988).

Disturbance and stability

Whether you see stability or chaos when you look at the world depends on your perspective. When you kneel on the ground and look very close, patches can seem random and chaotic, but when you stand miles away and see an entire **watershed**, those chaotic changes appear to even out, translating into a dynamic stability. In a forest, individual tree falls or forest fires can seem radically destabilizing and chaotic. But at the scale of watersheds, disturbances can create a shifting mosaic that may be stable. For centuries, roughly the same percentage of the watershed may contain stands of old-growth forest at any given time, even though the locations of those old growth stands will move as disturbances move through the forest, as trees age and die, and as new stands regenerate (Wu and Loucks 1996).

For thousands of years, people in western cultures have viewed disturbances—fires, windstorms, insects, disease—as destructive processes that destroy stability. But now many ecologists argue that from a long-term perspective, repeated disturbances may actually help maintain diversity and increase the chance that a given ecological system will persist. "Harmony is embedded in the patterns of fluctuation, and ecological **persistence** is 'order within disorder'" (Wu and Loucks, 1996). Patterns of disturbance can shape a diverse forest that over the long term may be much more stable than a forest protected from disturbance by people. For example, in western forests, fires of different intensities, along with storms, windthrows and insect attacks, together can create complex and shifting mosaics of forests across the landscape (Figure 2.16). Repeated light fires kept fuel loads (thick carpets of needles, branches, leaf litter, etc.) to a minimum, making huge, stand-replacing fires rare. When foresters tried to keep all fire out of the forest, the result was a community far more unstable in the long run, because it was far more vulnerable to intense fires that would remove the forest entirely from the site.

One of the critical insights of modern ecology is that human disturbances are now among the most important factors shaping ecosystem change. One hundred years ago, within mixed conifer communities in the American montane west, hot fires created gaps that let in pockets of sunlight where shade-intolerant trees such as ponderosa pine, larch, and lodgepole pine could establish in single-species stands. After a generation of human intervention in the form of fire suppression and logging, however, the effects of these intense fires have dramatically changed.

Think of a watershed as your hand, and the streams feeding into a watershed as the fingers on your hand (subwatersheds). Before logging and fire suppression, a catastrophic fire might have burned all the cover along one of those streams, but other streams with intact mature forests along them buffered the disturbance to the entire watershed. Twenty years later, cottonwoods and willows would have sprouted from the burned-out stumps, while ceanothus would have formed a thick low canopy that protected the soil from heavy rains. Across the stream, an old stand of ponderosa pine that had escaped the fire might have been ravaged by pine beetle. But a mile away, another patch of pine matured, another stand of spruce grew, and larch came in under a stand of lodgepole that had burst into flames. The trees shading another stream in

Figure 2.16. Fires, windthrow, and other disturbances shaped patchy forest such as these ponderosa pine stands in eastern Washington. From a photo courtesy of USDA Forest Service.

the watershed might have been blown over by a windstorm, but three streams still retained their old forest, and one stream had a recovered vegetation cover. Complexity buffered the disturbances.

If extensive clear-cutting and road building had denuded four of the streams, what would happen if a catastrophic fire destroyed cover along the stream with the only mature forest in the watershed? There would be nothing left to buffer the fire's effects across the land, no place for the animals whose habitats were destroyed to go, no mature trees left to seed in the burned site, and no pine forests nearby that could mature into old growth for another two hundred years. Whatever functions the mature forest naturally performed in the watershed would thus be missing from the system for centuries. This argument can also be extended to entire regions, especially when considering the effects of habitat loss on wide-ranging vertebrates such as spotted owls or pine martens (Perry 1995).

When ecologists point to evidence for dynamism in ecosystems, they focus on the most obvious (and directly economic) parts of the system: the plants. But large changes in the plant community, as in disturbance and succession, may be accompanied by much smaller changes in less obvious parts of the ecosystem—nutrients, soil structure, integrity of food webs. These relatively constant parts of the system may serve as legacies that maintain constraints on changes in plant communities. As these examples show, our attempts to say whether nature is balanced or dynamic depend on our perspective and the scale of our analysis. Many ecologists tend to dichotomize, calling nature either balanced or imbalanced, either stable or unstable—but nature works in many more than two modes (Perry 1995).

How Do Ecological Theories Change the Earth?

The science of ecology has dramatically shaped our current perceptions of nature. One good example of this is the way the term "ecology" can mean both the science and the political movement of nature preservation. Although one is a science and one is something very different, the political movement has borrowed deeply from the science. While many current ecologists may disavow their profession's links to the political movement, other ecologists now firmly believe that their research does and should aim to protect and restore the Earth's diversity and ecological functions. The science of ecology began, however, as a field devoted not to preserving natural relationships, but to better controlling them, and this goal affected the kinds of questions that were asked.

Foresters and succession: The Blue Mountains

Early ecologists, particularly in the United States, were motivated by the desire to use ecological knowledge to better manage ecosystems. For much of the twentieth century, scientists tried to use ecology to maximize outputs from public forests, hoping to minimize conflicts between different users who all wanted access to forest resources. They believed that maximizing production first required the liquidation of slow-growing ancient forests, so that rapidly growing forests could be grown instead. This management policy backfired, leading to devastating insect epidemics and fires across the West. Understanding how these ecological problems came about requires a look at the ways foresters have used ecological concepts of succession, competition, and climax communities, in their attempts to transform the forest (Langston 1995).

When Euro-Americans first came to the Blue Mountains of eastern Oregon and Washington in the early nineteenth century, they found a land of open forests full of large ponderosa pines—fire resistant, insect resistant forests that had been growing for centuries (Figure 2.17). But after less than ninety years of management by federal foresters, what had seemed like paradise was irrevocably lost. The great ponderosa pines vanished, and in their place were thickets of dying fir trees that were first attacked by defoliating insects (Figure 2.18), and then devastated by intense fires.

In part, the landscape changed for straightforward ecological reasons. When foresters suppressed fires in open, semi-arid, disturbance-prone forests dominated by ponderosa pine, firs grew faster than pines in the resultant shade—a successional change which left firs dominating the forests. Heavy grazing, which eliminated the grasses that had previously suppressed tree regeneration and kept forests relatively open, also contributed to the changes. High grading, or logging that removed old pine while leaving firs behind, also encouraged the replacement of open pine forests with dense fir stands. When droughts later hit, firs growing on dry sites succumbed to insect epidemics.

But the real story is much more complex than this. Changes in the land are never just ecological

Figure 2.17. Open forests dominated by ponderosa pine were common in the Blue Mountains when Euro-Americans first arrived in the 19th Century. From a photo courtesy of National Archives, #95-G-320935.

Figure 2.18. With the suppression of frequent, light fires, thickets of fir trees grew up in many forests. These dense stands were susceptible to attacks by defoliating insects, as shown in this illustration of the Yakima Reservation in eastern Washington, 1964. From a USDA Forest Service photo by R. G. Mitchell, Entomology Collection, PNW Research Station, La Grande laboratory, #PS-7990.

changes; people made the decisions that led to these ecological changes, and they made those decisions for a complex set of motives.

Federal foresters came to the Blue Mountains with the best of intentions: to save the forest from the scourges of industrial logging, fire, and decay. When they looked at the Blue Mountains, they saw two things: a "human" landscape in need of being saved because it had been ravaged by companies and the profit motive—and also a "natural" landscape that they felt needed saving because it was decadent, wasteful, and inefficient. Not only were federal foresters going to rescue the grand old western forests from the timber barons, they were going to make them better. Using the best possible science of the day, foresters felt they were going to make the best possible forests for the best of all possible societies—America in the brand-new twentieth century.

Industrial logging had been underway for less than a decade, and in that short time enormous changes were already beginning. Government foresters firmly believed that these industrial cutting practices produced sterile lands and, in particular, those practices destroyed the vegetation cover that protected the water supply in the arid west (Figure 2.19). Young trees were critical to the future of the forest, and industry seemed to have sacrificed them for short-term profits. Scientific forestry, the foresters felt, would change everything.

Two major interrelated tenets of scientific forestry developed. First, foresters felt they should encourage the growth of young trees by suppressing fire; and second, foresters felt they should replace old growth with regulated, rapidly growing forests. Fire seemed to threaten the forests by killing young trees, and since foresters were certain that young trees were the future of the forest, fire was clearly the enemy. The foresters thus decided that to protect the pine forests and the water supply, they needed to keep out fire and encourage reproduction. The Forest Service was convinced that the more young pines they had, the more merchantable pine would necessarily follow.

Although it was clear to early foresters that suppressing fire would lead to dense thickets of young trees, they felt that surely this would be a good thing, for reasons that show how their cultural beliefs affected their ecological reasoning. The early twentieth-century United States was a culture that glorified competition and masculinity. *Social Darwinism* promoted the belief that struggling with others for power and gain made men better men. Competition would eliminate the weak and favor the mighty. This ethos was fostered by emerging industrial capitalism and then projected onto the landscape. Foresters reasoned that dense stands of young trees would lead to intense competition for light and water, and that competition in the forest economy, just as in the industrial economy, would create vigorous individuals. Without competition, weaklings would

Figure 2.19. Early foresters believed that tree cover, as illustrated in this scene along a high-elevation stream in the forests of Eastern Oregon in 1929, protected the water supply in the semi-arid West. From a 1929 photo by N. J. Billings, U.S.D.A. Forest Service.

result—or so the foresters reasoned. The opposite turned out to be true, unfortunately. Western conifers do not self-suppress; without fires to thin them, what resulted was not a few big trees, but a thicket of stunted trees all the same age.

Replacing old growth with young trees was the second critical tenet of applied forest ecology. In 1905, the basic premise of the new Forest Service was simple: If the United States was running out of timber, the best way to meet future demands was to grow more timber. According to early Forest Service surveys, more than 70% of the Western forests were dominated by old growth. Foresters felt that meant western forests were losing as much wood to death and decay as they were gaining from growth.

Ecological theories of stability and predictability shaped the ways foresters viewed old-growth forests. Such forests, they thought, were at a stable, climax equilibrium, so the same amount of timber was lost to death and decay each year as was created through growth. Young forests would put on more wood volume per acre faster than old forests at equilibrium, since they were still growing rapidly. Therefore, foresters believed that young forests would create more wood for human use, so overmature forests needed to be cut down immediately. Scientific forestry seemed impossible until the old growth had been replaced with a regulated forest. These theories were not new; American foresters had borrowed them from European forestry, where old growth had long been eradicated. Trying to apply these European beliefs to a completely different set of ecological conditions in America shaped a Forest Service that, in order to protect the forest, believed it necessary to first cut it down.

When early foresters thought the forest would be improved if old growth were removed, their ideas about ecology were at the heart of their decision. These were ideas deeply shaped by their culture as well as by their science. Foresters seized on simple ecological theories—in particular, succession and competition theory—as a way of reducing the complexity of the forest to something they could hope to manage. Every species might be different, foresters reasoned, but with any luck they all followed the same simple rules. The alternative—that forest development might not follow orderly laws, and that nature might be so complex that people could never precisely predict the result of any action—was not something foresters wanted to contemplate.

Foresters interpreted succession theory in a particularly narrow way. Succession, they argued, was driven almost entirely by competition for light and moisture. In a struggle that approached warfare, each tree species tried to control limited resources. Forestry seemed to be simply a matter of manipulating nature's own competitive struggles, so as to tilt the balance towards economically useful species.

At the heart of forestry were two assumptions of competition theory: that the forest was a collection of resources, and that those resources were limited. There is only so much time in the world, so much energy, so much food. Tradeoffs are therefore inevitable: any time or energy spent on one activity is time taken from another activity. Any water you give out to the soil or to other plants is less water for you. Any nutrients another plant takes means less for you.

These assumptions stem from a vision of the world as a collection of separate, exchangeable parts. But there are other ways of seeing nature. One of the important insights of early ecological theory was that the natural world is not merely a collection of separate objects. Instead, indirect effects form an intricate series of relationships between the parts of a natural system. If a ponderosa pine releases water to surrounding shrubs, that does not always mean there is less for the pine. Those shrubs might in turn increase the local population of mycorrhizae, some of which might join with the pine and help it to better absorb nutrients from the soil.

Do the shrubs compete for resources with the tree? That question assumes a simple relationship of gain and loss; it is a human framework for seeing the forest, rather than the way the forest really is. Nor is the forest really an interconnected web of relationships—both competition and interconnection are human metaphors which help people work with the forest. But by focusing on competition theory and its applications to succession, early foresters discounted other ecological theories that were current at the time, particularly a recognition of the indirect effects that tied communities together. In their attempts to create an ideal, unchanging forest through competition, foresters had to discount these interconnected processes.

The assumptions of competition theory made it difficult for foresters to imagine that insects, waste, disease, and decadence might be *essential* for forest communities; indeed, that the economically productive part of the forest might depend on the economically unproductive part of the forest. But these semiarid, disturbance-prone forests were anything but efficient; their very inefficiency and redundancy was what allowed them to persist on dry, marginal sites. In trying to eliminate what they saw as wasteful insect enemies, foresters instead destroyed the habitat for the predators of those insects, making future outbreaks ever worse. In trying to eliminate what they saw as wasteful forest fires, they instead created conditions for ever more intense, catastrophic fires.

Scientific theories are important, but ideas alone do not shape the physical world. To transform the landscape, the right constellation of ideas, markets, ecological and economic conditions has to exist. At first, the Forest Service's dreams of transforming nature had little effect on the forest itself. Before the foresters could reshape the forests, they needed markets for the timber. And until the First World War, nobody was interested in buying federal timber. But after markets for Forest Service timber opened up, sales in the pinelands gained a momentum that quickly overwhelmed the conservative ideals of the foresters. All the sales were driven, not by the Forest Service's desire to make money, but by their firm conviction that scientific forestry would never be possible until old growth was eliminated.

To convert old growth to scientifically regulated forests, the Forest Service needed markets for that timber, and they needed railroads to get the timber out to the markets. Railroads were extraordinarily expensive, however. Financing them required capital, which meant attracting Midwestern lumber companies. But Midwestern corporations were only interested in spending money on railroads if they were promised sales rapid enough to finance those railroads. In the Blue Mountains, as across the West, this management policy devastated both the land and the local communities that depended on that land. Across the region foresters set up intensive harvests in the 1920s which would ultimately ensure that, by the late 1980s, harvests would drop by at least 60%, and mills would have little or no ponderosa pine left to harvest. This is exactly what happened—harvests collapsed and mills closed throughout the region. Ecologically, the effects were equally dramatic. Heavy logging, combined with fire suppression, high-grading, and grazing, led to the replacement of millions of acres of pine with thickets of drought-stressed, fire-susceptible firs.

The decisions of early scientists, while shaped by an ideology of efficiency and productive use, were not driven by individual greed or stupidity. Foresters destroyed the forests not in spite of their best intentions, but because of them—precisely because foresters' ideas of what was good for the forest were based on an ideal of deliberately transforming nature to serve industrial capitalism.

Early foresters had hoped that science would let them properly manage public forests for the public good, rather than for private gain. Only scientific experts, they felt, could solve public land conflicts. But what they did not realize was that their science was far from unbiased. It was deeply shaped by their culture, and this shaped a series of decisions that led foresters to attempt the transformation of old growth into productive, profitable forests. Anytime people use science in the formation of resource policy, they must grapple with these dilemmas. Scientific understanding is always simpler than the ecosystems; no one can ever understand everything about the ecosystems they are hoping to manage. Yet management is necessary, and ecology is one of the best tools for doing so, even if ecology can never offer a complete understanding of the entire world.

Does flux in nature justify transforming nature?

Fire suppression in the Blue Mountains illustrates a situation where beliefs in predictable succession, equilibrium, and climax communities helped justify intensive efforts to manipulate nature in order to increase production for human commodities. A belief in the balance of nature convinced many people that humans could never really upset that balance. Given enough time, people believed that succession would heal the wounds of human activity; people could therefore manipulate ecosystems without worry, since natural compensating mechanisms would eventually restore harmony (Bowler 1993, Christensen 1989).

Equilibrium models assumed that disturbances such as fire were external to the ecosystem. Many managers took this assumption one step further, attempting to remove these "external" disturbances to create a system they thought would be more balanced and stable. The Blue Mountains example is only one where this attempt to remove fire and increase stability badly backfired. Similar attempts at fire suppression around the globe have had equally dramatic and unintended effects (Pyne 1982, Goudie 1994). In Alaska, fire suppression in lowland sites has led to an increase in moss, which densely carpets the ground, raising the permafrost level, in turn encouraging the growth of black spruce—an unpopular species among foresters, for it has little timber or wildlife value. Ironically, fire suppression ultimately can magnify the effects of fire, as the geographer Carl Sauer (1969) pointed out nearly three decades ago:

> *The great fires we have come to fear are effects of our civilization.*
> *These are the crown fires of great depths and heat, notorious*
> *aftermaths of the pyres of slash left by lumbering. We also increase fire*
> *hazard by the very giving of fire protection which permits the*
> *indefinite accumulation of inflammable litter. Under the natural*
> *and primitive order, such holocausts, that leave a barren waste, even*
> *to the destruction of the organic soil, were not common.*

Trying to remove disturbances can ultimately change an entire community, by changing the ecological relationships within that community (Goudie 1994). In South Africa, when Kruger National Park was established, white park administrators removed native peoples, for human hunters did not fit with the image of a pristine wilderness park. The native Africans who had lived in the area had set frequent fires to improve hunting. With the removal of the Africans and their fires, bush encroached onto grassland, reducing grazing for wild animals. Removing people to create an undisturbed, stable wilderness ironically harmed what the Europeans were trying to protect (Goudie 1994).

While models of nature's balance and stability justified some attempts to control nature by removing disturbances, many equilibrium ecologists such as the Odums were quite explicit about how they wanted their work used—to preserve rather than control nature. In contrast, current non-equilibrial ecologists have sometimes been far less interested in conservation or preservation. Donald Worster (1993, 1994) argued that part of the cultural appeal of disequilibrium as the norm of the natural world comes from people who want to blunt some of the more radical implications of environmentalism. As Worster argues, some of the ecologists who developed the initial theories about instability in nature were also fairly hostile toward environmentalism, justifying their hostility with a nod toward natural instability, and implying that since nature does not have a holistic balance, any human interference is acceptable—indeed even scientifically validated. For example, companies that specialize in genetic manipulations have argued that the new paradigm of nature's flux means that they can ignore the ecological implications of introducing created species into the environment. An article in the April 26th, 1997 *Economist* states that, since nature is a dangerous and unruly place regardless of humankind, people do not need to worry about introducing genetically manipulated plants. The implication is that, if there is no stable climax community, then no ecological relationship needs our concern. The image of the natural world as a place of competition, flux, chaos and instability provides for some people a justification for unrestrained human intervention in nature.

Many foresters have seized upon arguments about flux in nature, seeing them as permission for continued intensive management and extraction. If incessant, unpredictable change is the way of nature, why then bother to preserve any natural systems? If forests always change, then why worry about a web of interrelationships and interconnected effects? Why even bother to preserve any ancient forests, some people wonder, if they are all what one anti-environmentalist calls "fakes"—artifacts of humans (Budiansky 1995). As forester Henry Alden of the Michigan-California Lumber Company said in a 1992 speech,

> *The impact of the fire was more intense than any clear-cut I have*
> *ever seen...If fire is a natural and essential part of the ecosystem we*
> *have been far too delicate in our attempts to simulate the natural*
> *cycles of the forest.*

Likewise, some people now justify their calls for logging old growth forests by saying that, since recent ecological theory shows there is no balance in nature and no stability in diversity, old growth has no special value and should be removed to make way for tree farms (Chase 1995, Budiansky 1995). Chase and Budiansky have argued that if people are part of nature, and all ecosystems change, then people can do whatever they like to the land, since any changes thus caused are not degradations but merely new directions for the ecosystem.

Much of the appeal of earlier equilibrial models of nature's balance came from cultural beliefs; the same is surely true of current dynamic models of nature. But when anti-environmentalists argue that new ecology proves their claims, they are not basing their arguments in science, nor does this show that new ecology creates anti-environmentalism. People have always looked to scientific theories to support their political claims; that does not mean the scientists who came up with the theories share the same political beliefs.

Although some anti-environmentalists interpret dynamic ecology in ways that radically devalue non-industrialized ecosystems, a belief in the importance of disturbance history in ecosystems does not necessarily foster disrespect for nature. Much of conservation biology is based on the belief that humans can have profound effects on nature; therefore, people need to make changes only with great caution. Theories of dynamic, changing ecosystems can lead as easily to a respect for nature's extraordinary complexity, as to the opposite. If people are connected to the land in intricate but poorly understood ways, there can be no stronger argument for living with ecological respect, as part of an interdependent community (Grumbine 1992). Respect for the complexity and interconnections of ecosystems need not imply a belief in static equilibrium conditions. Instead, such respect recognizes value not just in what the land can give, but in a larger set of communities and processes which include but are not controlled by humans.

The new field of **adaptive ecosystem management** attempts to use some of the findings of dynamic ecology to manage natural resources, not for maximum commodity production (a traditional industrial forest), or for preservation of current conditions (a traditional reserve), but for the perpetuation of patterns and processes that allow the ecosystem to persist. Adaptive ecosystem management rests on several critical principles. First, all ecosystems change, often in ways that are difficult to predict. Because humans have influenced ecological processes and patterns for thousands of years, understanding human disturbances is important for understanding current ecosystem functions. Management must therefore pay attention to the changing human framework as well as to a changing natural framework. Traditional management tried to manage forests by removing elements that seemed unproductive. Adaptive management recognizes that all species and processes may play important roles in the forest ecosystem—even if these functions are not yet understood. Therefore, instead of trying to make natural ecosystems more efficient by removing whatever is not a useful human resource, adaptive management tries to maintain, restore, and mimic natural processes.

Adaptive management is not a new idea; it is simply a way of applying the scientific method to management. Nearly a century ago, Frederick Ames (1910), who first worked for the Forest Service in the Blue Mountains and then became Chief of Silviculture for the nation, warned his fellow foresters that they had to practice something akin to adaptive management. Ames argued that before foresters could begin to manage the western forests, they needed to recognize that they did not understand the forests well enough to predict their response to management. Nevertheless, they had to manage, and even doing nothing at all was a form of management. Therefore, what they had to do was treat "all of sales as a vast experiment." Ames outlined an extremely ambitious monitoring plan: after each timber sale, foresters would go in every three years and record the response of the site to whatever experimental treatment—also known as logging—they had devised. Over the next 100 years, they could then compare the effects of different kinds of logging, fire exclusion, and grazing on different forest conditions. Ames called for close attention to both the forest and the effects of human actions on the forest. In modern terms, Ames was telling his foresters they needed to practice "adaptive management" that recognized foresters could not always predict the effects of their actions.

This was an excellent idea, but unfortunately it did not work at the time, for practical and political reasons. Even when conscientious foresters gathered all the data Ames called for, these reports accumulated dust, first on the top of the supervisor's desk, then in the office's filing cabinets, then in cardboard boxes in the storage attics. No one knew what to do with all this information, and it continued to multiply exponentially while managers tried to figure out a solution. When foresters did try to monitor the effects of their logging practices, superiors in the regional offices usually shied away from making recommended changes, often for political reasons. Caution seemed easier than adapting to uncertainties, given the pressures on foresters to make timber available for sale (Ames 1915, Langston 1995).

At best, adaptive management is a way of monitoring the results of tree cutting, burning, favoring pine or anything else. But what is most innovative and promising about adaptive management is the way it tries to meet head-on the challenges outlined in this chapter: how do you manage in a world where you know that your models of the forest are always much simpler than the forest itself? In the words of Jack Ward Thomas, the wildlife biologist who attempted to bring adaptive management back to the Forest Service in the 1990s:

> *The forest is an extremely complex place; in fact, it is*
> *too complex for us to ever hope to understand.*

Nevertheless, one still has to manage it; no neutral position is possible—doing nothing is also a management strategy. All attempts to manage are attempts to tell a story about how the land ought to be, and by definition, all these stories are simpler than the world itself.

As the first foresters in the Blue Mountains recognized, everything resource managers do is nothing more, and nothing less, than an experiment. The critical step for management, however, comes after the experiment: using all that information to change how you work with the land. Here the young Forest Service found itself unable to resist pressures to continue business as usual. Monitoring does not necessarily mean big government programs; what it means above all is people on the ground being responsive to what the land is telling them, and being responsible for acting on that knowledge. It means a dialogue between people and land; it means people knowing the place they log, and knowing the place they work.

Science, like management, works best when it is adaptive. Early foresters came to the natural world with a set of general models borrowed from ideas about competition for light and water, and these rules allowed them to try to make sense of ecological patterns. More importantly, they enabled ecologists to see these patterns in the first place. These models were simplifications, but they gave foresters a way of seeing so they could notice something more than just a lot of trees. Some philosophers of science have argued that scientific models determine what you notice, and therefore blind you to what does not fit your models (Foucault 1980). Nonetheless the alternative is even more blinding. If you go out without a set of rules and questions, you do not see anything at all: the differences are lost to you. Your eyes are closed to the fact that spruce never grows in the open, that larch stands change to fir stands after a time, that lodgepole changes to spruce, that ponderosa pine stands change to fir. You see a static mass of green, not a history of change or a forest in motion.

The underlying lesson of these stories about the difficulties faced by earlier ecologists is not that they were wrong and unenlightened. The lesson is more complicated: all hypotheses are only partial models, simplifications of the world that are influenced by the cultural lenses with which we view nature. Yet those models engage the scientist with the world in an important way. The scientific method requires that the scientist approach the world with an open mind; as a scientist, you are supposed to treat your own ideas with humility, modifying your hypotheses if the results do not support them. This process is never completely open-minded; initial ideas about how the world ought to work shape how you construct hypotheses, what you see when you set out to test those hypotheses, and what you think worth noting down. But there is an important ideal here: you allow the natural world to shape your ideas, not the other way around. The history of ecology has consisted of a long series of negotiations with the natural world, which have allowed better management and restoration of ecological systems. Ultimately, all models are wrong—but they are still useful. When new data refutes or complicates current models, the result is an increased understanding of the world that makes the new models better approximations of nature. Ecology cannot offer a pure view of the world, untinged by politics and uncertainty. But it is still essential as a tool to make human relationships to nature more sustainable.

TECHNIQUES SPECIFIC TO ECOLOGICAL HISTORY

Reconstructing past landscapes and understanding the patterns of change in those landscapes requires the diverse tools of a detective. Ecological historians use documentary sources including old maps, photographs, corporate records, travelers' accounts, local histories, early scientific reports, farm account books, land survey records, letters, diaries, tax records, and old court cases. They also use field sources, such as pollen records, tree rings, and stumps in a forest, which provide information preserved in the landscape instead of in an archive (see Whitney 1994 for an excellent review).

For North America, written materials are available since European settlement. Many explorers and settlers recorded their impressions of the landscape in copious notes, diaries, memoirs, and reports. Although these are usually qualitative impressions rather than quantitative sources, they are nonetheless valuable. They do need to be used with caution when attempting to reconstruct the preindustrial landscape, for many of the sources reflect as much about the authors as they do about the landscape. Promotional tracts from land companies trying to attract buyers tended to exaggerate the fertility of the soil and the size of the trees, while settlers accustomed to forests tended to exaggerate the poverty of Midwestern prairie landscapes in their letters and journals (Whitney 1994). Legal documents are also useful; bounties on wolves, regulations of tree cutting, laws about fire use, drainage laws, game control, flood control regulations and water law all provide information about changes in the landscape.

In North America, early scientists provided extensive records from the turn of the century on ecological conditions. These records contain a great deal of quantitative material as well as qualitative impressions, so they are particularly valuable. Map series depict the changes in forest cover for many townships in North America. In the West, photographs of the changing landscape provide clues to succession, deforestation, urbanization, and effects of grazing. Aerial photographs, beginning from the Second

World War, provide larger scale views of a changing landscape. Census reports detail changes in agriculture and forest cover and soil productivity.

Land survey records are particularly valuable for North America, which is one of the few places on Earth that possesses detailed survey records describing vegetation before major transformations by European settlement. Survey records provide a qualitative and quantitative record of the pre-settlement forest, making it possible to derive tree species composition, size-class structure, and density of the forest communities. Yet these surveys need to be used with caution, because fraud and incompetence were not unknown among the early surveyors. Certain surveyors did their surveys from a bar stool; others misidentified trees, and some seem to have been biased in favor of the easiest routes across the territory they were surveying. Nevertheless, the land survey records are our best source of information on the preindustrial forests of North America.

While many sources offer clues to the American landscape that greeted European settlers, it is critical to remember that this is not the natural or original landscape. Rather, it is a historical snapshot of one moment from the past—a moment that occurred before extensive industrial transformations, but came after a long history of interaction with people. Learning how different native peoples lived on the Earth presents another set of challenges. Interviews with tribal elders provide useful information about the twentieth century; for earlier periods, ecologists, anthropologists, and archaeologists reconstruct past practices with pollen histories, fire histories, the records of early explorers, and interpretation of cultural stories. Archaeological studies of changing resource use can be particularly fertile sources (for example see Grayson 1993).

Landscapes contain a great deal of information about their own disturbance histories. Box 2.2 explains the ways that pollen records can help understand the past. Observation of current plant communities can also give clues to the past. Walking through many North Carolina piedmont forests, it is easy to stumble across the furrows in the soil left by plows. These forests grew up in abandoned farm fields, and the marks of the plow are still visible decades later. Daffodils

Figure 2.20. Trees with wide crowns in a dense forest indicate that this woodland was once an open field.

growing in the forest usually indicate the site of an old farmhouse, where homesteaders planted flowers to cheer the view. When you come across an oak or a maple with a wide, spreading crown in the middle of a dense forest (Figure 2.20), you can be fairly certain that tree is a relic of a pasture or farmyard, since competition for light and space means that trees growing up in a forest tend to have narrow crowns. The effects of fire, windstorms, insect epidemics, logging, and farming can still be seen in forests, sometimes for centuries after the event occurred. Cores from trees tell a great deal about the forest's history, since each tree ring tells about the year it was formed. Narrow rings mean the tree grew only a little during those years, perhaps because of drought. Wide rings mean the tree was growing rapidly; fire scars indicate when ground fires swept through the forest (Figure 2.21). Plants, in other words, provide clues to the conditions in which they grew.

Figure 2.21. Tree rings offer clues to the history of the forest. Redrawn from Perry 1994.

SUMMARY

Understanding ecological systems requires understanding the ways people have altered those systems, and the reasons for those alterations.

- Ecosystems are shaped by people as well as by ecological processes. Human events, as well as natural processes, help shape an ecosystem's history and its current patterns and processes.
- Ecology is a set of questions people ask about the complexity of life on Earth, not just a set of facts. The ways that scientists formulate hypotheses reflect their cultures as well as, the natural world that they are studying. Ecological ideas emerge in cultural contexts, and understanding the history of those ideas enables better understanding of the natural world.
- Scientific ideas affect how people transform ecological systems. Wildlife biologists, foresters, and other applied ecologists use ecological theories to manage nature. Modern debates about environmental degradation can only be resolved by understanding the complex cultural and scientific factors that have created those problems.

BOX 2.2. RECONSTRUCTING PAST LANDSCAPES WITH POLLEN RECORDS

Pollen history has proven to be a useful tool for reconstructing some past landscapes. In the spring, some trees, grasses, and herbs produce massive quantities of wind-blown pollen, as any allergy sufferer knows. Most of this pollen is decomposed by soil microbes or else destroyed by oxidation. Some, however, is captured in waterlogged lake sediments, peat deposits, and acidic soils in bogs. Each year, sediments at the bottom of lakes accumulate, and those sediments preserve the pollen deposited during the year that the sed-

iment accumulated. Over thousands of years, these pollen samples fossilize; eventually the mud at the bottom of an old lake will contain the history of the region since the lake's formation, with the oldest material at the bottom and the youngest at the top. Researchers drill into the mud, extracting a vertical core of mud and fossilized pollen. Back in the lab, scientists called paleoecologists—ecologists of past eras—identify and count the pollen at different layers, documenting a history of vegetation change.

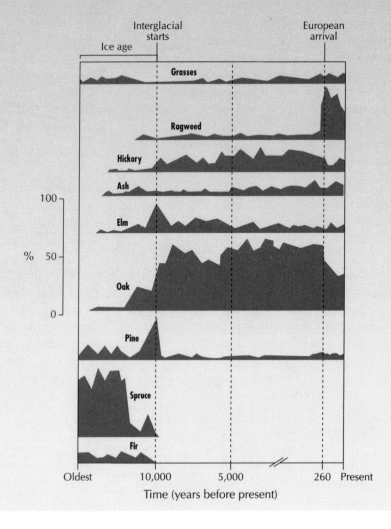

Thousands of pollen diagrams have been constructed from the United States; one from Silver Lake, Ohio, is shown on page 50 (redrawn from Ogden 1966). About 14,000 years ago, the lake formed after the retreat of the last glaciation and the pollen record started. Reading from the bottom of the diagram, at 14,000 years ago, the pollen from fir, pine, spruce and grass are common, indicating that a mixed coniferous woodlands existed in central Ohio at the end of the last ice age. The combination of these species suggests that the climate was cooler and drier than now. About 12,500 years ago, oak pollen begins to increase, suggesting a warming climate. Spruce and fir died out at the site, unable to compete in the altered climatic conditions. A cold period at 9800 years ago shows a brief peak in pine, and then it rapidly declines as the climate warms, and elm and hickory increase. At about 260 years before present we see a sharp decrease in forest species and an increase in ragweed, a now-common farmland weed. These changes correlate with the clearing of the forest for agriculture, tracking the invasion of European settlers and the tilling of the land.

Other material in lake sediments can also provide valuable clues to the past. Charcoal from ancient fires preserved within the mud can provide information about fire history and climate change. For example, researchers have traced increases in erosion (diagrammed below) and fires following European settlement and clearing of the land for agriculture.

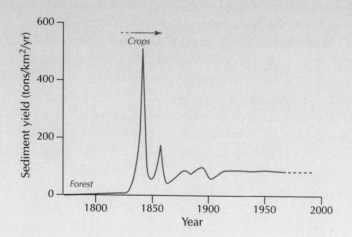

RECOMMENDED READING

PETER BOWLER. *The Norton History of the Environmental Sciences.* Norton, New York. 1993.
  A useful introduction to the history of ecology, geography, geology, and meteorology.

WILLIAM CRONON. *Changes in the Land: Indians, Colonists, and the Ecology of New England.* Hill and Wang, New York. 1983.
  A fascinating and readable classic on the environmental history of New England.

ANDREW GOUDIE. *The Human Impact on the Natural Environment.* MIT Press, Cambridge, Massachusetts,. 1994.
  An excellent overview of human transformations of the Earth.

EDWARD J. KORMONDY, ED. *Readings in Ecology.* Prentice Hall, Englewood Cliffs, New Jersey. 1965.
  A collection of classic ecology papers that shaped the development of the science.

NANCY LANGSTON. *Forest Dreams, Forest Nightmares.* University of Washington Press, Seattle. 1995.
  A study of the ways forests, ecologists, and human cultures have changed over the past century, leading to current environmental crises.

LESLIE A. REAL AND JAMES H. BROWN. *Foundations of Ecology.* Chicago University Press. Chicago, Illinois. 1991.
  This is a collection of classic discussions on ecology.

DONALD WORSTER. *Nature's Economy: A History of Ecological Ideas.* 2nd Edition. Cambridge University Press, New York. 1994.
  A very readable and provocative history of ecology and environmentalism.

Chapter 3

LANDSCAPE ECOLOGY
Living in a Mosaic

Monica G. Turner

A rural landscape which omits the city and an urban landscape which omits the country are radically incomplete portraits of their shared world.
—William Cronon

Nature keeps things controlled, but rarely keeps anything constant.
—Paul Schullery

WHAT IS LANDSCAPE ECOLOGY?

When you look across a landscape, have you ever wondered how the patterns you observe developed? Or how different animals might perceive that same landscape? Or whether the water quality in a nearby river or lake is affected by the patterns on the surrounding lands? Landscape ecologists study the causes and ecological consequences of *spatial patterns* in the environment, often over very large areas.

Even in places where the landscape may appear relatively undisturbed, chances are that the distribution and arrangements of **land cover**— the vegetation or habitat types— have changed greatly through time. Consider two examples. First, land-cover changes have been particularly profound since Europeans settled North America three centuries ago. Although Native Americans established settlements, practiced agriculture, hunted, and used fire to induce vegetation changes, European settlement led to much more dramatic and rapid landscape changes. Once-continuous forests were cleared for fuel, wood products, and crop cultivation during settlement of the eastern and central United States, initiating a widespread loss of forest cover that lasted through the early 1900s. Figure 3.1 illustrates changes in the abundance and arrangement of **forest patches**—small forested areas that are separated by some distance from other forested areas—in Petersham Township, Massachusetts (Foster 1992), and Cadiz Township, Wisconsin (Curtis 1956).

New England experienced the earliest deforestation by the European colonists, but subsequent abandonment of croplands beginning in the late 1800s resulted in widespread reforestation of this region (Figure 3.1A). Deforestation in the Midwest began somewhat later as settlers moved westward, but the suitability of these areas for agriculture kept the area of forest reduced (Figure 3.1B). These changes in the land have had tremendous implications for the distribution of resident plants and animals, the production and storage of biomass, and the cycling of essential chemical elements. For example, in Petersham Township, Massachusetts, regional forest communities in the 1700s were distributed along slopes and valleys much as they are today (Foster 1992). However, certain trees such as birch *(Betula populifolia)*, red maple *(Acer rubrum)*, and oaks *(Quercus* spp.) that are found in young forests or that can resprout from their stumps have increased in abundance during the past two centuries. At the same time, there has been a decline in long-lived, shade-tolerant species such as eastern hemlock *(Tsuga*

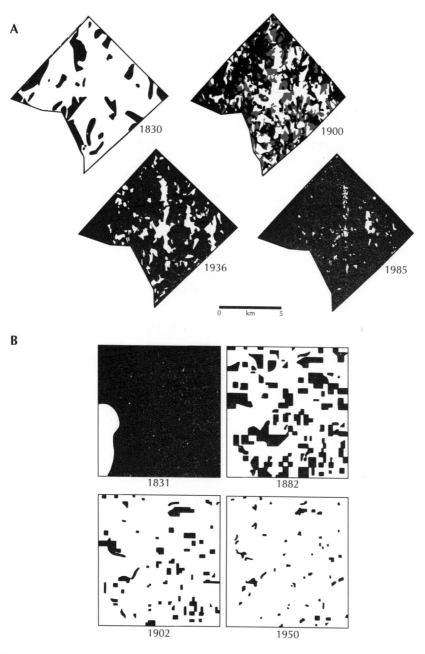

Figure 3.1. Changes in the amount and spatial arrangement of forested areas. Black areas indicate forest, white and colored areas are non-forest vegetation types or land covers. **A.** Petersham Township, Massachusetts. **B.** Cadiz Township, Wisconsin. Redrawn from from Curtis 1956 and Foster 1992.

canadensis) and sugar maple *(Acer saccharum).* Although the reforestation that occurred through most of the eastern United States suggests these forested landscapes are very **resilient**—that is, they are able to rapidly recover from past disturbance—substantial changes have nonetheless occurred in the assemblages of plants and animals. Understanding how landscapes change is an important objective of landscape ecology, and human activities are key pieces of this puzzle.

As a second example of how land cover changes through time, consider Yellowstone National Park, Wyoming, established in 1872 as the world's first national park, and an area in which natural processes rather than patterns of human settlement dominate. In 1988, wildfires burned through nearly 45% of the park, igniting tremendous public controversy about the appropriate role of natural disturbances on public lands. Rather than leaving behind a vast area of desolation, these fires instead created a mosaic in which islands of unburned or lightly burned forest were embedded within areas of severe fire (Plate 3.1) (Turner et al. 1994). Landscape ecologists were eager to explore the following questions:

- What caused the fires to burn in such a patchy manner?
- What are the effects of the post-fire mosaic on Yellowstone's plant and animal communities?
- Will the re-establishing vegetation differ in small and large burned areas?
- Will the re-establishing vegetation differ near the edge of the burned area and out in the middle of the burn?

Landscape ecologists continue to learn about the causes and consequences of natural disturbances by studying events like the Yellowstone fires (Box 1). Fires and other natural disturbances such as hurricanes, windstorms, or pest outbreaks, create patchiness in the landscape and provide excellent "natural experiments" to study the development of spatial pattern and its effects on many ecological processes.

Definition

Landscape ecology examines the interaction between **spatial pattern** (referring to the amount and configuration of something within an area) and ecological processes—that is, the causes and consequences of **spatial heterogeneity** across a range of scales. Spatial heterogeneity refers to the differences or dissimilarities observed across an area of land or water. Even in lakes and oceans, spatial pattern can be observed in nutrients and abiotic conditions, and in plants and animals. Note that landscape ecologists (and other ecologists as well) also examine differences or dissimilarities through time, which is referred to as **temporal heterogeneity**. The opposite of heterogeneity is **homogeneity**, which indicates sameness or similarity across an area. Various descriptions of landscape ecology reflect its emphasis on spatial patterning:

- Landscape ecology emphasizes broad spatial scales, large areas or regions, and the ecological effects of the spatial patterning of ecosystems (Turner 1989).
- Landscape ecology deals with the effects of the spatial configuration of mosaics on a wide variety of ecological phenomena (Wiens et al. 1993). Landscape ecologists often use the term "**mosaic**" to refer to the arrangement of habitats or vegetation because, when mapped or viewed from above, landscapes resemble the patterns in mosaic art created by small pieces of variously colored materials. Landscape ecology is the study of the reciprocal effects (that is, the effects act in both directions) of spatial pattern on ecological processes.
- Landscape ecology promotes the development of models and theories of spatial relationships, the collection of new types of data on spatial pattern and dynamics, and the examination of spatial scales rarely addressed elsewhere in ecology (Pickett and Cadenasso 1995).

Landscape ecologists often talk about three important aspects of the landscape: structure, function and change (Forman 1995). **Structure** refers to the spatial patterns—what is present and how it is arranged. **Function** characterizes the interactions among the spatial elements of the landscape—for example, movements of organisms, materials, or energy. **Change** recognizes that both the structure and function of landscapes are not fixed through time, but rather are very dynamic. Over very long time periods (such as hundreds to thousands of years), changes in geomorphology (the form and composition of the Earth's surface), climate and species' ranges lead to alterations in landscape structure and function. Our landscapes were very different when the glaciers last retreated than they are at present. Over shorter time periods such as years to centuries, disturbances and **succession**—change in the composition of vegetation through time—alter both structure and function of landscapes.

Landscape ecology has a very long history in Europe, where it developed out of traditions in geography and land-use planning. Indeed, landscape ecology provides the foundation for most land planning in northern and central Europe. In North America, where applications of landscape ecology are rapidly expanding, the discipline emerged largely from ecosystem ecology and developed with a strong focus on basic research. In part, these different traditions reflect differences in the landscapes and cultural histories of both continents. Europe has experienced greater population densities over a longer period of time than North America, and little of the European landscape remains as wilderness. The locations of many towns and farms have remained unchanged for centuries, and even the less-developed areas (e.g., forests) may be intensively managed. In contrast, North America contains larger expanses of land that are less influenced by human population pressures; this fostered a greater emphasis on understanding the dynamics of so-called "natural" landscapes. However, the demand for landscape-level land management strategies is now strongly influencing the science of landscape ecology in North America.

BOX 3.1. CAUSES AND CONSEQUENCES OF THE 1988 YELLOWSTONE FIRES

Stand-replacing fires, which kill mature trees and initiate succession, are an important component of the disturbance regime in many coniferous forest landscapes around the world. These fires tend to be very large and relatively infrequent, having *return intervals* ranging between 80 and 400 years. Although many fires start, often by lighting storms in summer, most burn out on their own. It is only during severe burning conditions—when there has been little precipitation for an extended time and the weather is very dry and windy—that the fires really take off; thus a very few fire events account for most of the area burned. When the big fires do occur, they generate a mosaic of different stand ages across the landscape. The spatial and temporal dynamics of these large, infrequent fires are in fact a dominant force structuring boreal and other coniferous forests. Yellowstone National Park offers a good case study of a fire-dominated landscape.

Yellowstone Park encompasses 9,000 km² in the northwest corner of Wyoming, and is primarily a high, forested plateau. Most of the park is covered with coniferous forests dominated by lodgepole pine *(Pinus contorta* var. *latifolia)*. Fire has long been an important component of this landscape, and has profoundly influenced the fauna, flora, and ecological processes of the Yellowstone area and most of the Rocky Mountains. The fire history of Yellowstone was reconstructed by obtaining narrow cores from living trees; dates of fires are revealed by counting the rings, marking which rings show burn scars. These data demonstrated that extensive fires occurred in the early 1700s, and that the 1970s landscape was a nonequilibrium mosaic of forest stands in differing successional stages (Romme 1982). Because large fires occurred periodically during the past few centuries at intervals of between 100 and 300 years, the area covered by young, middle-aged, and old forest stands was always changing (Romme and Despain 1989).

With the initiation of a natural fire program in Yellowstone, fires that began naturally were permitted to burn as long as human life and property were not in danger. Between 1972 and

1987, 235 lightning-caused fires burned without interference. As is typical in this kind of disturbance regime, most of these fires went out by themselves before burning more than a hectare; the largest fire burned about 3,100 ha in 1981. However, as a consequence of unusually prolonged drought and very strong winds, fires in 1988 affected more than 250,000 ha in Yellowstone and surrounding areas. The enormous extent and severity of the 1988 fires surprised many managers and researchers.

The 1988 fires created a mosaic of burn severities (i.e., effects on the vegetation) across the landscape. Even very large fires do not burn everything in their paths because of variations in wind, topography, vegetation, and time of burning. Some areas of Yellowstone experienced stand-replacing *crown fires*—burning only the upper leaves and branches that make up tree crowns—other areas experienced stand-replacing, severe-surface burns. Still other areas received light-surface burns in which trees were scorched but not killed. Understanding the effect of fire on landscape heterogeneity is important because the kinds, amounts, and spatial distribution of burned and unburned areas may influence the re-establishment of plant species on burned sites. Surprisingly, even though the fires were quite large, the majority of severely burned areas were within close proximity (50 to 200 meters) to unburned or lightly burned areas, suggesting that few burned sites are very far from potential sources of seeds for plant reestablishment.

How has the fire-created mosaic influenced the developing forest community? Burn severity and the size of the burned patch had significant effects on initial post-fire succession (Turner et al. 1997). Severely burned areas had higher cover and density of lodgepole pine seedlings and shrubs, a greater abundance of opportunistic species (species absent before the fires or present at very low abundance, but which were able to quickly colonize the areas opened up by the fires), and fewer species of vascular plants than less severely burned areas. Surprisingly, dispersal into the burned areas from the surrounding unburned forest has not been an important mechanism for reestablishment of forest species. Many plants survived the fires because their roots or rhizomes were not affected, even though their stems and leaves were burned. These survivors soon resprouted and flowered, and seedlings produced from these surviving plants filled in much of the burned area.

Patterns of initial post-fire succession were surprisingly more variable in space and time than current theory would have suggested (Turner et al. 1997). Although succession across much of Yellowstone appears to be moving toward plant communities similar to those that burned in 1988, there are some interesting differences in plant reestablishment. Regenerating forests show tremendous spatial variation in succession across the landscape. Lodgepole pine stands range from sparse to dense "doghair" stands, due in part to spatial variation in *serotiny* in lodgepole pine. **Serotiny** is an adaptation in pine cones whereby the cones remain closed and retain their seeds until heated. Fires open the serotinous cones to release their seeds, thus providing a tidy means of getting trees reestablished following a fire. In Yellowstone, however, some stands of lodgepole pine contain almost all serotinous trees, whereas other stands contain trees with almost no serotinous trees. Instead, these non-serotinous trees produce "open cones" that open and release seeds as they ripen. In burned areas where serotinous trees were abundant, dense thickets of pine seedlings now grow. On the other hand, pine seedlings are scarce in large burned areas that were occupied by old (>400 years) forests with low pre-fire serotiny (Turner et al. 1998), and forest reestablishment is questionable. Thus, the 1988 fires may have initiated multiple successional pathways related to differences in fire severity, fire size, and pre-fire community structure. These alternative pathways include development of non-forest communities in some areas previously characterized by coniferous forest.

Another surprise following the 1988 fires was the unanticipated recruitment of seedling aspen (*Populus tremuloides*) in areas of Yellowstone previously dominated by lodgepole pine. Aspen occupied only ~1% of the park prior to the 1988 fires, occurring almost exclusively on the low-elevation, sagebrush-grasslands in northern Yellowstone, and tree-sized aspen have not regenerated since park establishment in 1872. Many ecologists, including this author, assumed that aspen had not reproduced by seed in the northern Rocky Mountains since the Pleistocene—the last Ice Age, about 10,000 years ago! Indeed, aspen plants can persist for centuries because they develop clonal structure and can send up new sprouts—trees—from their roots as the older stems die. However, abundant aspen seedlings were observed in 1989 across the expansive burned areas in Yellowstone, and these seedlings, though browsed by elk, were still persisting eight years after the fire. The flush of post-fire aspen seedling establishment may enhance the long-term ability of this species to persist in the park. Because the trees reproduced sexually, generating many new genetic combinations in the offspring, the seedling aspen population now contains much greater genetic diversity than the mature clonal stands. This increased genetic diversity in the seedling populations may enhance the ability of aspen to withstand current climate conditions, levels of interspecific competition, and ungulate browsing.

Future forest conditions in the regions with very low tree seedling densities are least predictable across the Yellowstone landscape. Ecologists continue to study these plant communities to determine whether succession is simply proceeding at a much slower rate toward a coniferous forest, or whether a non-forest community—a meadow—will persist. Little is known about the long-term implications of a fire-generated landscape mosaic for indicators of ecosystem function at landscape scales. The 1988 Yellowstone fires will likely lead to new insights into the dynamics of large landscapes for many years to come.

Figure 3.2. Spatial heterogeneity at different scales: **A,** across a large landscape illustrating how humans often perceive landscapes, and **B,** over a fine scale such as a few square meters, as a small organism might perceive its landscape. From photos by John Wiens.

Landscape patterns are observed and studied across the large areas that humans typically call landscapes (Figure 3.2,A), but landscapes can also be rather small in size (Figure 3.2, B). The "correct" scale depends on what is being studied. For example, the landscape mosaic important for a beetle or a crab may consist of bare and vegetated patches within a relatively small area. Understanding the dynamics of landscape mosaics at many spatial scales, how these mosaics influence ecological processes, and what landscapes may look like in the future are within the purview of landscape ecology.

Causes of Landscape Patterns

What causes landscape pattern? All landscapes have a history. Today's landscapes have resulted from many factors, including species interactions that can generate spatial patterning even under homogeneous conditions; variability in abiotic conditions such as climate, topography and soils; past and present patterns of human settlement and land use; dynamics of natural disturbance and succession; and through "engineering" of the landscape by dominant organisms such as beaver. **Climate**, the composite weather across a region, often controls what plant and animal species can occur within a given landscape and where they occur spatially. Climatic conditions are modified by topography, or landform (which ranges from nearly flat plains to rolling, irregular plains, to hills, to low mountains, finally to high mountains). For example, temperatures decrease as you move up a mountain, and south-facing slopes are warmer and drier than north-facing slopes at the same latitude in the northern hemisphere. Climate, topography, soils, and the biogeographic ranges of plants and animals all contribute to the landscape mosaic.

Earth's climate has changed dramatically during the history of the planet, and these changes have contributed to the composition of present-day landscapes. Groupings of plants and animals seen today are relatively recent assemblages. **Paleoecology** is the study of individuals, populations, and communities of the plants and animals that lived in the past, and their interactions with and responses to changing environments. Paleoecologists expand the view of ecology to consider changes over the long time scale of the Earth's history; insights derived from their data help landscape ecologists to understand how spatial patterns of plants and animals developed through time. Consider some differences in **biotic communities** (the local assemblages of species) of eastern North America during the 10,000 years since the end of the last glaciation. As the glaciers retreated to the north, tree species were able to migrate from their southern refuges to the gradually warming areas to the north. However, entire groups of plants and animals did not move together as units. Sequences of fossil pollen (see description of this technique in Chapter Two) have shown that different species moved in different directions and at varying rates (Figure 3.3). White pines began from the mid-Atlantic coast, and spread to the northwest. Hemlocks also began from the mid-Atlantic region and spread north and west, but more rapidly than white pine. Oaks migrated from the southeastern United States to the north and east. Elms spread rapidly to the northeast from the southeastern

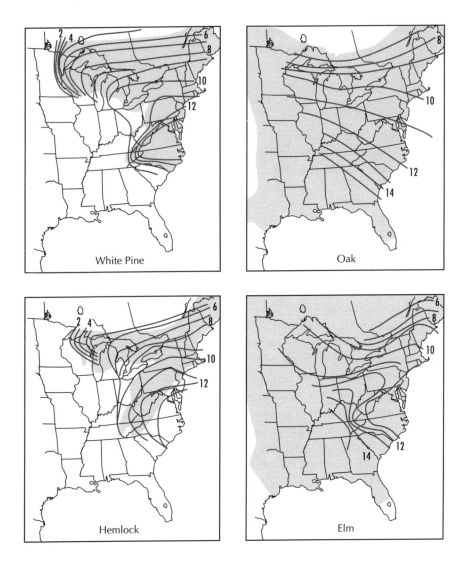

Figure 3.3. Changes in the northern and western range limits for tree species common in eastern North America, based on radiocarbon-dated pollen records from lakes. Numbered lines indicate the range limits at times in the past in thousands of years before the present. Colored areas represent the modern range of the tree taxa. Redrawn from Davis 1983.

United States, then more gradually moved to the northwest. These documented changes in species' distributions are important because they demonstrate clearly that our landscapes have been changing gradually but continually over very, very long time scales. Past changes also provide information that may help ecologists predict how landscapes might change in the future if the climate of the Earth changes. Climate change will be discussed further at the end of this chapter.

Human activities, from prehistoric times through the present, have also contributed to the landscapes we observe today. Over long time periods, humans have influenced landscapes in five major ways. Humans have:

- Changed the relative abundances of plants, especially by modifying the dominant tree species within forests;
- Both extended and truncated the distributional ranges of plant and animal species;
- Provided opportunities for weedy species to invade into disturbed areas;
- Altered the nutrient status of soils;
- Changed the pattern of the landscape mosaic by settlement patterns and land-use practices.

It is important to recognize that the terrestrial and aquatic landscapes we observe today have a substantial history that reflects changes in climate and in human activities.

Importance of Scale

Landscape ecologists often grapple with issues associated with spatial scale. Two important but different aspects of spatial scale are grain and extent (Figure 3.4). **Grain** refers to the spatial resolution of a study or data set; for example, are the data or samples obtained at a resolution of 1 m^2, 100 m^2, 1 ha, or 1 km^2? **Extent** refers to the size of the study area being considered; for example, is the landscape small or large? No fixed scale, either grain or extent, unambiguously identifies landscape ecology. This is because although landscape ecologists often focus on landscapes as humans perceive them (and as discussed in Chapter One), a landscape ecological approach can also be applied to a very small area at fine resolution if warranted by the organisms or process.

Scale is sometimes considered to be problematic because the answers obtained for a question are influenced by the scale of the study. Comparing data collected at different scales is like comparing apples and oranges, although rules explaining how to extrapolate across scales (i.e., take information obtained at one scale and apply it at another) can sometimes be developed. One example of the practical importance of scale relates to the use of data obtained from satellites. The earliest Landsat Multispectral Scanner satellites launched in the 1970s collected data at a grain size of approximately 90 m $\times$ 90 m. The more recent Landsat Thematic Mapper satellites have a grain size of approximately 30 m $\times$ 30 m, and the French SPOT satellite has even finer resolution, 10 m $\times$

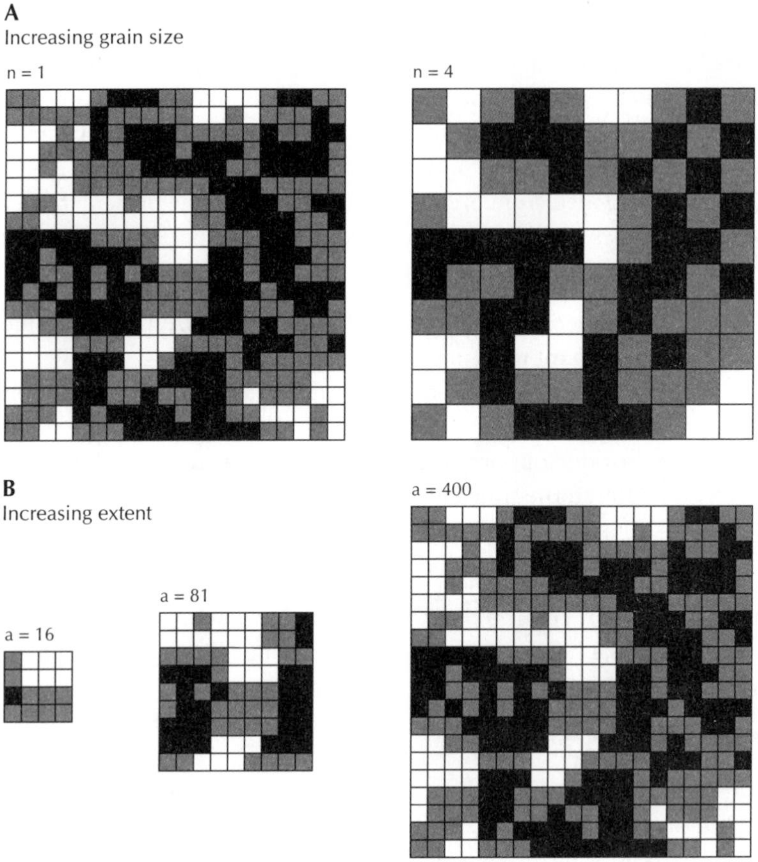

A
Increasing grain size

Figure 3.4. Schematic illustration of changes two components of spatial scale. **A.** The grain size of the data is increased, leading to more coarse resolution. **B.** The extent, or area, of the landscape is increased. Redrawn from Turner et al. 1989.

B
Increasing extent

10 m. Smaller grain sizes allow more resolution of features on the landscape; with larger grain sizes, smaller elements are easily missed, or at best underestimated. The newer satellites also have greater spectral resolution, meaning that they measure more wavelengths of light, and this also contributes to their improved accuracies. Thus, to describe changes in a landscape during the past 20 to 30 years using satellite data, we must account for the changes in grain size and spectral resolution between the satellites and make sure that our conclusions reflect actual changes in the landscape—not just differences in scale.

How is the "right" scale for studying landscape patterns or processes to be determined? Patterns and processes are observed at characteristic scales in both space and time, and these scales are often positively correlated. For example, consider the temporal and spatial scales of variation in the environment, which contributes to landscape patterns, and some important processes in forested landscapes (Figure 3.5). To understand variation in fine-scale topography, one would study an area of 10–100 m^2

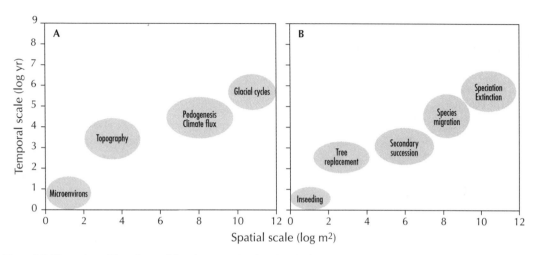

Figure 3.5. Illustration of how the spatial and temporal scales of ecological phenomena are often correlated: **A.** Variability in the environment. **B.** Processes that occur in forests. Redrawn from Urban et al. 1987.

over perhaps 1–15 years (Figure 3.5, A). However, to understand the development of soils (**pedogenesis**) or variation in regional climate, one would study a larger area over a longer time period, perhaps thousands of years. Similarly, to examine the production of seed and establishment of new seedlings in a forest, one would study a small area for one to a few years (Figure 3.5, B). To examine the migration of different tree species across a landscape, however, both the temporal and spatial scale would need to be expanded. In general, ecological events have a characteristic natural frequency and a corresponding spatial scale, and an ecological study of the landscape must scaled accordingly.

How are Patterns Measured on Landscapes?

Techniques used in landscape ecology are discussed later in this chapter, but it is important to have a sense of the kinds of data commonly used in landscape studies. Many analyses of landscape pattern are conducted on land use and land cover data that have been digitized and stored within a **geographic information system (GIS)** (Figure 3.6). A GIS refers to a powerful set of computer tools for collecting, storing, retrieving, transforming, and displaying spatial data from the real world for a particular set of purposes. Geographical data describe phenomena in several terms: their location with respect to a known coordinate system (latitude and longitude is a familiar one, but there are many others); their attributes that are unrelated to their position (such as soil type, vegetation cover, or presence of a species of interest); and their spatial interrelations with each other. Developing a familiarity with the use of GIS is becoming an important part of the training for ecologists studying landscape patterns and processes.

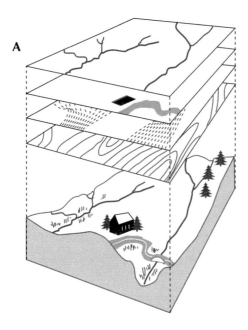

A

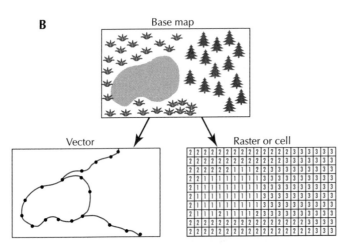

B

Base map

Vector

Raster or cell

Figure 3.6. A. In a Graphic Information System, different types or themes of data (e.g., land cover, topography, mean annual temperature, population density) that are aligned on a common coordinate system are stored and manipulated. **B.** Illustration of raster and vector formats, two common ways in which data are represented in a geographic information system. Redrawn from Coulson et al. 1991 and Burrough 1982.

Three main types of data are often used for landscape analyses. **Aerial photography** (photographs taken of the Earth's surface from airplanes) is an important data source for landscape studies, particularly for detecting changes in a landscape during the twentieth century. Aerial photos are generally available back through the 1930s. *Digital remote sensing* (images of the Earth's surface obtained from satellites or from airborne sensors) is now widely used and is accessible to many researchers. The Landsat and Spot satellites have provided frequent and spatially extensive coverage worldwide and are a very useful source of digital data. *Airborne imaging scanners* may also be used to provide fine resolution data for a particular locale. *Published data and censuses* provide a valuable source of landscape data, particularly for temporal comparisons that extend back beyond the record of aerial photography. For example, the United States General Land Office Survey data have been used extensively to describe vegetation in the United States prior to European settlement. In addition to these three sources, field-mapped data may be used for smaller landscapes in which the investigator might map the spatial patterns of particular vegetation classes or landscape elements of interest in a relatively small area. Field mapping is not generally feasible for studies that cover a large area.

Most researchers store their landscape data in a GIS for ease of manipulation and display. Many computer programs that have been written for landscape analyses were developed for use with raster, or grid cell, data, but vector-based versions are sometimes available. In **raster format**, a landscape is usually divided into a fixed grid of square or hexagonal cells of equal size. Irregularly shaped landscapes can be represented within a rectangular perimeter larger than the landscape itself. In **vector format**, lines are defined by sets of coordinates that define the boundaries of polygons (Figure 3.6, B). The polygons are of variable size and shape, but a minimum mapping unit (i.e., level of resolution) is specified. Raster data are more commonly used in landscape analyses largely because the computer programming of the analyses is somewhat easier, and most satellite imagery is in raster format.

QUESTIONS ASKED BY LANDSCAPE ECOLOGISTS
How Does the Spatial Arrangement of Habitat Influence the Presence and Abundance of Species?

Organisms move, eat, reproduce, live and die in spatially heterogeneous landscapes. The question of how they are influenced by the spatial patterning of their habitat has long stimulated active research in landscape ecology and clearly has important implications for conservation. Indeed, the literature is filled with studies of animal responses to the arrangement and distribution of habitat patches on land and in water. Population ecologists also have found it increasingly necessary to take a landscape perspective when studying or interpreting the dynamics of single populations (see Chapter Seven, Population Ecology). This landscape perspective includes consideration of patch size, habitat arrangement, the identification of suitable habitat, and connectivity.

Effects of patch size

One well-studied aspect of population responses to habitat arrangement is the effect of patch size. A **patch** is defined as a relatively homogenous nonlinear area that differs from its surroundings (Forman 1995). In general, larger patches of habitat contain more species and often a greater number of individuals than smaller patches of the same habitat. This occurs for several reasons. First, the larger the habitat patch, the more local environmental variability is contained within it; for example, differences in microclimate, structural variation in the plants, and diversity of topographic positions. This variability provides more opportunities for organisms with different requirements and tolerances to find suitable sites within the patch. Moreover, the edges and interiors of patches may have quite different conditions that favor some species but not others, and the relative abundance of edge versus interior habitats varies with patch size.

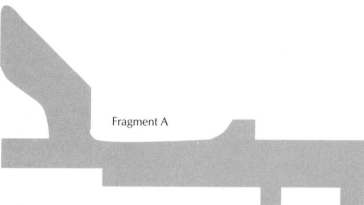

Fragment A

Fragment B

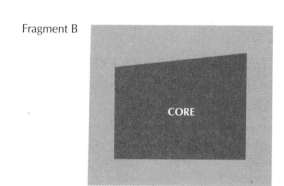

Figure 3.7. Comparison of the effect of patch size and shape on the amount of edge and interior habitat in two forest fragments. Fragment A is 39 ha in size but is entirely edge habitat. Fragment B is 47 ha in size and contains 20 ha of interior habitat. When the breeding success of 16 bird species sensitive to fragmentation was studied in these two forest patches, six bred in Fragment B and none bred in Fragment A. Redrawn from Temple 1986.

If you were to walk all around a small and a large patch, measuring the perimeter, you would immediately observe that the larger patch had a longer perimeter, or edge. However, if you computed the ratio of the length of perimeter to the area of the patch, you would find that the smaller patch has a greater perimeter to area ratio than the larger patch. This means that the smaller patch will have a greater proportion of *edge habitat* and the larger patch will have a greater proportion of *interior habitat.* Consider these differences for forested patches. At the edge of a forested patch, there is generally more light, a warmer, drier microclimate, and greater access to organisms that frequent open habitats. In contrast, the interior of the patch tends to be more shady, cooler and more moist, and often off-limits to the organisms of open habitat. When a patch gets sufficiently small or elongated in shape, all interior habitat is lost (Figure 3.7), leading to a loss of interior species and dominance by edge species. Large patches typically include both edge and interior species.

Effects of habitat arrangement

A series of studies of small mammals in an agricultural landscape mosaic in Canada, conducted by Gray Merriam and his colleagues, nicely illustrates important effects of habitat arrangements (how patches are arranged relative to each other) on small mammals as well as the integration of field studies and modeling. Their study area is a landscape containing crop fields along with scattered woodlands. Sometimes these woodlands are isolated from one another by being completely surrounded by crop fields, and sometimes the woodlands are connected by fencerows—narrow **corridors** of trees and shrubs that grow up along the borders of the crop fields. Early studies of this landscape demonstrated that chipmunks and white-footed mice traveled frequently along the fencerows but seldom moved between the woodlands and the field or across open fields (Wegner and Merriam 1979). Birds also seldom flew directly over the open fields between the woods and preferred to fly along the fencerows. Similar results have been observed in many locations in Europe, where fencerows between woodlands have developed over many centuries.

Merriam and colleagues next explored what happened to the small mammal populations if they were **extirpated**—that is, became locally extinct—within the woodlands (Henderson et al. 1985). Local extinctions are a frequent natural occurrence each year because relatively small numbers of animals survive the winter. When chipmunks were live-trapped and removed from woodland patches, the rate of recolonization of these patches depended upon their connectivity to other woods. Recolonization occurred more rapidly in patches that were connected to other wooded areas by fencerows than in the isolated woodlands. This study also suggested that an area of at least 4 km^2 and containing at least five woodlands and interconnecting fencerows would be required for the populations in this mosaic to persist through the years.

A simulation modeling study was used to further explore the effects of alternative patterns of connectivity among patches (Lefkovitch and Fahrig 1985). Simulation modeling involves the use of computers and mathematical equations to imitate a process of interest. Thirty-four different arrangements of connections among five patches (Figure 3.8) were simulated to determine which spatial characteristics of groups of habitat patches were important predictors of the survival of a resident animal population, like the chipmunks or white-footed mice. Results indicated that populations in isolated patches died out much earlier and had lower population sizes than did the populations in the connected patches. Furthermore, woodlots that formed part of a square or pentagonal arrangement had higher population sizes and were more likely to persist than those that formed parts of a line or triangle. The line and triangle arrangements offered fewer opportunities of exchanges of organisms than did the others. These results were also supported by field data in which four interconnected woodlands had higher mean population sizes of white-footed mice than two isolated woodlands of similar size. New organisms could disperse into the connected woodlands and augment or replenish the local populations, whereas the isolated woodland received fewer new mice.

A key insight that resulted from these and many other studies of interconnected populations is that the birth and death rates of populations may vary significantly among different patches across the landscape. Within a landscape mosaic, each patch may have

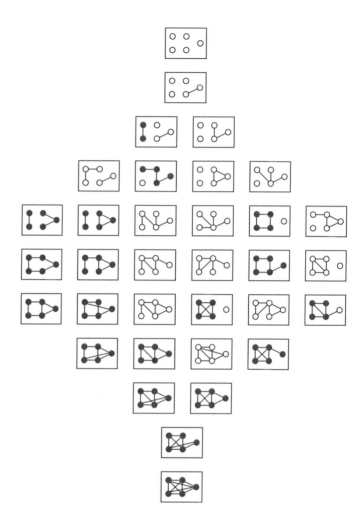

Figure 3.8. Illustration of the unique alternative arrangements of habitat patches used in a simulation model (Lefkovitch and Fahrig 1985). The circles are patches and the lines are connections between patches. Populations in the patches indicated by white circles died out during the 100–year simulation; populations in the patches indicated by colored circles persisted.

its own **demographics**—the rates of birth, death, immigration and emigration. These demographic differences can determine whether a given patch is a source or sink for the population (Pulliam 1988). In **source patches,** the local reproductive rate exceeds the local mortality rate, leading to a surplus of individuals that are potentially available to colonize other patches of suitable habitat across the landscape. These patches provide a "source" of new colonists. In **sink patches,** local mortality exceeds local reproductive success, and the population in such patches will be extirpated unless it is regularly colonized by new individuals. Individuals arriving in these patches land in a "sink" that effectively drains the source population. However, colonization from source to sink patches can maintain the overall population on the landscape at an apparent **equilibrium**, or constant, size. An important characteristic of populations structured in this source-sink manner is that the overall population can decline dramatically if key source patches are eliminated. The field of conservation biology, which is related to this aspect of landscape ecology, often has a goal of identifying particular subpopulations, habitat patches, and links among patches that are critical to allowing the entire population of organisms to persist (Meffe and Carroll 1994).

Identifying suitable habitat

In contrast to the relative ease of identifying habitat for small mammals in an agricultural landscape, it is often difficult to determine the distribution of **suitable habitat**—areas having the conditions required for a given species to meet its needs for resources, shelter and reproduction—across a landscape for different species. What constitutes suitable habitat for a particular species depends on a variety of factors, including some characteristics that might be surprising. Consider habitat perceived as suitable or favorable by eastern timber wolves in the upper midwestern United States. Although driven nearly to extinction during the early part of the century, wolves have been gradually expanding their range during the past 15 years, moving eastward from Minnesota into northern Wisconsin and the Upper Peninsula of Michigan. To determine where the wolves were going and what habitats they were using, individual wolves were fitted with radio collars that transmit a signal to determine their locations. Analyses of the data from these radiocollared wolves revealed that suitable habitat was a function not only of vegetation type and deer density (deer are commonly preyed upon by the wolves) but also of land ownership class, road density, and human population density across the landscape (Mladenoff et al. 1995). Although wolves were moving throughout the landscape and crossing unsuitable areas, successful establishment of a wolf pack was restricted to the high-quality habitat.

Once the important relationships that define suitable habitat for a target plant or animal are known, one very useful step is to map the distribution of suitable habitat across a large region. In the wolf study mentioned above, a statistical model developed to describe wolf habitat was combined with spatial data and used to predict the likelihood of suitable habitat occurring across the upper Midwest (Plate 3.2). This allowed ecologists and land managers to look at the landscape from the viewpoint of a wolf and to identify the areas that would most likely be important for the expanding wolf population.

Habitat connectivity

Once suitable habitat for a species of interest is characterized, determining whether the habitat is connected spatially or not is often of interest. Both plants and animals need suitable areas in which movement and dispersal can occur to maintain the populations. Plants, of course, usually move at the seed stage rather than as mature organisms, but both plants and animals have varying degrees of mobility. Therefore, the connectivity of suitable habitat can constrain the spatial distribution of the species by making some areas accessible and others inaccessible. Habitat connectivity is a threshold dynamic that depends on two factors: (1) the abundance and spatial arrangement of the suitable habitat, and (2) the movement or dispersal capabilities of the organism. A **threshold** refers to a point at which the habitat suddenly becomes either connected or disconnected, depending on point of view. Let's consider separately each of the two factors—habitat arrangement and organism mobility—and imagine a landscape on which we've overlain a grid, like that on graph paper, to identify different locations across the landscape.

First, think about a set of landscapes in which suitable habitat for an organism is distributed randomly, but may occupy different fractions of the landscape (e.g., 10%, 40%, 90%, etc.). Assume the organism can only move short distances, such as only to its surrounding four neighboring grid cells (up, down, left and right in our grid). When the suitable habitat occupies less than 60% of the landscape, the habitat is distributed in a number of relatively small disconnected patches; however, when suitable habitat occupies more than 60% of the landscape, the habitat is distributed in a few, large, well-connected patches (Plate 3.3, A & B). Under these conditions, the threshold of connectivity is about 60%. If the habitat is not distributed at random but has a greater degree of "clumping," as observed in most landscapes, then the threshold of connectivity usually moves down. In other words, the habitat would be connected at some lower fraction of occupancy on the landscape.

The values of these thresholds are obtained by using computer simulations to generate many random maps on which the spatial patterns of habitat patches are analyzed. The notion of critical thresholds of connectivity in random or structured maps developed from **percolation theory**, a branch of physical science (Stauffer 1985). Historically, the theory goes back to the 1940s, when it was used to describe how small, branching molecules form larger and larger macromolecules if more and more chemical bonds are formed between the original molecules. This process may lead to the formation of a network of chemical bonds spanning the whole system. A large part of percolation theory has to do with the changes in pattern that occur near the threshold at which the chemical bonds are connected. Percolation theory was adapted for studies in landscape ecology to evaluate changes in the connectivity of habitats rather than molecules.

Second, consider the different movement abilities of an organism on our same set of landscapes where the suitable habitat is distributed at random. If the organism is a bit better at getting around—say it can move to its eight nearest neighbors, i.e., the adjacent and diagonal neighbors—then the threshold of connectivity decreases to about 40%. If the organism can "jump" across a single cell of unsuitable habitat and essentially ignore that interruption, the threshold of connectivity decreases again, and a landscape containing only 25% suitable habitat could be traversed by the organism. Changes in the clumping of the habitat will also interact with the movement capability of the organism to determine the threshold of connectivity. Thus, connectivity is greatly influenced by the behavior of the organism (see Chapter Six), which is difficult to quantify across heterogeneous landscapes.

The important idea about habitat connectivity is that habitat is either connected or disconnected, and that the change between these two states occurs at a threshold of habitat abundance. For an organism, this threshold means the difference between being able to move about the landscape to locate suitable sites for foraging, nesting, and dispersal, and being unable to do so. Thresholds have important implications for conservation because suitable habitat might be lost for a while, with no apparent negative effect on a plant or animal of interest until the threshold is passed. Negative effects may then occur suddenly as the organisms can no longer meet their needs on the fragmented landscape. The exact location of the threshold depends on the organism, the amount of habitat, and the spatial clustering of the habitat. Different species might perceive different thresholds in the same landscape. Studies to date suggest that conservation actions, such as adding habitat or protecting key locations, are most likely to have substantial effects on habitat connectivity when the suitable habitat is low to intermediate in abundance. In this range, small changes in habitat abundance are likely to cause the threshold of connectivity to be passed.

The combination of field studies, experiments, and theory development has clearly demonstrated important effects of spatial patterning on a variety of organisms. Patch size, patch arrangement, and habitat connectivity have strong influences on the abundance and persistence of populations. However, "habitat" and "patch" must be defined based on the characteristics and requirements of the organism of interest, rather than on a human's view of the habitat. Suitable habitat for an elephant and a dung beetle living in the same region would be quite different. The same landscape may be very, very different for different species, and an organism-centered view of the landscape is required to understand the response of populations to spatial patterning.

Does the Surrounding Landscape Influence Local Populations?

When we see plants and animals in an area, are they there just because their habitat patch itself is suitable, or does the surrounding area—the **landscape context**—also have an effect? The abundance of organisms at a given location or sampling point may sometimes be better explained by attributes of the surrounding landscape than by characteristics of the immediate locale. For example, Pearson (1993) studied wintering birds in powerline rights of way in the Georgia piedmont. A right of way is a corridor in which the vegetation is maintained in an open state, usually by mowing, so that shrubs and herbaceous plants dominate. The areas surrounding the right of way may be open, forested, or in cultivation. Pearson recorded the abundance of bird species and such vegetation characteristics as height, density, and species composition within each right of way, and quantified the types of habitats in the surrounding landscape based on aerial photography (Figure 3.9). He found that variability in the presence and abundance of certain wintering birds (e.g., titmice and rufous-sided towhees) was best explained by the habitats in the surrounding landscape. Other species (northern cardinals and white-throated spar-

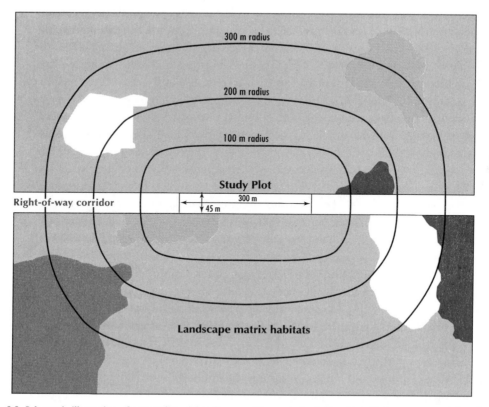

Figure 3.9. Schematic illustration of a powerline right-of-way corridor, a study plot in which bird presence and abundance was recorded, and the 100-m radius bands used to evaluate landscape context. Only three bands are shown here, but five were used on the original maps. Colors indicate differing habitats. Redrawn from Pearson 1993.

rows) responded only to the characteristics of the local habitat. Yet other species (Carolina wrens) responded both to local conditions and to the landscape context.

Winter foraging patterns of elk and bison in northern Yellowstone National Park clearly differ with spatial scale and provide another example of the importance of landscape context (Pearson et al. 1995). Elk and bison are large mobile ungulates (hoofed animals) that can select feeding areas from extensive landscapes. Winter snows cover their forage, and they must dig through the accumulated snow to find food, which is typically dried grasses and forbs. How do they select their feeding areas? Do they choose very locally, or do they pay some attention to the surrounding landscape? Feeding activities of elk and bison were monitored across approximately 7,500 hectares in Yellowstone during two winters, and related to both fine-scale (within a single hectare) and broad-scale (up to 225 ha) heterogeneity. The ungulates selected winter feeding areas based on environmental heterogeneity at broad scales (81 to 225 ha), rather than at fine scale. In other words, their cues about where to search for food appeared to be related to topography, the composition of the grassland vegetation, and whether a site had been recently burned. Feeding occurred more often on south-facing slopes where snow accumulation was less, in habitats that produced moderate to high amounts of biomass, and on sites that had burned two to three years earlier and subsequently contained more biomass than unburned sites. Once the animals were positioned within snow-covered feeding patches, however, selection of where to actually dig through the snow and feed occurred at random. This random pattern is typically expected when animals are eating a low-quality food, as is the case for the dried grasses and leaves that make up the ungulates' herbaceous winter forage.

These examples demonstrate that it is important to consider landscape context along with local site attributes when trying to explain local ecological processes. This insight has substantial implications for land management because it suggests that what happens in small local areas may be influenced considerably by the surrounding landscape. An example of landscape effects on aquatic systems is discussed below. Ecologists are conducting research to understand better when the overall landscape influences are important or may even dominate, and when the local conditions are more important.

Do Landscape Patterns Affect the Transport of Materials from Land to Water?

Landscape patterns affect not only plants and animals, but also exert tremendously important effects on interactions between terrestrial and aquatic ecosystems. Consider a **watershed** (a naturally bounded land area that is drained by a stream or river) containing croplands or pastures. Farmers often apply fertilizers high in nitrogen and phosphorus to their fields, but not all of the added nitrogen and phosphorus is taken up by the plants. When it rains, some of these nutrients are leached from the soil and transported through the watershed and into the stream by both surface and subsurface water flow. However, the length and width of areas of natural vegetation along the stream (called the **riparian zone**) can reduce the transport of these nutrients from upland agricultural areas to streams (Figure 3.10). For example, substantial quantities of particulate materials, organic nitrogen, ammonium, nitrate, and particulate phosphorus were removed in an agricultural watershed when waters flowing from a corn field passed across approximately 50 meters of riparian-zone forest (Peterjohn and Correll 1984). Here a natural element of the landscape—the riparian vegetation—reduced the amount of nutrients being transported to the stream from another landscape element, the crop field. This process of nutrient removal is ecologically important, because the excess nutrients that unintentionally end up in lakes and streams are a major cause of **cultural eutrophication**, the overfertilization of fresh water systems (see Chapter Four, Ecosystem Ecology), which has detrimental effects on the aquatic ecosystem. The presence and location of particular vegetation types can strongly affect the movements of materials across the landscape and help to regulate the quality of surface waters within the landscape.

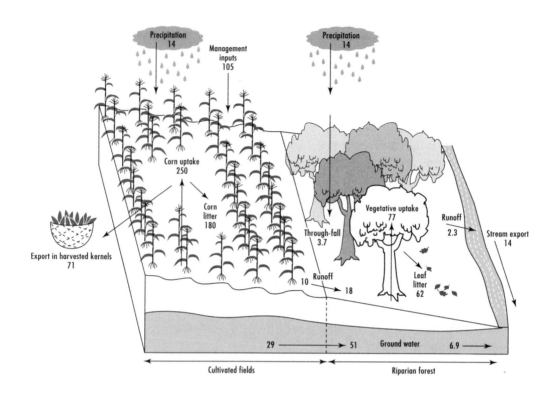

Figure 3.10. A study watershed in Maryland illustrates the flux and cycling of total nitrogen in water moving from upland agriculture, through riparian forest, and into a stream. All values are kilograms per hectare within each habitat. Redrawn from Peterjohn and Correll 1984.

Landscape patterns are also important in urban areas, as illustrated for Minneapolis-St. Paul by Detenbeck et al. (1993). Like agricultural areas, cities and suburbs are important contributors to **nonpoint source pollution**—pollution that does not come from a single source, like a pipe, but rather is delivered from widespread areas of the landscape (see Figure 4.1). Homeowners often apply as much fertilizer and pesticides per unit area to their lawns as farmers do to their crop fields! In 33 lake watersheds in the Minneapolis–St. Paul area, landscape and vegetation patterns were obtained from aerial photographs and then compared with measured lake water quality. When lakes were dominated by forested lands in the surrounding watershed, they tended to be less eutrophic and have lower levels of chloride and lead. Lakes with substantial agricultural land uses in their watersheds were more eutrophic. When wetlands remained intact in the watersheds, less lead was present in the lake water. Percent urban land use has also been found to be positively correlated with the export of phosphorus (a nutrient that limits algae growth in lakes) to lakes in the Minneapolis-St. Paul area.

Thus, components of the landscape surrounding a lake, stream or river can have a strong influence on water quality. Elements of the landscape may serve as sources, sinks, or transformers for nutrient, sediment, and pollution loads. The type of land cover—such as agricultural or urban—is only part of the equation; actual practices employed within a type of land cover can have quite strong effects. In addition, the topography of the region influences the rate of delivery from landscape components to water bodies. When watersheds are steeply sloped and soils are highly erodible, the flux or export of nutrients and sediments to surface waters will increase. In both urban and agricultural landscapes, native riparian vegetation, whether wetland or forest, can reduce nonpoint pollution and help maintain satisfactory quality of surface waters. Improving our understanding of the complex relationships between the land and water is an important goal of both basic and applied research in landscape ecology.

How Do Ecosystem Processes Vary Spatially?

An **ecosystem** is a spatially explicit unit of the Earth that includes all of the organisms, along with all components of the abiotic environment, within its boundaries. The **productivity** of ecosystems (how much biotic matter is produced, including plants, animals and microbes) and the processes that lead to the cycling, retention or loss of essential elements are fundamental to ecosystem function and the production of goods and services upon which humans depend, such as food and fiber (see Chapter Four). **Primary production**—the total amount of plant material produced within an ecosystem in some amount of time, often a year—forms the basis for the food webs in ecosystems. **Food webs** describe the relationships of who eats whom, and are the vehicle through which energy flows through an ecosystem . Elements such as nitrogen, phosphorus, calcium, potassium, and others essential for life, cycle between the living and nonliving components of the ecosystem. Ecologists refer to this continuous flow and transformation of essential elements as **biogeochemical cycling**. The accumulation and decomposition of organic matter in the soil (e.g., when trees shed their leaves, or when plants and animals die) and the subsequent uptake of the nutrients released through **decomposition** (the breakdown of biotic matter into its inorganic components) are very important parts of the functioning of ecosystems. These ecosystem processes vary spatially across broad regions, and this variability is often attributable to differences in physical conditions like slope, elevation, and soils, and to land-use activities.

Can we really explain and predict the variation in ecosystem function across landscapes? Extensive studies have been done on regional patterns of primary production, the accumulation of soil organic matter, and biogeochemical cycling in the Great Plains region of North America. Across the Great Plains and other broad geographic regions, general trends in processes like net primary productivity and decomposition may be predicted reasonably well by broad-scale variability in temperature, precipitation, and soils. For example, plant productivity tends to be higher in the more eastern, moister

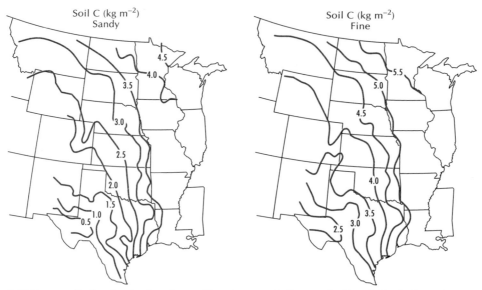

Figure 3.11. Patterns of soil carbon levels in the top 20 centimeters of soil across the Great Plains landscape in sandy soils (left) and fine-textured soils (right). Contour lines were developed from statistical methods and represent the potential values for soils of specified textures. Actual soil carbon could be lower. Redrawn from Parton et al. 1987.

areas of the Great Plains than in the western areas. The accumulation of organic matter in soils across the Great Plains depends on four important variables: temperature, moisture, soil texture, and plant lignin content (Parton et al. 1987). **Lignin** is a structural compound that is difficult for most microbes to decompose, and thus plant tissues with higher lignin contents tend to decompose more slowly. Across the dry landscape of the Great Plains, soil carbon increases from southwest to northeast, and is greater on fine-textured soils than on sandy soils (Figure 3.11). General patterns across the landscape are reasonably well known; however, to predict ecosystem processes at finer scales, additional variables like plant species composition, past land use, and landscape position are also required. A recent study conducted in agroecosystems illustrates some of these patterns.

Burke et al. (1995) studied the patterns of soil organic matter, microbial biomass, and the availability of carbon and nitrogen in soils in two geographic locations, then at different landscape positions (i.e., at the bottom, or toe, of a slope; on a slope; or at the top of a ridge), and under different types of land management. They found that these indicators of ecosystem functioning (soil organic matter, microbial biomass, and soil carbon and nitrogen) were highest at the more northern of two study sites in eastern Colorado, and at **toeslope** landscape positions—the bases of slopes where materials tend to accumulate. The geographic differences between sites reflect broad-scale influences of temperature and soil texture on the accumulation of organic matter; the northern site had cooler temperatures and finer soils. The effects of landscape position reflect the long-term transport of fine organic matter downslope as well as gradients in available moisture and production. Ridgetops tend to be exposed and dry, toeslopes more sheltered and moist. Thus, there are natural gradients in soil organic matter and nutrient supply capacity across the landscape. Interestingly, the effects of the different agricultural management practices on ecosystem processes were strongly dependent on climate and landscape position. In relatively unproductive sites, management practices such as eliminating tilling that reduced soil disturbances had little effect during a period of five years.

The activities of animals in the landscape can have important effects on the spatial dynamics of biogeochemical cycling. Some animals, like beaver, can completely transform portions of a landscape from upland to aquatic communities, conspicuously altering landscape structure and substantially changing nutrient dynamics. When beaver dams are abandoned, the ponds gradually undergo succession to upland communities (see Chapter Two for the surprising history of beaver impact on ecosystems). In other cases, actions of animals on the landscape are much more subtle and may not be detected by the casual observer. Studies of the effects of moose-browsing on nitrogen cycling on Isle Royale, Michigan, illustrate how these subtle interactions provide an important linkage between species and ecosystems (MacInnes et al. 1992). Moose prefer to browse on balsam fir and deciduous tree species such as birch and aspen, rather than on spruce. In areas of Isle Royale where fences (exclosures) were built to prevent moose from browsing, the deciduous trees have persisted and grown larger. Outside the exclosures, white spruce was the only tree species that could grow above the browsing height of a moose. Moose browsing on balsam fir and the deciduous trees thus prevents saplings of preferred species from growing into full-sized trees. Such browsing also opens up the forest canopy and reduces tree biomass, allowing more light to reach the forest floor and stimulating more production of shrubs and herbaceous species.

By selectively foraging on specific plant species, moose and other large herbivores thus can influence ecosystem dynamics—changing plant community composition, biomass, production, and nutrient cycling (MacInnes et al. 1992). The soils in the areas dominated by spruce no longer received as much **litter**—the leaves dropped each fall—and the nutritional quality of the litter, especially its nitrogen content, declined for the decomposers. This decrease in both the quantity and quality of litter leads to a decline in the microbial processes that in turn determine nitrogen availability for the living plants. Moose are highly mobile and can forage all around the landscape, so the interactions between moose and the vegetation creates a mosaic of nutrient cycling regimes in these boreal forests. Moose respond to spatial heterogeneity in the plant community by their selective foraging, and they in turn contribute to patterning in ecosystem processes and in plant community composition—e.g., a shift toward dominance by the unbrowsed conifers in areas where moose were foraging. Understanding the spatial heterogeneity of ecosystem processes in this system requires forging a linkage between the feeding ecology and population dynamics of moose and ecosystem function, all within the context of a landscape. Determining the patterns, causes, and effects of ecosystem function across a landscape remains an important topic of contemporary research in both ecosystem and landscape ecology.

How Are Disturbances an Integral Part of Landscapes?

People are often shocked by the dramatic changes that natural disturbances can cause in landscapes, but disturbance is actually a very important and integral part of many ecosystems and landscapes. A **disturbance** may be defined as a relatively discrete event that disrupts the structure of an ecosystem, community, or population, and changes resource availability or the physical environment. Disturbances happen over relatively short intervals of time: hurricanes or windstorms occur over hours to days, fires occur over hours to months, and volcanoes erupt over periods of days or weeks. Ecologists often distinguish between a particular disturbance event—like an individual storm or fire—and the disturbance regime that characterizes a landscape. The **disturbance regime** of a landscape refers to the spatial and temporal dynamics of disturbances over a longer time period. It includes such characteristics as spatial distribution of the disturbances, disturbance frequency (i.e., number of disturbance events in a time interval, or the probability of a disturbance occurring), return interval (mean time between disturbances), rotation period (how long would it be until an area equivalent to the size of the study area was disturbed), disturbance size, and the magnitude, or force, of the disturbance.

Disturbance is a major agent of pattern formation, and it may even be required for the maintenance of ecosystem function. For example, hurricanes contribute to the maintenance of species diversity in many tropical forests, while prairie landscapes may be maintained by regular fires. The diversity of species in rocky intertidal regions such as the northeastern and northwestern coasts of the United States result in part from storm and wave disturbance. Effects of natural disturbances range in size from small "gaps" in a forest canopy or rocky intertidal region created by the death of one or a few individuals, to larger patches created by a severe windstorm, to extensive areas burned by fires.

Natural disturbances both create and respond to landscape pattern. Certain topographic positions across a landscape may be more or less susceptible to wind damage by a hurricane, and the spread of fires may be limited by the distribution of vegetation types or the presence of natural fire breaks such as rivers. These disturbances also result in a new mosaic of disturbed and undisturbed patches across the landscape to which the biota must respond; indeed, disturbances can sometimes set the stage for ecological processes for years or centuries to come (Box 3.1).

Because disturbance is often responsible for creating and maintaining the patterns we observe, intentional or unintentional shifts in the disturbance regime may dramatically alter the landscape. Baker's (1992) study of changing fire regimes in the Boundary Water Canoe Area of northern Minnesota demonstrated how landscape structure varied with fire frequency. Prior to European settlement, fires were relatively large and infrequent. As the upper midwest was settled by Europeans, fire frequency increased substantially because of indiscriminate burning by early settlers, land speculators and prospectors. Subsequently, there was an extensive period of active human fire suppression. The periods of settlement and fire suppression, which represented substantial shifts from the presettlement disturbance regime, produced significant effects on landscape structure. However, sometimes the effects on landscape structure lagged in time—that is, the effects were not apparent for years after the disturbance regime changed. When fire size and frequency both declined (as with the introduction of fire suppression following settlement), some changes in landscape structure occurred immediately, but some were delayed for decades or even centuries. Effects on landscape structure were more likely to be immediate when fire size declined but frequency increased, as with the change from presettlement to settlement regimes (Figure 3.12).

Given the importance of disturbance, how should ecologists conceptualize the changes that are constantly playing out on most landscapes? Do landscapes remain in any kind of **equilibrium**? These types of questions are of particular importance when we consider the goals of land management in national parks, for example. Should we attempt to maintain landscapes in a particular state, even in the face of natural disturbances, or do we allow them to change? How much change is to be expected over the course of one or more human generations? The properties used to evaluate equilibrium on a landscape fall into two general categories. **Persistence** refers simply to the continued presence of key elements of the

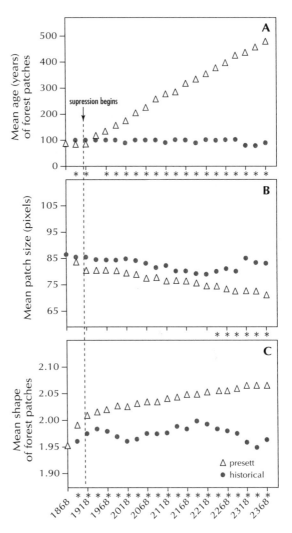

Figure 3.12. Simulated changes in landscape pattern measurements associated with changes in the fire regime in the Boundary Water Canoe Area, Minnesota. **A.** Mean age. **B.** Mean size. **C.** Mean shape of forest patches. A * below the X-axis indicates years for which the simulated presettlement and historical (which included fire suppression) fire regimes were statistically different. Redrawn from Baker 1992.

landscape, which might include species, habitats or successional stages, or age classes of the vegetation. **Constancy** refers to a lack of change or minimal change in the numbers, densities, or relative proportions of the elements of interest. For example, this might be applied to the size (total number) or density (number per unit area) of populations, the amount of **biomass** (the mass of living plants, animals and microbes in an area) present, or the fraction of the landscape covered by different successional stages. Given that disturbances can result in dramatic changes in the landscape, are there conditions under which either persistence or constancy might be anticipated? Maintaining such constancy or persistence is likely to be a desirable goal for many land managers.

Whether a landscape is perceived to be in equilibrium or not depends on the size and frequency of the disturbances relative to the size of the study landscape and the life span, or recovery period, of the biota (Turner et al. 1993). If the disturbance is small relative to the size of the landscape, and if the rate of recovery from disturbance is faster than the return time of the disturbance (Figure 3.13), then the landscape will appear to be a steady-state mosaic. In a **steady-state mosaic**, the vegetation at any given location would be changing through disturbance and succession, but the proportion of the landscape in various stages of succession would remain relatively constant. This dynamic characterizes much of the deciduous forest of eastern North America, where the primary natural disturbance is the creation of small "gaps" in the forest by the death of one or a few trees. This "**shifting mosaic**" is found on some landscapes under specified conditions: when the size of individual disturbance events is small relative to the size of the landscape, and when disturbed areas usually recover before they are again disturbed. Thus, the shifting mosaic concept is very much dependent upon the scale of observation and is not a general property of all landscapes.

If the disturbances are large relative to the size of the landscape and if they are relatively infrequent, a steady-state mosaic will not be observed. Two fire-influenced landscapes have provided similar examples of constant fluctuation in the patch mosaic during the past several centuries. Romme (1982) described wide fluctuations in landscape composition and diversity within a 7,300-ha watershed of Yellowstone National Park, suggesting a landscape characterized by continual change. Even when the study landscape was expanded to encompass nearly 130,000 ha, constant change in the patch mosaic during the past 250 years was still the rule. Baker (1989) also failed to find equilibrium conditions in the Boundary Waters Canoe Area of northern Minnesota, even when the study landscape was 87 times larger than the mean disturbance-patch size. These landscapes affected by large but infrequent disturbances may appropriately be viewed as systems characterized by cyclic, long-term changes in structure and function. Although variation through time and space is substantial, these disturbance-driven systems can be considered stable because the same general sequence of vegetation development recurs following each disturbance event. Thus, landscapes can exhibit a variety of dynamics under different disturbance regimes, and conclusions regarding landscape equilibrium are only appropriate for a specified spatial and temporal scale.

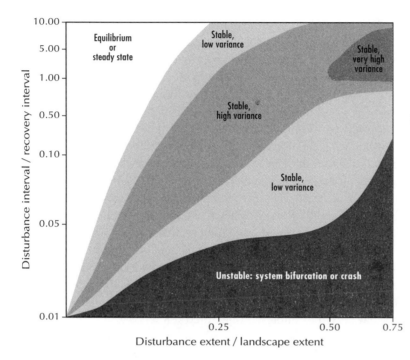

Figure 3.13. State space diagram illustrating where the temporal and spatial characteristics of a disturbance regime lead to qualitatively different types of landscape dynamics. Redrawn from Turner et al. 1993.

THE ROLE OF MODELS IN LANDSCAPE ECOLOGY

As in other areas of ecological research, **models** (representations of reality which are also by necessity simplifications) are important in landscape ecology as a means to formalize our understanding about a landscape of interest. Models provide a mechanism for generating hypotheses that can be tested **empirically** (that is, by collecting data and comparing those observations with the model's prediction), and suggest general insights into the relationships between landscape patterns and ecological processes. Models complement other approaches used in landscape ecology. Landscape ecological studies generally utilize one of five approaches for obtaining empirical data from the real world:

- Whole landscapes may be manipulated experimentally, although this approach is relatively uncommon because it is logistically difficult, hard to replicate, and very expensive.
- Comparative data can be obtained from similar regions that vary in some key parameter of interest. For example, one might select from locations within the same biome a set of study areas that differ in the amount or arrangement of a particular habitat, then relate population movements or density to these patterns.

- Micro-landscapes (relatively small plots that contain spatial variability) might be manipulated experimentally, providing opportunities for controlled testing of hypotheses. This approach has been effectively used to study the responses of small animals to spatial patterning by using insects (e.g., beetles, ants, and grasshoppers) as the study organisms.
- The effects of broad-scale land management activities such as timber harvests or prescribed fires can be studied within an experimental framework. Because planned experiments at landscape scales are so difficult, management actions can provide an alternative source of information about the effects of pattern on ecological processes.
- So-called "natural experiments" that contribute to the development of landscape structure can also provide valuable opportunities for empirical study. The 1988 fires in Yellowstone and the 1980 eruption of Mount St. Helens are examples of such events.

In addition to their overall uses in ecology, models play two additional very important roles in landscape studies. First, because the opportunity to manipulate large landscapes is very rare, models may play a part in suggesting what the consequences of various landscape changes may be. For example, consider the 1988 Yellowstone fires. Ecologists clearly cannot conduct experiments in Yellowstone in which such proportions of the landscape are burned to study different patterns of burning. A model, however, can be used to **simulate** such experiments. Second, models may be used to conduct simulation experiments over a much broader range of conditions—climates, for example—than a field-based study of only a few years can capture. Simulation studies were done to explore the effects of fire size and pattern on elk and bison in Yellowstone under winter conditions ranging from very severe (a lot of snow) to very mild (minimal snow). The results demonstrated an important interaction between winter weather and the amount of the landscape that was burned. Incorporating a wide range of environmental variability is important for projecting how landscapes might be affected by changes in the global climate, or even how observed interactions between patterns and processes—such as elk feeding in a burned landscape—might be influenced by winters of varying severity.

Most models used in landscape ecology are **spatially explicit**—that is, they represent the spatial locations of the quantities being modeled. This makes sense, given landscape ecologists' emphasis on asking questions about spatial patterns and processes. Most models involve computer simulation. However, the models used may range from very simple and abstract representations of space, such as random patterns of one kind of habitat, to very detailed maps that include many variables. Spatial models are useful when one is interested in how spatial pattern affects some ecological variable or process. Consider the following questions in which spatial pattern is used as an "input"—you have a map that sets up the initial conditions for the model and which may or may not change through time.

- What is the effect of different arrangements of the same amount of habitat on a species?
- How does the input of nitrogen to a river or lake vary with positioning of vegetation/cover types across an area?
- How does the patterning of resources influence foraging by an animal?
- How is the spread of fire influenced by spatial pattern of vegetation? In this case, spatial pattern is used as an "input"—you have a map that sets up the initial conditions for the model, and the map may or may not change through time.

Spatial models are also useful when predicting changes in spatial pattern through time is an objective. For example, how will a landscape change in the face of future disturbance or land-use activities? In this case, abundance, configuration, and location are desired "outputs" from the model. An initial spatial pattern is used as an input, but then the model represents the processes that cause the pattern to change through time. A somewhat related use of spatial modeling is determining how a set of processes or biotic interactions can generate a pattern. For example, how does competition for space among species contribute to how the species are distributed across a landscape? Pattern is again an output from the model, but the simulation would typically begin with a homogeneous (i.e., unpatterned) area. We'll consider here a few of the many examples of spatial models used in landscape ecology.

It is important to note that spatial models are not without limitations. These models often require accurate and up-to-date spatial data bases, generally within a GIS, and this is an expensive undertaking if the data are not readily available. Extensive field studies may be required to estimate the parameters used in the model for a given species or process in a given landscape, and to monitor the success of the model in making predictions.

Spatially Explicit Population Models

Spatially explicit population models have yielded interesting insights into the interactions between single populations and their habitats. These models often simulate the birth, death, foraging, and movement patterns of animals or plants across a landscape, and are used to understand the effects of landscape change on a population. Thus, these models can be used to address questions of changing resource distributions, habitat fragmentation, isolation, habitat shape, and patch size. An example of this type of spatial model is BACHMAP, a model designed to elucidate the effects of habitat arrangement on the size and extinction probability of a Bachman's sparrow (*Aimophila aestivalis*) population in a region managed for timber production (Pulliam et al. 1992). Bachman's sparrow is a small bird found in pine woods of the southeastern United States, and the species is of concern to land managers because its population has declined since the 1930s over much of its range. This model includes life-history characteristics (such as

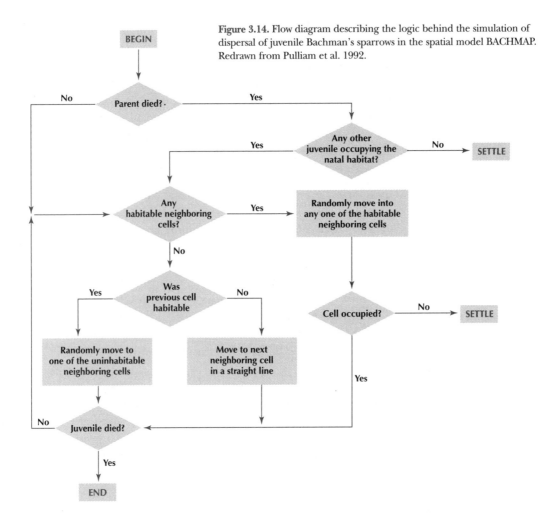

Figure 3.14. Flow diagram describing the logic behind the simulation of dispersal of juvenile Bachman's sparrows in the spatial model BACHMAP. Redrawn from Pulliam et al. 1992.

dispersal, survivorship, and reproductive success) of Bachman's sparrow and the landscape characteristics of the Savannah River Site, South Carolina (Figure 3.14). An interesting aspect of the population is that the sparrows occur primarily in very young pine stands (<10 years after planting) and in mature pine stands (>50 years old), but they rarely use stands of intermediate age. Thus, as trees are harvested and regrow, the distribution of suitable habitat for Bachman's sparrow across the landscape changes considerably. Simulations indicated that the demographic parameters (having to do with birth and death rates) had a greater impact on population size than the dispersal parameters (having to do with moving across the landscape to find suitable habitat), but many of the simulated population sizes were small—much less than the amount of suitable habitat might theoretically support. Other studies have suggested that suitable habitat may remain unoccupied if it is isolated from other suitable habitat occupied by a population. At the actual Savannah River Site, a stand of young pines suitable for Bachman's sparrow was initiated by a tornado. However, the tornado occurred in an area far away from the existing population, and the birds did not colonize the young pine stand, presumably because it was isolated from their occupied habitat.

Land Management: An Example of Timber Harvest Models

Landscape ecological models have also been used to explore the implications of different ways of harvesting timber from forested landscapes. These models typically take an area like a watershed or a national forest and simulate different sizes and arrangements of harvest areas, as well as how much time elapses until the next harvest. For example, small dispersed cuts as well as large aggregated cuts might be compared. Similarly, effects of varying the time between successive harvests—sometimes called **return interval**—from 50 to 100 to 200 years might be studied. In addition to projecting the configuration of forests of different age on the landscape, the models might also examine the effects of each scenario on the potential distribution of suitable habitat for various wildlife populations.

Some important insights for forest management have emerged from studies using spatially explicit forest harvesting models. The deleterious effects of small dispersed cutting patterns for habitat connectivity are readily apparent from simulation studies. The small dispersed cuts create a highly modified forest landscape reminiscent of Swiss cheese, resulting in very little interior forest conditions across the landscape (Figure 3.15). For the same total area cut, fewer but larger aggregated cuts actually can maintain greater connectivity of forest habitats. Another important insight gained from these models is an estimate of the amount of time required for the patterns established by a cutting regime to be erased from the landscape. For example, dispersed cutting has been practiced on federal lands in the Pacific Northwest for more than 40 years, but alternative cutting plans are now being considered. Simulation modeling studies demonstrated that, once established, the landscape pattern created by dispersed disturbances is difficult to erase unless the rate of cutting is substantially reduced or the rotation period is increased (Wallin et al. 1994). This occurs because the cut areas are not eligible for cutting again until the trees regrow (100 years in this example), and subsequent management has to work around these early cuts.

Modeling Land-Use Alternatives

Models have also been used to examine how different land-use policies or decisions may be expressed on the landscape. For example, studies of land-use change in central Rondonia, Brazil used a simulation model to represent the typical sequence of events that occur as new colonists arrive in the rain forest to begin cultivation of the land (Dale et al. 1993). Colonists often burn the tropical forest, plant annual and perennial crops,

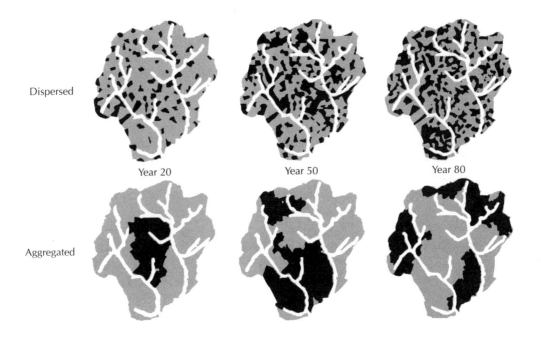

Figure 3.15. Simulated landscape pattern development under two different forest cutting strategies that both clear the same amount of forest area. Black represents areas recently harvested, colored areas represent closed-canopy forest, and white areas are riparian zones along streams. Upper series: the forest is cut in small patches dispersed throughout the watershed; Lower series: harvests are larger and more aggregated within the watershed. Redrawn from Wallin et al. 1994.

then use the land for pasture, and finally abandon their lots (Plate 2.1). The model was then used to compare the effects of this sequence on the landscape with an alternative sustainable scenario in which the farmers planted a diversity of crops and allowed some of their farm land to grow into perennial tree crops, from which products like rubber and cocoa could be harvested. Both scenarios begin with identical conditions in which 294 lots ranging in size from 53 ha to 120 ha are available for settlement. The simulations demonstrated that nearly all the forest would be cleared by the eighteenth year in the typical land-use scenario, whereas the sustainable scenario stabilized in the twenty-fifth year of the simulation with about 40% of the forest being cleared.

Understanding the causes and consequences of land-use changes remains an important area of current research in landscape ecology. When the ecological function of large regions is of interest, the actions of the human population become extremely important. There is a great need to link what has been learned in geography and the social sciences about what humans do and why they do it, with what we've learned in the natural sciences about the consequences of human activities. Without this linkage, attempts to project future ecological conditions across landscapes are bound to be unsuccessful.

The Importance of Linking Models and Data

As with other types of ecology, landscape ecological studies are most powerful when they combine modeling or theory with empirical study. Models provide for broader context and require a formal expression of the understanding of the system. They represent our best understanding of the important components and interactions within a system and often identify useful testable hypotheses. Empirical studies provide the necessary "ground truth" that suggests what relationships between elements of the landscape are important, and which hypotheses may in fact be reasonable representations of ecological processes.

The evaluation of a model is a very important step in the model development process. This is when the following questions are answered:

- How well does the model work?
- Does it agree with observations?
- How well does the model meet its objectives?
- How accurate are the underlying assumptions?

Comparison of model predictions and empirical data is done in many different ways: by graphical comparison, by tabular comparison, and by statistical methods. Regardless of what method of comparison is used, it is important to always include the range of variation in the data (and also the range of variation in the model prediction, if the prediction contains any element of chance) in the comparison, because data are just a sampling of reality and thus also a simplification of the real world.

TECHNIQUES SPECIFIC TO LANDSCAPE ECOLOGY

Landscape ecology is often computer intensive because of its use of spatial data and GIS or simulation modeling. Landscape ecology may also employ broad-scale field studies or experiments. Spatial data—obtained from the field, maps, aerial photography, and satellite imagery, as described earlier—provide the basis for many of the analyses. A GIS is often used to store and manipulate the data, and may be used in conjunction with models to predict patterns and processes across a landscape.

A landscape is typically represented as a grid of cells, and this grid then provides the basis for the quantitative analysis. Each grid cell has a numerical value assigned to it representing some characteristic such as vegetation or habitat type (as depicted in Figure 3.6B). For example, vegetation might be categorized as being deciduous forest, coniferous forest, or grassland. Each grid cell in the landscape would then be assigned to one of these three categories. The spatial scale of these data, both their grain and extent, strongly influences the numerical results of any pattern analysis. Recent analyses suggest that the grain of the data used for spatial analyses should be two to five times smaller than the spatial features of interest. In addition, the spatial extent should be two to five times larger than landscape patches to avoid bias in calculating landscape metrics.

The choice of what categories to include in a spatial analysis is also critical, as the classification scheme strongly influences the numerical results. When describing a forested landscape, one might classify areas within the forest based on the dominant species, or by how old the forest is regardless of species. The patches identified by these two schemes would be very different, and the metrics describing them would also be dissimilar. The classes must be selected for the particular question or objective. For example, general categories (deciduous vs. coniferous forest) might be appropriate to study landscape patterns in the eastern United States, but various forest community classes (birch-hemlock, oak-hickory, basswood-birch, tulip tree) would be needed to study patterns within a particular landscape such as the Great Smoky Mountains National Park in the southeastern United States.

Because patterns must be quantified in order to relate them to processes or to detect changes through time, a variety of **metrics**, or measurements, for the analysis of spatial pattern are used in landscape ecology. Overviews of the metrics that are readily available for quantifying spatial pattern are available elsewhere (e.g., McGarigal and Marks 1995) and will be discussed here only generally. Landscape pattern metrics fall within two general categories. First, the metric may reflect overall characteristics of the landscape by summing or averaging across the entire study area. Examples include the area or fraction of the landscape that is occupied by a particular habitat type, the total number of habitat types present, landscape diversity that combines information about both the number of habitat types and their relative abundances, or the total amount of edge between habitat types. Second, a variety of metrics apply to patches of habitat across the landscape. On the gridded map, a **patch** is defined as a group of contiguous cells of the same habitat type. Once patches are identified on the landscape, many of their characteristics can be computed, including mean patch size, distribution of patch sizes, length of perimeter around the patches, perimeter to area ratio, and patch shape complexity.

Choosing which metrics to use for different purposes, understanding their sensitivities, and knowing how they relate to particular ecological processes, are key to the successful use of these techniques. No single metric by itself is sufficient for quantifying spatial pattern. The choice of which metrics are "best" must be based upon the question at hand. Many metrics of spatial pattern are in fact strongly correlated with one another, containing much redundant information. In one study, correlations were examined among 55 different landscape metrics, and only five independent factors were identified among them (Riitters et al. 1995). Thus, many typical landscape metrics (e.g., number of patches, mean patch size, perimeter to area ratios, patch complexity, and indices like contagion that measure the interspersion among categories) are *not* measuring different qualities of spatial pattern, and the set of metrics to be used in concert should be carefully selected. Ideally, one would want to choose metrics that relate well to an ecological response of interest and that capture a different aspect of landscape structure. However, very simplistic descriptions of the landscape

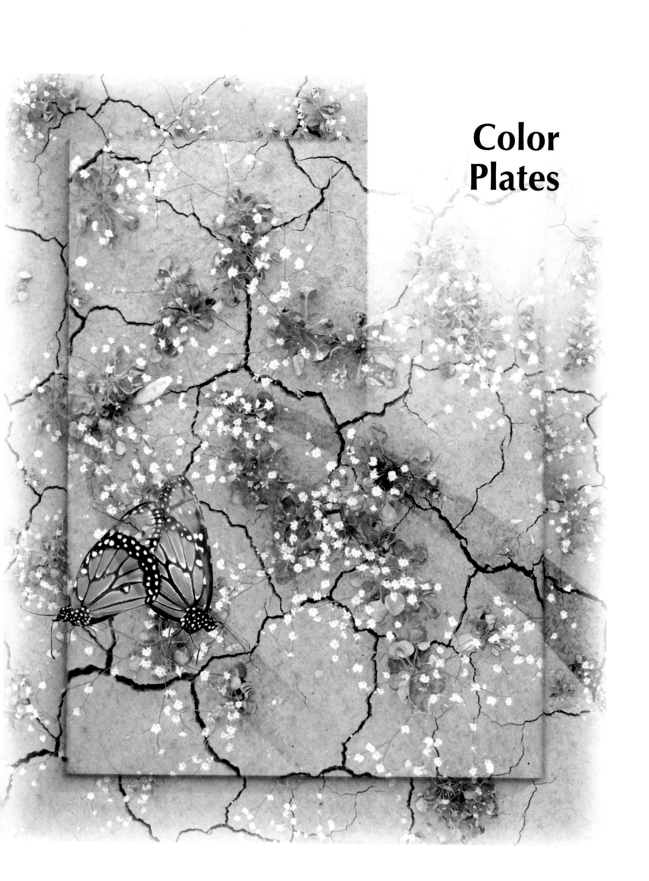

**Color
Plates**

Plate 1.1. A rural Wisconsin scene near Black Earth, Wisconsin. From a photo by Zane Williams.

Plate 2.1. A. In *la selva* at La Estación Biología Los Tuxtlas, a biopreserve in Veracruz state, Mexico. Tropical rain forests such as this one are threatened by human impacts. **B.** Burning the logged edge of the same forest, just beyond the preserve boundary. **C.** The day after the burn. **D.** Across the road from B and C; an area previously slashed, burned, cultivated and pastured for a decade. **E.** An area cultivated and pastured for several decades, now under intensifying development. Note the gully erosion on the slopes between a patchwork of cultivation, pasture, and other uncautious land use. While ecologists can explain the many ecological effects of such practices, the root causes of degradation lie in human culture. From photos by Kandis Elliot.

Plate 2.2. Most people tend to assume the moors of upland Scotland are naturally treeless, but they once were forested. Burning by prehistoric hunter-gatherers helped transform them from forested landscapes to treeless moors. From photos by William Feeny.

Plate 2.3. Ponds created by beaver dams provide habitat for fish and their ever-hopeful human predators. From a photo by Stanley Dodson.

Plate 3.1. The landscape mosaic of burn severities created by the 1988 fires in Yellowstone National Park, Wyoming. Black areas were affected by the most severe crown fires, brown areas had severe surface fires that also killed the trees, and green areas were unburned or just lightly burned. From a photo by M. G. Turner.

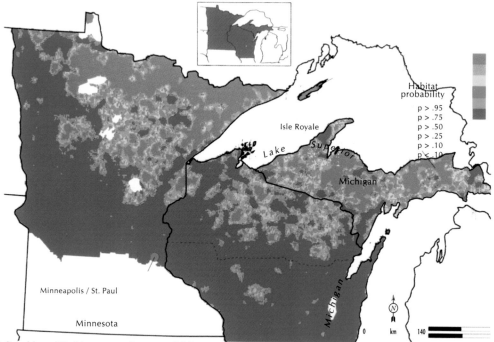

Plate 3.2. Suitable wolf habitat across the upper Midwest, as predicted by the statistical model developed by Mladenoff et al. 1995.

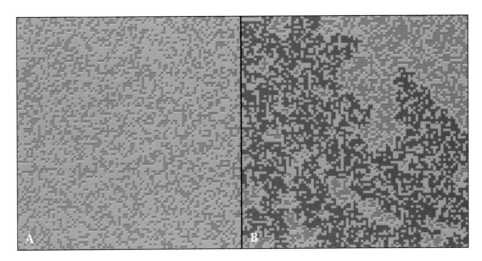

Plate 3.3. Gridded random maps illustrating critical threshold of connectivity. **A.** The habitat of interest, shown in blue, occupies 40% of the landscape and is distributed in small disconnected patches. **B.** When occupying about 60% of the landscape, the habitat becomes connected, with a single patch (changed from blue to red) spanning the entire map.

Plate 4.1. Paul (foreground) and Peter lakes at the University of Notre Dame Environmental Research Center. Peter Lake was treated with lime while Paul Lake served as a reference or control ecosystem (Hasler et al. 1951). This was the first ecosystem experiment to use a control. More recent ecosystem experiments have manipulated the food webs of these lakes (Carpenter and Kitchell 1993). From a photo by S. R. Carpenter.

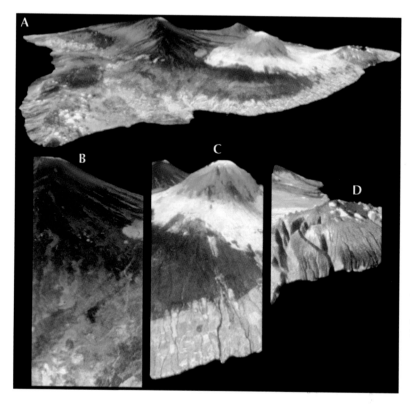

Plate 4.2. A. False-colored perspective of the island of Hawai'i. **B, C, D.** Close-up views of Mauna Loa, Mauna Kea, and Kohala volcanoes, in order of youngest to oldest. As soils develop, infiltration of rainwater to the underlying lava is slowed, leading to more deeply eroded surfaces on progressively older volcanoes. Comparison of volcanoes of different ages has shown that plant production is limited by nitrogen supply in young sites and phosphorus supply in the oldest sites (Vitousek et al. 1997). From an image courtesy of S. Adams, Jet Propulsion Laboratory.

Plate 4.3. Experimental watershed at Hubbard Brook Experimental Forest, New Hampshire. Harvested watersheds are compared with reference watersheds to investigate effects of forest management practices on nutrient cycling. Likens (1992) summarizes a few of the major findings of these studies. From a photo by G. E. Likens.

Plate 4.4. A. Fire is essential for restoring and maintaining prairie vegetation. **B.** On the day after a prairie fire, the ground is covered with ash. Invading woody plants have been killed. Mounds in the foreground are ant nests. **C.** Three weeks after a prairie fire, the vegetation is resprouting from rootstocks. Most of the plant biomass of prairies is underground. From photos by S. R. Carpenter.

Plate 5.1. A baobab tree in Tsavo National Park, Kenya. The bole of the tree has been gouged by elephants seeking water stored in the tree tissues. From a photo by J. Kitchell.

Plate 5.2. Examples of substrates preferred by zebra mussels in lakes of the Great Lakes region. **A.** Substrate found in Lake Wawasee, an inland lake in Indiana. **B, C.** Substrates found in Milwaukee Harbor and Green Bay, respectively, both of Lake Michigan. **C** does not infer that zebra mussels prefer Coke, only that the can was probably in the water longer. From photos courtesy of Cliff Kraft, University of Wisconsin Sea Grant Institute.

Plate 6.1. Territorial advertisement in songbirds. This male song sparrow *(Melospiza melodia)* is singing from a prominent perch in his territory to advertise to conspecific males that this territory is occupied. From a photo by Lukas Keller.

Plate 6.2. Reproductive behavior in damselflies. Left pair: copulation. The male, above, clasps the female with his abdominal claspers while the female receives sperm via the male's sperm transfer organ on his second abdominal segment. Right pair, foreground: oviposition and mate guarding. Following copulation, the female lays her eggs, while the male continues to clasp her to prevent other males from copulating with her and displacing his sperm from her spermatheca.

Plate 6.3. Example of a low threshold for mating attempts by a male: an Australian buprestid beetle attempting to copulate with a beer bottle with female-mimicking features. Inserts show comparative dimpling on (upper) the beetle elytra (wing covers) and (lower) the glass bottle. From photos by Darryl Gwynne.

Plate 6.4. The male peacock has an elaborate tail (actually his tail coverts) used to display to the female during courtship.

Plate 7.1. Carrion flies on white mouse cadavers. Left, the green-bottle *Phaenicia coeruleiviridis;* right, *Sarcophaga bullata.* From photos by T. Ives.

Plate 7.2. Left, common groundsel, *Senecio vulgaris;* right, bluegrass, *Poa annua.* Aggregation figures prominently in competition among many kinds of species, including carrion flies (Plate 7.1) and familiar backyard plants. From photos by T. Ives.

Plate 7.3. A. The rocky intertidal area of Tatoosh Island on the Pacific coast of Washington. **B.** Close-up of the community of goose barnacles, acorn barnacles, mussels and snails. **C.** The plot on the right has an exclosure apparatus that prevented bird predation on the intertidal community. The plot on the left was not protected against bird predation. (It is covered by a grid used for counting densities of species.) Notice how different the community looks after being protected from bird predation. From photos by Tim Wootton.

Plate 8.1. The rocky intertidal of the Oregon coast. Note red starfish, green algae and numerous mussels and other attached organisms in this habitat. From a photo by M. K. Wilcox.

Plate 8.2. Vegetation of the Great Smoky Mountains. Cove (valley) forests, dominated by tulip poplars, butt up against pine-dominated forest on low ridges; in the higher valleys they give way to hemlock draws. High ridges are occupied by rhododendron thickets, "balds" (open grassy areas), or spruce-beech associations on the highest ridges. From a photo by T. F. H. Allen.

Plate 8.3. A. Rolling prairie at the site of Custer's Last Stand near the Little Bighorn River, Montana. **B.** Prairie settlers' sod and earth dwelling. **C.** The plow of the settlers.

Born and raised in Nebraska, early ecologist F. E. Clements was 3 years old at the time of Custer's defeat at the Little Bighorn. Clements entered the University of Nebraska at the time of the massacre at Wounded Knee, just over the border in South Dakota. He began the recording of natural places as the steel plow broke the plains for the first time. Toward the end of his career, Clements watched the Dust Bowl consume the whole landscape that he knew as a boy and student.
From photos by T. F. H. Allen.

are also likely to be misleading. The issues associated with how to measure landscape pattern and detect ecologically meaningful changes in landscapes through time remain important research questions.

Landscape ecologists have also used **model systems,** in which spatial patterns can be manipulated experimentally at manageable scales and replicated. This experimental approach is an important complement both to models and to field studies where patterns and processes are observed in real landscapes but cannot be replicated. The model system approach has been especially useful in the study of the movement patterns of organisms in response to spatial structure of habitat. For example, the movements of beetles have been studied in model systems in which the arrangement of bare ground and clumps of vegetation was varied (Wiens and Milne 1989). Results show that the movement rates of these ground insects are indeed affected by changes in the vegetation structure, and the observed movements are faster and less complex in their pathway than those predicted by simulation models. However, the complexity of the pathways appears to be similar for different species, even in different vegetation types. If this similarity applies to various species, then ecologists may be able to establish some generalizations about movement and spatial pattern that will hold across various species and scales.

APPLICATIONS OF LANDSCAPE ECOLOGY

Landscape ecology offers a perspective to applied questions about our natural environment that complements the perspectives emerging from the other subdisciplines of ecology discussed in this book. By linking patterns and processes, landscape ecology may provide insight into many practical problems regarding the land, how it is managed, and how it will change.

Ecosystem Management and Land-Use Planning

Landscape ecological concepts form an important basis for ecosystem management and land-use planning. **Ecosystem management** has emerged as a new paradigm for land and resource management that emphasizes ecological systems as functional units, and stresses the long-term sustainability of those systems (Christensen et al. 1996). Resource management has undergone an important conceptual shift from emphasizing short-term yield and economic gain in resource decisions, to the recognition that ecosystem-provided commodities required by humans, and services such as clean air and water, must be assured over the long term. Landscape ecology plays an important role in the development and application of ecosystem management because many of these decisions must be made for whole national forests, entire river basins, and other spatially diverse areas.

By offering theory and methods for understanding ecological dynamics over larger areas, landscape ecology can provide an important conceptual foundation for resource management. Nearly all land management agencies have recognized that informed resource management decisions cannot be made exclusively at the site level. Plans for managing national forests, for example, now include large-scale landscape considerations. The resource is managed by managing the large-scale environmental context. Decisions about the size, shape, distribution and timing of resource management actions can be placed within a landscape-ecology framework. One of the easiest places to see this in action is in forest management. The decision about where to cut and in what size units is, by definition, spatial. Many of the federally owned forested lands are obliged to produce timber resources while maintaining the variety of plants and animals within the forest. In addition to timber harvest, forest managers must also consider the role of natural disturbances such as pest outbreaks or wildfires in shaping the forest and maintaining its diversity. Under direction from President Clinton, the Forest Ecosystem Management Assessment Team that evaluated the forest cutting alternatives for the Pacific Northwest of the United States worked out the implications of alternative scenarios within a landscape framework.

Landscape ecology offers techniques and approaches for evaluating alternative scenarios of land use. A comparison among options may provide insights into their pros and cons, and the likelihood of both desirable and undesirable outcomes. Changes in habitat connectivity or water quality could be projected under different patterns of development, clumped together versus dispersed across the landscape, or with varying densities of roads. Knowledge gained about the patterns generated by natural disturbances might be used as a guide for positioning exploitative activities across the landscape.

Habitat Fragmentation and the Conservation of Biodiversity

The maintenance of **biodiversity** (the abundance, variety, and genetic constitution of native animals and plants) also requires a landscape perspective (Franklin et al. 1993). **Habitat fragmentation** has two components (Meffe and Carroll 1994): the *reduction in the total amount of a habitat type* or perhaps all natural habitat in a landscape, and the *apportionment of the remaining habitat* into smaller, more isolated patches. Earlier in this chapter, we discussed the effects of patch size, arrangement and connectivity on populations within the patches, along with the importance of landscape context. When applied to conservation of biodiversity, landscape ecology encourages a broad-scale perspective. Developing a plan for maintaining biodiversity in fragmented landscapes requires an analysis of the pattern of habitats and connections, and how these match up to the needs of native species. The condition of the areas between primary habitat locations should not be ignored, because these areas may help maintain diversity, influence the effectiveness of reserves, and control landscape connectivity. It is thus critical to look beyond the boundaries of a park or nature reserve when considering the long-term future of the species and processes within the reserve. Landscape ecology suggests that ecosystem- or landscape-based approaches may be more desirable than species-based approaches if we want to maintain existing biodiversity.

Land-use patterns have important influences on biodiversity (Box 3.2). Land-use activities may alter the relative abundances of natural habitats and result in the establishment of new land cover types, thereby changing landscape diversity. Introduction of new cover types can increase the biodiversity by providing more habitats, but natural habitats may be compromised, leaving less area available for native species. Land-use activities can alter the spatial pattern of habitats, resulting in fragmentation of once-continuous habitat. Just as when a china plate is dropped and breaks into many small pieces, fragmentation of habitat results in large areas being broken into smaller, separated habitat patches. Natural communities are often islands in a sea of development. Habitat fragmentation exerts numerous effects on native species, almost always reducing biodiversity. Land use also modifies natural patterns of environmental variation, especially if natural disturbance regimes are changed. When fire control and logging alter the frequency and extent of natural fires, the environment may be changed directly. In general, altering the patterns of natural habits increases the likelihood of losing native species and disrupting ecological functions. The effects on landscape structure should be considered when making decisions about development locations, densities, and uses of the land.

Determining the size, shape and arrangement of nature reserves is another important area for the application of landscape ecology. The theory of **island biogeography** (MacArthur and Wilson 1963, 1967, MacArthur 1972) has been an important influence on how ecologists think about reserves. Island biogeography, developed as a general theory to predict the number of species found on oceanic islands, is based on two attributes of an island: its size and its distance from the mainland. The theory predicts that the number of species on an island will reach an equilibrium that is directly proportional to island size (meaning larger islands would contain more species) and inversely proportional to distance of the island from the mainland (meaning fewer species are on islands far away from the mainland, which is the source of new colonists).

Island biogeography theory has been important in highlighting effects of the size and isolation of natural areas on their effectiveness in meeting conservation objectives; indeed, size/distance factors are now important considerations in conservation planning. The theory was applied to the design of nature preserves in terrestrial landscapes, generating a long debate among ecologists about whether a single large preserve would be better than having several smaller preserves spaced such that organisms could move among them. The argument centered around the fact that a single preserve might hold more total species, but it could be wiped out with a single catastrophic event. In contrast, the smaller preserves would each contain fewer species, but some preserves would be likely to survive any particular catastrophe.

General guidelines for preserve design have emerged from many different studies:

- Large preserves are superior to small ones that encompass the same area.
- More circular shapes that reduce the ratio of edge to interior area are preferred over elongated shapes of the same size.
- Corridors connecting isolated areas are desirable.

BOX 3.2. TRENDS AND IMPLICATIONS OF LAND-USE CHANGE IN THE UNITED STATES

Land-use change in the United States is a vast, uncontrolled experiment in how human modifications of the landscape influence the nation's biota. Human influences on the flora and fauna of North America have occurred over several millennia. Native Americans established settlements, practiced agriculture, hunted, and used fire to induce vegetation changes. However, the land-use changes during the past three centuries since European settlement have been particularly profound. At the time of European settlement, forest covered about half of the present lower 48 states. Most of the forest was in the more moist eastern and northwest regions, and it had already been altered by Native American land-use practices. However, the clearing of forests for fuel, timber, other wood products, and cropland led to a widespread loss of forest cover that lasted through the early 1900s. In fact, the rates of deforestation during the mid 1800s were as fast or faster than the much publicized rates of deforestation in the tropics today. By 1920, the area of virgin forest remaining in the conterminous United States was but a tiny fraction of that present in 1620.

Cropland increased at the expense of other land covers through much of American history, reaching a peak in the 1920s and 30s. Between the 1780s and the 1980s, 53% of America's wetlands were converted to other uses; between the 1950s and 1970s alone, nearly 4.5 million ha were lost. Upland grasslands, including the native vegetation of the prairies and plains, also experienced a net decline. Developed land in the United States has increased as the population has grown. Present-day settlement patterns occupy more land per person than in the past, and homes and subdivisions are more dispersed across the landscape. Trends in urban land use are unusual in that they move in only one direction—they generally do not revert back to other categories. The distribution of developed lands across the nation will leave a long-lasting footprint on our landscapes.

What have been the ecological implications of these dramatic changes in land use? Existing information suggests that many states have lost between 1 and 5% of their flora and between 2 and 20% of their fauna. Although relatively few species have become extinct, many species were lost from portions of their ranges and have undergone huge changes in their abundance. The large animals, especially predators like wolves and mountain lions, disappeared throughout much of the United States soon after settlement. Numbers of furbearing animals diminished as they were trapped for their pelts. As forests were cleared and prairies plowed, species that require interior conditions in forests or grasslands were replaced by species that can live along edges or adapt well to disturbed conditions. Even in areas where forest has grown back, the relative abundance of the different species is often quite changed from the presettlement forests. In many states of the arid West, the Midwest, and the lower Mississippi Valley, the riparian forests have been reduced by more than 80% and have not recovered.

Past, present, and future patterns of land use continue to be a dominant influence on ecological processes in several ways. Continued habitat fragmentation is one of the most important results of recent land-use changes. Even though clearing of land for agriculture has slowed, urban and suburban development continues at a rapid pace. The length and area of roads increase with land development, and roads can be effective barriers to animal movement. The effects of habitat fragmentation on plants and animals will continue to be pervasive. Because landscape context is often important for local ecological processes, regional land-use changes are likely to affect many terrestrial and aquatic ecosystems. Although individual changes in land use may appear to have only local significance, the cumulative effects of local changes are transforming the landscape in the United States and elsewhere. Knowledge gained from landscape ecological studies could be used to improve land-use planning, so that negative ecological effects like habitat loss and fragmentation might be reduced.

Global Climate Change

An important applied problem facing ecologists of all kinds is determining the effects of potential changes in the Earth's climate. The **greenhouse effect** is a well-established natural phenomenon in which the Earth's atmosphere retains heat by partially absorbing and re-emitting long-wave radiation, trapping energy and warming the surface of the Earth and its atmosphere. Of concern to scientists and policy makers worldwide, however, is the **enhanced greenhouse effect** resulting from human-induced increases in greenhouse gases in the atmosphere. **Greenhouse gases** (including carbon dioxide, methane, and nitrous oxide) absorb infrared radiation emitted from the Earth's surface and from clouds, then emit this radiation back to the Earth. Since 1750, the atmospheric concentrations of these gases has increased significantly (carbon dioxide by 30%, methane by 145%, and nitrous oxide by 15%), primarily because of the use of fossil fuels (e.g., oil, coal and natural gas) and the conversion of natural vegetation to agriculture. Although still debated among governments, the scientific evidence now suggests a detectible human influence on global climate (Houghton et al. 1996). For example, global mean surface air temperature has increased by between about 0.3 and 0.6°C since the late 1800s, the global sea level has risen by between 10 and 25 cm over the past century, and recent years have been among the warmest since 1860 (the period of instrumental record). Furthermore, climate is expected to change in the future. The best estimate of future climate change suggests an increase in global mean surface air temperature of 2°C relative to 1990 by the year 2100 (Houghton et al. 1996).

Predicting the responses of landscapes to projected changes in global climate remains an important and active area of research and one which has very important implications for public policy. Broad-scale changes in today's landscapes would affect agriculture, forestry, and possibly even human settlement patterns. As discussed early in this chapter, the distributions of plants and animals on the Earth changed considerably during the last 10,000 years as climate warmed and the glaciers retreated. Future changes in climate may result in changes in the geographic distribution of **biomes** (major types of vegetation communities, such as grasslands, boreal forest, or deciduous forest), and the composition of natural and managed vegetation types. Obviously, this would dramatically alter the patterns of today's landscapes, and would likely result in new assemblages of plants and animals. Species presumably could migrate to new areas of suitable habitat at least as quickly as they did in the past (e.g., after the last glaciation), but extensive areas of agricultural and urban lands may act as barriers to migration.

Disturbance regimes may be very sensitive to climate change, and changes in disturbance regimes may rapidly transform the landscapes we observe today. Fire regimes are particularly responsive to climate, and paleoecological studies have demonstrated past changes in fire regimes under very slight shifts in climate. Large regions of the Earth may be subjected to more frequent fires if the climate becomes warmer and drier. Landscape patterns in such regions would also change, possibly becoming dominated by more continuous younger forests, or even shifting from forest to grassland. However, if more frequent fires occur in areas of human settlement rather than wilderness areas, risks to property and life will increase. Similarly, changes in the frequency and intensity of floods will affect many people. Landscape ecologists have an important role to play in understanding and predicting the effects of climate change on the Earth's biota. However, this very complicated endeavor involves many branches of ecology.

LINKS TO OTHER KINDS OF ECOLOGY

Landscape ecology offers a conceptual framework for linking many kinds of ecology because spatial patterning is important in exploring many different ecological questions. The branch of population ecology (Chapter Seven) that considers the dynamics of **metapopulations** (a collection of subpopulations of a species, each occupying a patch of habitat in a landscape of otherwise unsuitable habitat, but that function together as a demographic unit) overlaps with how landscape ecologists study the responses of organisms to spatial pattern of suitable habitat. When the movements or foraging patterns of individual organisms are considered in a spatial context, landscape ecology and population ecology again overlap, and behavioral ecology (Chapter Six) also joins in the mix. Landscape ecology is linked to community ecology (Chapter Eight) when spatial influences on the interactions among populations are considered. When ecosystem ecologists consider how matter or energy moves through spatially heterogeneous systems, landscape and ecosystem ecology intersect (Chapter Four). Because landscape ecology deals with spatial dynamics of all kinds, it naturally grades into many of the subdisciplines within ecology.

SUMMARY

Revealing the causes and consequences of spatial heterogeneity in natural systems has emerged as a fundamental challenge for contemporary ecologists, and this is the emphasis of landscape ecology. Spatial patterns result from complex interactions between the biota and the environment as well as human activities.

Landscape ecology addresses questions that relate spatial patterning in the environment to ecological processes:

- How does the spatial arrangement of habitat influence the presence and abundance of species?
- How does the surrounding landscape influence local populations?
- How do land-use patterns affect the transport of materials from land to water?
- How and why do ecosystem processes vary spatially?
- How are disturbances an integral part of landscapes?

Studies to date have clearly demonstrated that spatial pattern can indeed have a strong effect on many ecological processes. Species respond to patch size, the spatial arrangement of patches, and habitat connectivity. However, different species perceive the landscape quite differently, and habitat studies must always be conducted from an organism-based perspective. **Landscape context**, or the composition and arrangement of the surrounding landscape, can have a strong effect on the presence, abundance, and activities of local populations. Landscape elements are also important for land-water interactions: landscape elements may serve as sources, sinks, or transformers for nutrient, sediment and pollution loads. Other important processes that maintain ecosystem function also vary spatially in the landscape. Whether natural or caused by human activity, disturbance is an important agent of pattern formation in landscapes, and the "footprint" of a disturbance may persist for a very long time. When disturbance regimes are altered either naturally (e.g., as climate changes) or by human activities, shifts in landscape structure will often result.

Modeling plays an important role in landscape ecology because it is difficult to conduct or replicate experiments over large areas. The type of model most often used in landscape ecological studies involves computer simulations. A combined approach that includes empirical study as well as models or theory is the most powerful approach to the study of landscapes. In addition to the use of computers in modeling, landscape ecology employs many computer-based techniques designed to store and analyze spatial data. Geographic information systems (GIS) are often used for data storage and manipulation, and many metrics are available for quantifying spatial patterns in landscape data.

A landscape perspective is becoming increasingly important in applications such as understanding and planning for land-use; developing strategies to conserve biodiversity in fragmented landscapes, and predicting ecological effects of global climate change. The emerging paradigm of ecosystem management includes a strong emphasis on concepts and techniques of landscape ecology. By linking patterns and processes, landscape ecology may provide insights into many practical problems regarding the land, how it is managed, and how it is likely to change in the future.

RECOMMENDED READINGS

RICHARD T. T. FORMAN. 1995. *Land mosaics.* Cambridge University Press, Cambridge, Great Britain.

> *Professor Forman is one of the founders of the discipline of landscape ecology, and his book provides an excellent overview of ecological dynamics in landscapes.*

STEWARD T. A. PICKETT AND MARY L. CADENASSO. 1995. "Landscape Ecology: Spatial Heterogeneity in Ecological Systems." *Science* 269:331–334.

> *A relatively short article that concisely summarizes the focus of landscape ecology.*

H. RONALD PULLIAM, JOHN B. DUNNING AND JIANGUO LIU. 1992. "Population Dynamics in Complex Landscapes: A Case Study." *Ecological Applications* 2:165–177.

> *This article provides an example of how models are used to study population responses to landscape pattern by using Bachman's sparrow as the study organism.*

WILLIAM H. ROMME AND DENNIS H. KNIGHT. 1982. "Landscape Diversity: The Concept Applied to Yellowstone Park." *BioScience* 32:664–70.

> *Romme and Knight examine the history of fire in Yellowstone National Park and examine how the diversity of the landscape has changed over the past few centuries.*

MONICA G. TURNER AND ROBERT H. GARDNER, EDS. 1991. *Quantitative Methods in Landscape Ecology.* Springer-Verlag, New York.

> *This book provides an overview of breadth of methods used to analyze landscapes and the use of computer models to simulate landscape change.*

DEAN L. URBAN, ROBERT V. O'NEILL AND HERMAN H. SHUGART. 1987. "Landscape Ecology." *BioScience* 37:119-27.

> *This article emphasizes the position of landscape ecology within the hierarchy of scales in space and time considered by ecologists.*

Chapter 4

ECOSYSTEM ECOLOGY
Integrated Physical, Chemical and Biological Processes

Stephen R. Carpenter

X had marked time in the limestone ledge since the Paleozoic seas covered the land. Time, to an atom locked in a rock, does not pass. The break came when a bur-oak root nosed down a crack and began prying and sucking. In the flash of a century the rock decayed, and X was pulled out and up into the world of living things. He helped build a flower, which became an acorn, which fattened a deer, which fed an Indian, all in a single year.

—*Aldo Leopold*

WHAT IS ECOSYSTEM ECOLOGY?

How much time does a phosphorus atom spend in organisms during its long journey from eroding hills to the bottom of the sea? What happens to the fertilizers we use in our gardens? How much of the Earth's plant production is used by animals? How much by people? Why does the amount of oxygen in the air we breathe stay constant? These are some of the questions asked by ecosystem ecologists.

The goal of ecosystem ecology is to understand the *flow of energy and matter through organisms and their environment.* Energy from sunlight is captured by plants, which can be consumed by animals or microbes. Ecosystem ecology explores questions about this production of organic energy and its transfer among organisms; other questions address nutrient cycling—essential nutrients such as phosphorus and nitrogen can cycle repeatedly among organisms and the nonliving parts of ecosystems such as soil, sediment, water or the atmosphere. Studies of energy flow and nutrient cycling combine principles of physics, chemistry and biology. While many of the fundamental questions of ecosystem ecology are concerned with production and nutrient cycling, problems of environmental management have stimulated a great deal of work on ecosystems. Ecosystem ecology has been applied to understand the production by living resources that is used by people, the cycling of persistent toxic chemicals, the environmental impacts of human activities, and many other interactions of people and nature.

What Are Ecosystems?

"...A spatially explicit unit of the Earth that includes all of the organisms, along with all components of the abiotic environment within its boundaries." —Likens, 1992

Arthur Tansley introduced the term *ecosystem* in 1935. In doing so, he used the term *component* to refer to some part of an ecosystem. For example, "soil" and "trees" are ecosystem components. Tansley pointed out that it was very difficult to study the interactions

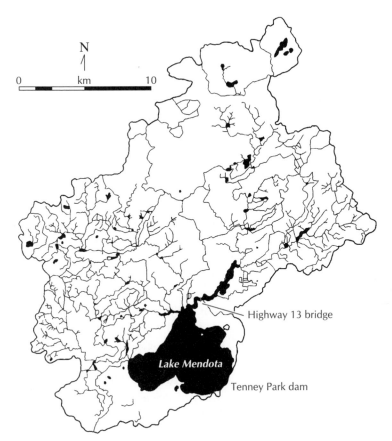

N

0   km   10

Figure 4.1. Lake Mendota and its watershed. Ecologists studying Lake Mendota have defined boundaries between the lake and the Yahara River at the Highway 113 bridge and the Tenney Park dam. Because the lake receives no significant groundwater input, the lower boundary is the maximum depth in the sediments inhabited by organisms, about 10 cm. The watershed boundary delimits the land from which rainwater runs into the lake. Subwatershed boundaries divide the flow of water between two streams that feed into the lake.

Highway 13 bridge

Lake Mendota

Tenney Park dam

of organisms (the biotic part of the ecosystem) separately from the physical and chemical features of their environment (the abiotic part of the ecosystem). Tansley proposed that organisms and their physical-chemical environment could be studied in an integrated way as ecosystems, "…the basic units of nature on the face of the Earth."

A lake is a superb illustration of an ecosystem (Figure 4.1). The lake shore is the *horizontal boundary*—although its exact location may require an arbitrary decision about where the lake ends and a stream begins. The lake's surface is the *upper boundary*. The *lower boundary* extends into the sediments to the maximum depth from which chemicals can flow, or organisms can move, into the lake water. In some lakes, the lower boundary is the maximum depth inhabited by organisms. In other lakes, the lower boundary extends to groundwater that flows into the lake. These boundaries define the part of Earth that is the ecosystem. The ecosystem contains sediments; water with many kinds of dissolved salts, organic chemicals and gases; and hundreds of kinds of organisms including bacteria, protozoa, fungi, algae, higher plants, worms, clams, insects, **zooplankton** (microscopic animals suspended or swimming in waters of lakes or oceans), fishes and many more.

Different ecosystem boundaries are appropriate for different questions. The lake ecosystem boundaries are suited to such questions as:

- How much oxygen flows in or out of the lake each day?
- How many kilograms of fish does the lake produce each year?

These questions are related to the metabolism of the lake and its potential yield of fish. But other questions may require different ecosystem boundaries. For example:

- Where does the phosphorus come from that enters the lake each year?

This question arises because phosphorus controls the production of plants in lakes. It requires us to define the **watershed** of the lake, the boundary of the land area from which water flows into the lake. Most of the phosphorus that enters the lake will be water-borne, arising from erosion of uplands, fertilizers or animal wastes in the watershed. Questions about the nitrogen sources to the lake expand the boundaries still farther. Some of the nitrogen will come from the watershed (soils and fertilizers), but some will originate from burning coal or oil far upwind of the lake. The appropriate boundaries include all areas that contribute significantly to the airborne nitrogen input to the lake. Still other questions require a global perspective. The biosphere, which consists of all the organisms on the planet plus the soil, water and air with which they interact, is the largest and most inclusive of Earth's ecosystems.

While ecosystem boundaries may seem arbitrary, they are no more problematic than other boundaries considered by biologists (Allen and Hoekstra 1992). For example, a nerve cell may be a convenient unit of study, but in practice a neuron may be difficult to discern and separate from other cells. Its intimate connections with other cells may mean that certain questions are best answered by experiments that consider a larger unit than the neuron itself. Every entity studied by biologists—cells, tissues, organs, populations, communities, ecosystems, landscapes—requires that scientists consider scaling questions and define boundaries.

What Are the Appropriate Scales for Studying Ecosystems?

Ecosystem boundaries are quite flexible, and in principle ecologists can define an infinite variety of ecosystems. The task of choosing boundaries for an ecosystem is part of the more general issue of choosing the **scale** for an ecosystem study (Allen and Hoekstra 1992). Scale refers to the scope chosen for an ecological study by the investigator. Several important kinds of scale are commonly considered in ecosystem ecology (Levin 1992, Chapters 3 and 8).

- *Spatial scale* involves choices of ecosystem boundaries and the degree of spatial resolution of the entities within the ecosystem. How large an area was studied? How big is the ecosystem we need to study? What subunits of the ecosystem are important, and how big are they?

- *Temporal scale* asks: What period of time was studied? How fast is the ecosystem changing? What is the duration of the study, and how frequently will samples be taken?
- *System components* addresses such questions as: What organisms or processes were studied? Is it sufficient to know only the total fish production of a lake? Or are we interested in the production of a particular species of fish, such as walleye?

Ecologists agree that there is no single "correct" scale for all ecosystem studies (Allen and Hoekstra 1992, Levin 1992). However, certain scales of study have been more insightful than others. Ecosystems with tangible boundaries such as lakes and watersheds have been especially useful in ecology. Often the appropriate scales for an ecosystem study follow from the question being asked. Ecologist Stuart Fisher and colleagues, for example, have investigated the processing of materials like nitrate in rivers (Fisher et al. 1998). They described different zones of the river (open water, water-saturated sediments, surrounding terrestrial vegetation) in terms of their impact on downstream nitrogen cycles and response to floods. The zones were distinctive in spatial scale (extent of their effects downstream) and temporal scale (length of time required to recover from flood).

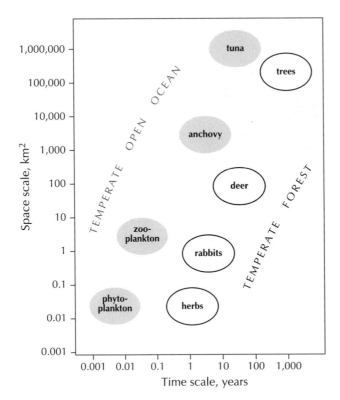

Figure 4.2. Scales appropriate for selected components of temperate open ocean and forest ecosystems. The ellipses are the scales at which the biomass is most variable, suggesting scales that would be useful for studying the processes that control the biomass.

Ecologists can vary the scale deliberately to determine the effects of choosing different scales. In this approach, scale is viewed as an independent variable. For example, the space and time scales of ecosystem components are often related (Figure 4.2). Fluctuations in **phytoplankton** (microscopic algae suspended in lake or ocean water) can be understood by studying a jar of water for a few days. To understand the fluctuations in populations of tuna, however, we must study thousands of square kilometers of ocean for years. Components of terrestrial ecosystems also have characteristic scales. Fluctuations in rabbit populations occur at scales of about a square kilometer from year to year, while fluctuations in tree populations occur over thousands of square kilometers and several centuries. The scales at which a population is most variable provide clues to possible controlling factors and to the scales of studies that would be necessary to understand those factors.

Applied questions of **ecosystem management** focus ecologists' attention on particular scales. Ecologists often work at scales that concern society, but these scales may not correspond to useful physical, chemical or biological boundaries. For example, what will be the effects of improved technology for removing mercury from coal smoke (Figure 4.3)? This question pertains to several scales: the zone of mercury deposition from burning of coal in the U.S., and the cycling of mercury in several different types of terrestrial and aquatic ecosystems. But the zone of mercury deposition does not map neatly onto the aquatic ecosystems which transport, transform and cycle mercury once it leaves the

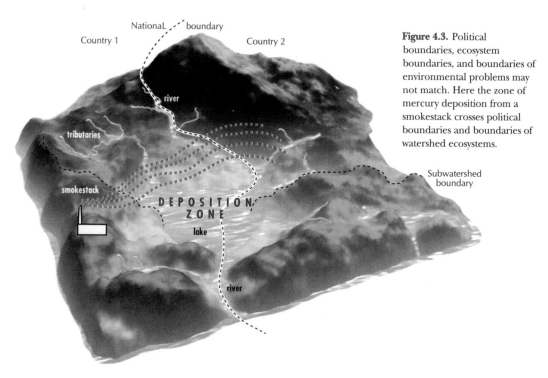

Figure 4.3. Political boundaries, ecosystem boundaries, and boundaries of environmental problems may not match. Here the zone of mercury deposition from a smokestack crosses political boundaries and boundaries of watershed ecosystems.

atmosphere. And the ecological boundaries (the deposition zone and the watersheds) do not match the national and state boundaries of management jurisdictions. Ecologists must tackle such questions at more than one scale. A global picture can be built from component studies of ecosystems like watersheds, streams and lakes. Cross-scale comparisons can be used to determine how processes change as a function of scale choice. Such questions are an important research frontier in ecosystem ecology.

QUESTIONS ASKED BY ECOSYSTEM ECOLOGISTS

Ecosystem ecologists study a remarkable diversity of systems and ask a wide range of questions. This chapter will attempt to represent the breadth of the field by considering just a few of the questions asked by ecosystem ecologists.

Ecologists have traditionally divided Earth's ecosystems into categories. The categories provide a basis for comparison, and each ecosystem type has some distinctive methods and questions. Forests, grasslands, **savannas** (grasslands with sparsely distributed trees), deserts, **tundra** (alpine or polar ecosystems with low-growing plants, no trees, and perennially frozen soils) and cultivated lands are commonly recognized types of *terrestrial ecosystems. Freshwater ecosystems* are often classified as lakes, rivers, or **wetlands** (ecosystems in which the soil is saturated with water part of the year, like marshes or bogs). Open oceans, continental shelves, and **estuaries** (tidal ecosystems where rivers enter the sea) are types of *marine ecosystems.* Ecosystems are often further classified by latitude; for example, tropical forest is distinguished from temperate forest, and polar oceans are distinguished from tropical ones.

How Productive Are Ecosystems?

Life on Earth is supported by the primary production of ecosystems (Box 4.1). The highest productivities occur in wetlands, tropical rain forests, coral reefs and coastal estuaries (1800 to 2500 g dry biomass $m^{-2} y^{-1}$). Low productivities occur in barren sand, rock and ice fields (3 g dry biomass $m^{-2} y^{-1}$), desert shrub ecosystems (70 g dry biomass $m^{-2} y^{-1}$), and open ocean (260 g dry biomass $m^{-2} y^{-1}$). The contribution to the biosphere's production by a particular ecosystem type depends on the total area of the ecosystem as well as the **productivity**. Most of the new biomass production of Earth occurs in tropical rain forest (high productivity, small area) and the open ocean (low productivity, large area).

The biomass of living plants on Earth (about 830 x 10^9 metric tons) is exceeded by the mass of **detritus**, or dead organic matter (about 1500 x 10^9 metric tons). However, the ratio of detritus mass to plant biomass varies considerably among ecosystems. In tropical rain forest, where decomposition is rapid, the detritus:plant biomass ratio is about 0.5. The living biomass is far greater than the mass of dead and decaying biomass plus soil organic matter. In temperate grassland, where decomposition is slowed by lack of

BOX 4.1. CONCEPTS OF PRODUCTIVITY

Biomass is the mass of an ecosystem component per unit area or volume of the ecosystem. Mass can be measured as wet mass (the mass of the organisms as they are collected in the field), dry mass (the mass of the organisms after drying at a standard temperature for a standard period of time), carbon, organic energy (calories), or a limiting nutrient. Biomass is usually expressed per unit area in terrestrial ecology. In aquatic ecology and soil ecology, biomass may be expressed per unit area or volume. The choice of units depends on the question and the ecosystem under study.

Productivity is the instantaneous rate of generation of new biomass. It has units of biomass/(area × time) or biomass/(volume × time). Production is the cumulative generation of new biomass over a specified period of time. In order to compare ecosystems, productivity is often integrated over a year to calculate annual production, the generation of new biomass per unit area or volume in a year. Productivity is sometimes expressed in energy units, or calories. A calorie is the amount of energy needed to heat 1 gram of water 1° C. One gram of dry weight biomass is equivalent to about 4,500 calories. This number varies depending on the relative proportions of fat, carbohydrate and protein in the biomass.

Primary productivity is productivity by plants or bacteria that generate new biomass using energy and inorganic chemicals (Figure 4.4). Energy from the sun drives primary productivity by higher plants (trees, shrubs, grasses and so forth), algae and some bacteria. Energy from reduced chemical compounds (such as hydrogen gas, methane, or hydrogen sulfide) drives primary productivity by some bacteria. *Gross primary productivity* is the total rate of new biomass generation, before any losses are subtracted. In carbon units, it is the rate of carbon fixation per unit area or volume. Net primary productivity is the rate of new biomass gener-

ation minus the rate of respiration. Respiration is a cost of maintaining the primary producer biomass. The *net primary production* is the new biomass available for growth or reproduction of the primary producers. Net primary production can be lost to grazing or death. Death is the loss of living organisms (e.g. whole plants) or parts of organisms (e.g. deciduous leaves shed in autumn, or the organic compounds excreted by some plants). Death contributes biomass to the detritus of the ecosystem. Detritus is dead plant or animal matter that is not yet completely decomposed.

Secondary productivity is productivity by organisms that consume biomass of other ecosystem components. Consumers include animals that ingest biomass, as well as bacteria and fungi that absorb nutrients and organic compounds dissolved from detritus. Animals that ingest biomass are often classified as herbivores (consumers of plants, such as bison and mice), decomposers or detritivores (consumers of detritus, such as vultures and fungi), predators (consumers of other animals, such as lions and marlins), or omnivores (consumers of two or more types of food). *Gross secondary production* is the assimilation of ingested biomass by a consumer. For an herbivore, detritivore or predator, gross secondary production is the rate of ingestion minus the rate of fecal loss. For bacteria or fungi, gross secondary production is the rate of uptake of organic carbon from detritus. *Net secondary production* is the secondary production available for growth and reproduction of the consumer. Net secondary production can also be lost to predation or death. For animals, net secondary production is the rate of ingestion minus the rates of loss as feces, urine and respiration. For bacteria and fungi, net secondary production is the rate of uptake of detrital carbon minus the rate of respiration.

Ecological efficiency is the ratio of the biomass incorporated by a consumer trophic level to the biomass of its prey trophic level.

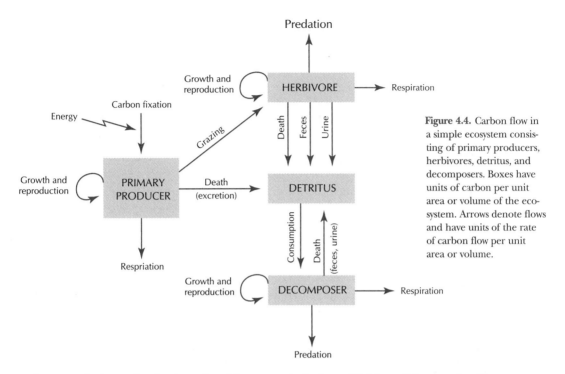

Figure 4.4. Carbon flow in a simple ecosystem consisting of primary producers, herbivores, detritus, and decomposers. Boxes have units of carbon per unit area or volume of the ecosystem. Arrows denote flows and have units of the rate of carbon flow per unit area or volume.

moisture and winter, the detritus:plant biomass ratio is about 30. Most of the grassland's organic matter is in dead and decaying plants, and soil. In most ecosystems, detritus represents a substantial reservoir of organic carbon and nutrients, and consumption of detritus supports a significant biomass of microbes and animals.

What controls primary production?

Differences in **primary productivity** rate among types of ecosystems can often be explained by moisture and temperature. High primary productivities occur in warm, moist environments (e.g. tropical forests, wetlands, coastal tropical marine ecosystems). Low primary productivities occur in dry (e.g. desert) or cold (e.g. tundra) ecosystems.

Within an ecosystem type, differences in primary productivity depend on moisture and temperature, but additional factors also become important. Light intensity affects productivity, and can vary considerably with latitude and cloud cover. Nutrient availability also has strong effects on primary productivity. Open ocean ecosystems have low nutrient supplies and low primary productivity per unit area.

Productivity changes with time after **disturbance.** Disturbances are occasional physical events that destroy biomass; examples are forest fires, volcanic eruptions, and severe storms. Immediately after the plant biomass of a forest ecosystem is destroyed, productivity is low. Net primary productivity increases quickly, however, as the site is colonized by fast-growing, opportunistic plants. Later, the net primary productivity declines as the site becomes occupied by slower-growing, mature trees. At this stage, most of the gross primary production is respired (used for maintenance of biomass) and relatively little remains for net primary production. This cycle of change in net primary productivity with time since disturbance is known from many ecosystem types including forests, grasslands and streams.

Primary productivity also depends on the concentration of carbon dioxide (CO_2) in the air. The concentration of CO_2 in the atmosphere is increasing due to the burning of coal, oil and natural gas (Figure 4.5). This increase is causing the surface temperature of Earth to rise, because CO_2 and other gases trap infrared radiation from the sun (Holland and Peterson 1995). Warmer temperatures may increase the primary productivity of temperate and polar ecosystems by increasing the length of the growing season. Higher concentrations of CO_2 may increase rates of photosynthesis and thereby cause primary productivity of ecosystems to increase. Alternatively, primary productivity may remain about constant while water-use efficiency increases. When plants open the pores in their leaves to take up CO_2, they lose water through evaporation. In ecosystems where water is scarce, increased availability of CO_2 may allow plants to maintain the same rate of photosynthesis while losing less water. How will increased CO_2 affect Earth's primary productivity? We don't know, but the experiment is under way. If human emissions of CO_2 are stabilized at 1994 levels, the steady-state concentration of CO_2 in the atmosphere will stabilize at about twice the preindustrial concentration (Intergov-

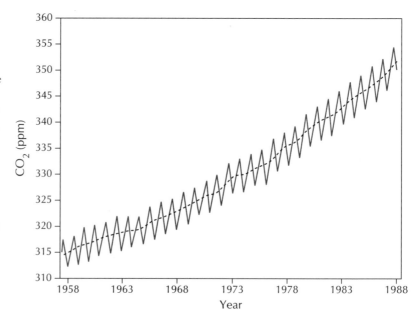

Figure 4.5. Carbon dioxide concentration versus time at the peak of Mauna Loa, Hawaii. The increasing trend is due to burning of coal and oil. The annual cycle is due to ecosystem primary production and respiration of ecosystems upwind of Mauna Loa. Through the year, the lowest CO_2 concentrations occur in summer, and the highest in winter.

BOX 4.2. CONTROL OF PRIMARY PRODUCTIVITY IN LAKES

Lakes have provided some of the best studies of processes that control ecosystem primary productivity. Lakes are well-bounded ecosystems in which rigorous measurements of primary productivity are possible. Also, concern about eutrophication has prompted research on lake primary production. Eutrophication is a state of excessive primary productivity caused by fertilization. Extensive plant growth, including potentially toxic cyanobacteria ("blue-green algae") dominate the lake ecosystem. This excess primary production is not consumed effectively by herbivores. When it decomposes, oxygen is stripped from the water, and animals, including fishes, are killed. The water has foul odors and tastes, and may be toxic to humans and livestock. Use of the water for drinking, irrigation and industry is compromised, and the poor water quality also interferes with recreational uses of the lake such as fishing, swimming and boating. Eutrophication is the most widespread water quality problem in the United States (National Research Council 1992). Concern about eutrophication became acute in North America and Western Europe in the late 1940s, although symptoms of the problem appeared at least a century earlier.

By the 1960s, ecologists had showed that phosphorus input was closely related to the concentration of chlorophyll, a measure of the biomass of algae (Figure 4.6). However, the input of phosphorus was associated with inputs of other nutrients, such as nitrogen. Furthermore, eutrophication was correlated with other characteristics of lakes such as inorganic and organic carbon concentrations and loss of game fish and grazers, which could have causal links to eutrophication. The debate about the chemical cause of eutrophication was settled in the early 1970s by David Schindler and colleagues (Schindler 1977).

They enriched two adjacent lakes, one with inorganic nitrogen and carbon, the other with inorganic nitrogen, carbon and phosphorus. Algal outbreaks developed immediately in the lake with additional phosphorus, but did not occur in the other lake. Given enough phosphorus, the algae obtained carbon and nitrogen from the atmosphere. This work convinced ecologists and lake managers that reductions of phosphorus input were essential to solving the problem of eutrophication.

Phosphorus inputs to lakes come from sewage, certain industrial processes, agricultural fertilizers and erosion, and urban sources (eroding construction sites, lawn fertilizers, pet wastes and so forth). "End of pipe" sources (sewage and industrial inputs) have now been reduced in some areas. Diffuse ("nonpoint") inputs from agriculture and urban areas are still a significant cause of eutrophication in all countries.

Food web structure and humic substances can also regulate lake productivity (Plate 4.1). The grazers in lake food webs are strongly regulated by fish predation. Large fish-eating fish, such as bass, regulate smaller zooplankton-eating fish, such as shiners and sticklebacks. The zooplankton-eating fish control the size and biomass of herbivorous zooplankton. This trophic cascade causes primary production to be lower when grazers are abundant, zooplankton-eating fish are rare, and fish-eating fish are abundant (Carpenter and Kitchell 1993).

Humic substances are dissolved organic compounds formed by the incomplete decay of detritus derived from plants, especially wetland plants and submerged aquatic plants. They stain the lake water, reduce the transmission of light, and decrease primary production. Humic compounds also affect the availability of nutrients to algae and bacteria in lake water.

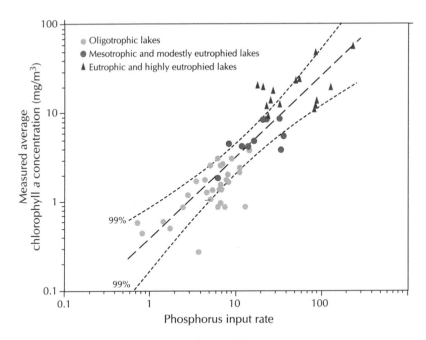

Figure 4.6.
Concentration of chlorophyll in lake water versus phosphorus input rate for temperate lakes. The x-axis is adjusted for differences among lakes in flushing rate. Redrawn from Vollenweider 1976.

ernmental Panel on Climate Change, 1995). On the other hand, it is likely that emissions will not be stabilized and atmospheric CO_2 concentrations will rise even higher. Likely consequences include a warmer Earth and higher sea levels. How will the distribution of ecosystem types be affected by these changes, and what are the implications for people and wildlife that depend on these ecosystems? Ecologists are just beginning to consider this important question.

How are primary and secondary production related?

Consumers affect primary production by removing plant biomass and recycling nutrients. Plant production in turn supports herbivore production. Interactions of producers and consumers are highly variable in time and space and rarely approach **steady state**. Shifts in top predators cause **trophic cascades** that pass through food webs and change primary production (Box 4.2). Despite the negative effects of herbivores on plants, herbivore biomass increases as primary production increases. The relationship of plant production to herbivory is important because it affects the production and abundance of animals, including people.

Herbivory rates are faster in aquatic ecosystems than terrestrial ecosystems, for a given rate of primary production (Figure 4.7). Herbivore biomass and herbivory rates increase with primary production with similar slopes in both terrestrial and aquatic

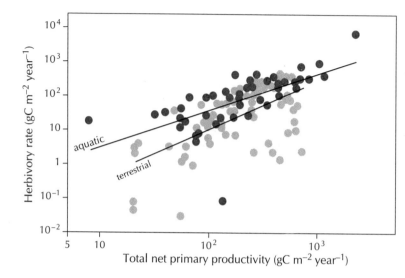

Figure 4.7. Annual rate of herbivory versus annual net primary productivity rate in aquatic (dark circles) and terrestrial (light circles) ecosystems. Redrawn from Cyr and Pace 1993.

ecosystems (Cyr and Pace 1993). However, aquatic herbivores remove about 51% of the primary production, three times the amount removed by terrestrial herbivores. These high herbivory rates suggest that aquatic herbivores leave a smaller amount of net primary production to decomposers than their terrestrial counterparts.

Differences in herbivory rate help explain the contrasting ratio of herbivore to producer biomass in aquatic and terrestrial systems. In terrestrial ecosystems, primary producers attain higher biomass than herbivores. Herbivore biomass is constrained by low herbivory and consequently low secondary production. In aquatic ecosystems, herbivory is faster, secondary production is higher, and consequently herbivores attain higher biomass than producers. How can consumer biomass exceed plant biomass in aquatic ecosystems? The aquatic plant biomass has a fast growth rate (replacing its biomass several times per year, or even several times per week), so the relatively high consumer biomass is maintained by rapid growth and grazing of a relatively small plant biomass.

Are trophic cascades more common in aquatic than terrestrial ecosystems (Polis and Winemiller 1996)? The high relative biomass of consumers relative to producers in aquatic ecosystems suggests this could be the case. On the other hand, diet choices by consumers may be more important than their relative biomass. In lakes, consumers with broad diets, such as the water-flea *Daphnia* and largemouth bass, are crucial for trophic cascades (Carpenter and Kitchell 1993). In terrestrial ecosystems, the analogs of *Daphnia* are large herbivores like deer, caribou and wildebeest, and the analogs of bass are wide-ranging opportunistic predators like lions and wolves. Will the return of the wolf to the northern United States cause a trophic cascade through deer to affect plant diversity and production? Ecologists may have the opportunity to answer this question someday.

How Are Nutrients Transformed and Cycled in Ecosystems?

About two dozen chemical elements are known to be essential for the growth, development and reproduction of organisms. A shortage of any of these elements could reduce ecosystem productivity. The most limiting nutrient for primary productivity is the one least available in the environment, relative to demands for plant growth, because any atoms that become available are quickly taken up by primary producers. If a small amount of the limiting nutrient is added to an ecosystem, primary productivity will increase. Adding a bit of the limiting nutrient often makes a different element the most limiting. The most limiting nutrients for primary production are usually phosphorus and nitrogen.

Biogeochemistry is the study of the movement of elements or compounds through living organisms and the nonliving environment. Box 4.3 presents some of the terminology and concepts that ecologists use to study biogeochemical cycles.

The phosphorus cycle

New phosphorus becomes available to the biosphere through the weathering of rock. Phosphorus is lost from the biosphere when it sinks to permanent sediments of lakes or oceans. Geologic changes that expose sedimentary rocks can make phosphorus accessible to weathering or mining. Such deposits are the source of "new" phosphorus to ecosystems.

Once phosphorus enters ecosystems, it has limited mobility. Phosphorus moves through soil pores and aquatic ecosystems in the form of phosphate (PO_4) salts or organic phosphate compounds dissolved in water. Phosphate dissolves readily in acidic or anoxic water, but if oxygen is present and the pH is neutral or alkaline, phosphate forms complexes (mostly with calcium or iron) and becomes immobile. Because phosphorus is usually in short supply for organisms, any phosphate that becomes available in soils or aquatic ecosystems is quickly taken up by plants. Consequently, phosphate concentrations in groundwater and aquatic ecosystems are usually quite low.

In organisms, phosphorus is used mainly for energy transformation, nucleic acids, phospholipids and bones. Phosphorus is recycled in ecosystems by excretion and decomposition of detritus. Animals excrete phosphorus (Box 4.4). Part of the phosphorus in plant detritus is soluble and becomes available within days after the plant dies. Some detrital phosphorus, such as animal bones, is insoluble and takes many years to recycle.

People have increased the global recycling rate of phosphorus from sedimentary deposits to ecosystems by mining the deposits for phosphate fertilizer. Phosphate added to farmland has several possible fates. Some is bound in the soil; some is harvested in crops, meat, eggs and dairy products; and some is washed into aquatic ecosystems. Some of the phosphate in food is assimilated by people. Phosphorus egested and excreted by people is either captured by sewage treatment plants or added to aquatic ecosystems. Increased inputs of phosphorus to aquatic ecosystems cause **eutrophication** (Box 4.2).

BOX 4.3. CONCEPTS OF MATERIAL CYCLING

The concepts of residence time and turnover rate are often used to compare material cycles (DeAngelis 1992). Residence time is the average amount of time that a unit of material (atom, molecule, particle etc.) spends in an ecosystem or a compartment of an ecosystem. It is equal to the amount of material in the ecosystem (units of mass) divided by the flow rate through the ecosystem (units of mass/time). Turnover rate is the proportion of material in the ecosystem that is replaced by flowthrough per unit time. It is the reciprocal of residence time.

Consider the oxygen in the atmosphere (Figure 4.8). Oxygen is produced in photosynthesis, and lost in oxidative respiration and reactions with reduced chemicals in Earth's crust (for example when coal is burned, or sulfides vented from a volcano are oxidized to sulfate in the atmosphere). The residence time of oxygen in the atmosphere is 38,000,000 / 8440 or about 4,500 years, and the turnover rate is about 0.00022 year-1 or 0.022% per year. Over a human lifetime, the amount of oxygen in the atmosphere appears constant because photosynthesis roughly balances respiration, the residence time is long, and the turnover rate is slow.

Phosphate in lake water illustrates the opposite situation. Phosphate concentrations are low, and uptake rates by planktonic algae and bacteria are fast. Thus the residence time of phosphate atoms in lake water is a few seconds, and the turnover rate is about 25% per second.

At steady state, the rate of change of the mass of material in an ecosystem is zero. In other words, the mass of material is constant, and inflows balance outflows. No ecosystem is strictly at steady state. However, over certain time scales, ecosystems are close enough to steady state that ecologists assume steady-state conditions hold. During a human lifetime, the amount of oxygen in the atmosphere is close to steady state because the turnover rate is slow; during a daily primary production measurement, the phosphate concentration of lake water is close to steady state because the turnover rate is fast. At steady state, the turnover rate is equal to the return rate of the system from a small perturbation. If the amount of oxygen in the atmosphere were suddenly reduced by 10%, it would take a long time (about 450 years) to return to the present level. If the amount of phosphate in lake water was reduced by 10%, it would take less than a second to return to the steady-state level. Return rate can be used to compare the capacity of different ecosystems to recover from disturbance.

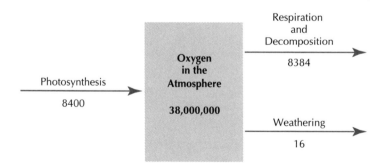

Figure 4.8. The oxygen cycle of the atmosphere. The mass of oxygen in the atmosphere is given in trillions of moles O_2, and the rates are trillions of moles of O_2 per year. Redrawn from Walker 1984.

BOX 4.4. HOW DO ANIMALS AFFECT NUTRIENTS AVAILABLE TO PRIMARY PRODUCERS?

Excretion by herbivores can be a major source of recycled nutrients to plants. Some of the carbon and nutrients assimilated by herbivores is used for secondary production. However, much of the assimilated carbon is consumed in respiration, and the nutrients not needed for secondary production are excreted.

Herbivores differ in their requirements for nitrogen and phosphorus (Sterner et al. 1992). Herbivores that require N:P ratios higher than those found in the plants they consume recycle N and P at a low ratio that promotes N limitation of primary producers (Figure 4.9). Herbivores that require N:P ratios lower than those found in the plants they consume recycle N

and P at a high ratio that promotes P limitation of primary producers.

The best examples of herbivore control of nutrient limitations to primary producers comes from plankton. Zooplankton are often the largest source of nutrients to phytoplankton. Zooplankton differ widely in their nutrient requirements, with some species' N:P ratio as low as 7 and other species' N:P ratio as high as 50. Also, a high percentage of aquatic primary production is consumed by herbivores (Figure 4.7). Thus rates of biomass turnover and nutrient recycling are high, and herbivores can quickly alter the N:P ratios available to producers.

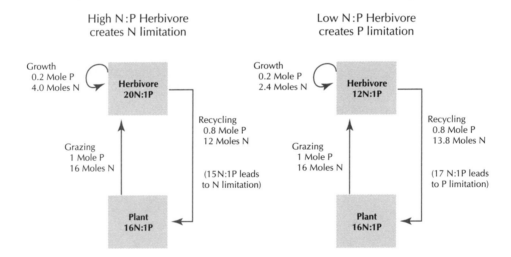

Figure 4.9. Contrasts in nutrient limitation of plants produced by herbivores with different requirements for nitrogen (N) and phosphorus (P). In both cases, the plant N:P ratio is 16:1, the herbivore ingests an amount of plant material equivalent to 1 mole P and 16 moles N, and 20% of the ingested P is assimilated. The herbivore with a high N:P requirement assimilates 0.2 mole P and 4 moles N, and therefore excretes 0.8 mole P and 12 moles N. The excreted N:P ratio of 15:1 would cause the plant to be N-limited if the herbivore was its only source of nutrients. The herbivore with a low N:P requirement also assimilates 0.2 mole of P, but needs only 1.4 moles of N. It therefore excretes 0.8 mole P and 13.8 moles of N. The excreted N:P ratio of 17:1 would cause the plant to be P-limited if the herbivore was its only source of nutrients.

Will phosphorus added to the biosphere by humans build up in soils, sediments or aquatic ecosystems? Questions about the fate of phosphorus and control of phosphorus pollution continue to be important ones for ecologists.

Is the phosphorus cycle responsible for the oxygen balance of the atmosphere (Holland 1995)? Oceanic primary production is a major source of O_2 to the atmosphere, and oceanic primary production depends on the supply of phosphate to surface waters of the oceans. Phosphate tends to be immobilized under oxygenated conditions, and mobilized in the absence of oxygen. If O_2 concentrations were lower than the present levels, would phosphorus be released from marine sediments, thereby increasing oceanic production and the O_2 content of the atmosphere? If O_2 concentrations were higher than at present, would phosphorus be immobilized in sediments, thereby decreasing production and the O_2 content of the atmosphere? Phosphorus inputs to the oceans are now increasing because of increases in fertilizer use and erosion (Howarth et al. 1995). This inadvertent experiment will probably teach us more about the interactions of phosphorus in marine ecosystems.

The nitrogen cycle

Atmospheric nitrogen, N_2, is not chemically useful to most organisms. To enter the biosphere, N_2 must be fixed as ammonia (NH_4), nitrate (NO_3), nitrite (NO_2) or an organic compound of nitrogen. Some nitrogen is fixed by lightning, and a small amount of ammonia is introduced to the biosphere by volcanoes, but most of the nitrogen that cycles in the biosphere is fixed by organisms. Atmospheric nitrogen can be fixed by cyanobacteria in aquatic ecosystems, and by symbiotic bacteria associated with the roots of certain plants (legumes) in terrestrial ecosystems.

Nitrate and ammonia are relatively soluble in water and relatively mobile in groundwater and aquatic ecosystems. In organisms, nitrogen is used for many constituents of cells, including protein and nucleic acids. Nitrogen enters the detrital pool through excretion and death. Nitrogen is excreted as ammonia or urea (Box 4.4). In dead organic matter, most of the nitrogen is in the amino acids that make up protein. **Ammonification** is the process by which bacteria and fungi convert urea and amino acids to ammonia. Ammonia can be lost to the atmosphere, taken up by plants, or nitrified. *Nitrification* is the process by which certain bacteria convert ammonia to nitrate, a process that consumes oxygen and yields energy to the bacteria. The nitrate can be readily transported in groundwater or surface water, taken up by plants, or denitrified. *Denitrification* closes the nitrogen cycle by transforming nitrogen to the atmospheric form, N_2. Denitrification is carried out by bacteria in anoxic environments. In the respiration of these bacteria, nitrate (not oxygen) is the electron acceptor. Carbohydrate and nitrate are converted to CO_2, water and N_2, yielding energy to the bacteria.

The most limiting nutrient to annual productivity of oceans and terrestrial ecosystems is often nitrogen (Vitousek and Howarth 1991). Why don't nitrogen-fixing organisms become more abundant, and fix enough nitrogen to make phosphorus limiting? In comparison with phosphorus, nitrogen tends to be lost rapidly from ecosystems (by washout, vaporization of ammonia, or denitrification), but released slowly from detritus. Also, nitrogen fixation costs organisms a great deal of hard-won organic energy, and can itself be limited by other nutrients including phosphorus. Thus, nitrogen becomes limiting in marine and terrestrial ecosystems because it is lost relatively rapidly, yet is recycled and fixed relatively slowly (Plate 4.2). Limitation of primary production in marine and terrestrial ecosystems may be shifting because human activity is increasing the inputs of nitrogen more than inputs of phosphorus.

Humans have roughly doubled the input rate of fixed nitrogen to the biosphere, and these inputs are still increasing (Vitousek et al. 1997). Combustion of **fossil fuels** (coal, oil, natural gas) forms nitrogen oxides which are later oxidized to NO_2 and other compounds that may be taken up by organisms. By planting nitrogen-fixing crops, people increase global nitrogen fixation. Nitrogen is mobilized from wood, soil and sediments by clearing land and draining wetlands. Most importantly, an enormous amount of nitrogen is fixed industrially for use as fertilizer. Industrial nitrogen fixation now exceeds natural nitrogen fixation! More than half of the nitrogen fixed industrially in the course of human history has been fixed since 1980. The momentum of human population growth will lead to high rates of industrial nitrogen fixation for decades.

The nitrogen fixed by people is mobile in air and water, and has significant effects on ecosystems. Concentrations of nitrous oxide (N_2O) and nitric oxide (NO) are increasing in the atmosphere. Nitrous oxide is one of the gases that absorb infrared radiation and contribute to global warming. Nitric oxide is involved in reactions that create peroxyacyl nitrate, one of the chemicals in **smog** that irritates the eyes and lungs. Smog (a word formed by combining "smoke" and "fog") is a mixture of by-product gases from

burning fossil fuels. As one might expect, smog has negative effects both on primary productivity and human health. Nitric oxide is also the source of atmospheric nitric acid, a principal cause of "acid rain." Acid rain causes long-term declines in soil fertility by depleting calcium and potassium from soil (Likens et al. 1996). Streams and lakes in several regions of Earth are severely acidified (Likens 1992). Nitrate in groundwater is a human health hazard in some regions. Nitrogen washed into estuaries and coastal oceans has caused eutrophication, blooms of toxic algae, and die-offs of fish (Vitousek et al. 1997).

The problems caused by human mobilization of nitrogen have no simple solution. Industrial nitrogen fixation supports food supplies and human survival. Combustion of coal, oil and natural gas will increase for the foreseeable future, although improved technologies may decrease the release of fixed nitrogen from this source. It may be possible to increase the efficiency of fertilizer use, increase the efficiency of nitrogen removal from industrial effluents, decrease the transport of nitrate and ammonia to aquatic ecosystems, and increase denitrification by restoring wetlands and submerged aquatic plant beds. How much nitrogen should be fixed? How can we used fixed nitrogen as efficiently as possible? How can the impacts of fixed nitrogen on air and water be minimized? Questions about managing the nitrogen cycle will be a priority for ecosystem scientists and society in the future.

THE ROLE OF EVOLUTION IN ECOSYSTEM ECOLOGY

Evolution and ecosystem ecology are linked because evolution occurs in ecosystems, and the species that perform ecosystem processes were shaped by evolution and continue to evolve. Perhaps the most spectacular event linking evolution and ecosystems was the rise of photosynthesis and creation of an oxidizing atmosphere about 2.1 billion years ago. The shift from a reducing atmosphere to an oxygen-rich one completely changed ecosystem processes and the context of later evolution. Even today, however, more subtle signs of the linkage between evolution and ecosystems are evident.

Because ecosystem processes determine, in part, the reproductive success of organisms, they are a factor in natural selection. Consequently, ecosystem processes may account for some of the regularities found in community structure. Fishes, for example, tend to feed in a few distinct ways: eating fish, zooplankton, or invertebrates from rocky surfaces or soft sediments. Each feeding type has a number of morphological attributes, and species of diverse lineages have converged on these attributes through evolutionary time. In Lake Thingvallavatn, Iceland, all four feeding types of fish occur within a single species, arctic char (Figure 4.10). In just 10,000 years since the lake was formed, four morphologically distinct races of char have evolved to the same feeding types found in far older and more species-rich lake ecosystems. The chars of Thingvallavatn suggest that features of the fish body plan interact with characteristics of lake ecosystems to structure fish feeding types.

C. S. Holling (1992) has discovered that body size distributions of terrestrial birds and mammals from a given ecosystem type tend to have clumps. That is, there tend to be groups of species that have similar adult body size, with substantial differences in body size between groups. This finding suggests that certain body sizes (the ones in the gaps between groups) have not been successful in these ecosystems, and that body sizes tend to cluster in certain ranges where populations can succeed. Holling suggests that the clumped size structure of the animals reflects the spatial organization of the habitat.

Evolution affects ecosystem processes through the diversity and adaptations of organisms. Some of the most spectacular examples arise in situations where only one species is capable of performing an ecosystem process (Jones and Lawton 1995). Some

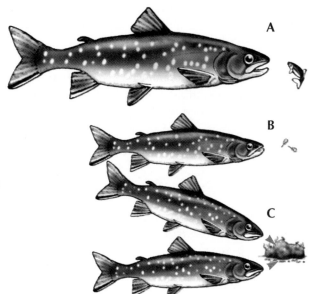

Figure 4.10. Three feeding types of arctic char in Lake Thingvallavatn, Iceland, and their typical prey. **A.** Fish-eating char; **B.** Plankton-eating char; **C.** bottom-feeding char (two behavioral types; one grazes the upper side of stones, and the other forages beneath. Redrawn from Campbell 1996.

trophic cascades, for example, depend on the capabilities of a single predator species, and no other species can substitute. When only one species can perform a particular process, the ecosystem is vulnerable to stresses that displace that one species. Acidification of lakes knocks out the bacteria that convert ammonia to nitrate (Rudd et al. 1988). This loss of a particular group of organisms blocks the entire nitrogen cycle of the ecosystem.

The more important role of evolution may be to create diversity that buffers ecosystems against stress. Often, many species perform a given ecosystem process. If fluctuations in climate or inputs to the ecosystem should eliminate some species, others remain to sustain ecosystem functions. For example, as lakes are acidified, zooplankton biomass remains about the same even though many species are lost and other species become abundant (Frost et al. 1995). The processes performed by zooplankton —grazing, secondary production, nutrient recycling—may change little despite wholesale shifts in species composition (Figure 4.11).

This "functional complementarity" of species is sometimes misunderstood to mean that species are redundant and therefore expendable. This is not true; complementary species have similar but not identical roles in ecosystem processes. *Complementarity* allows ecosystem structure to be renewed, and ecosystem processes to recover, following many different kinds of disturbances (Peterson et al. 1998). The mechanisms that people use for temperature regulation are analogous to the roles of complementary species in ecosystem processes (Gunderson et al. 1995). As your environment becomes hotter or colder, you can regulate your body temperature by changing your location, adding or removing clothing, perspiring, changing blood flow to your body surface, or altering your metab-

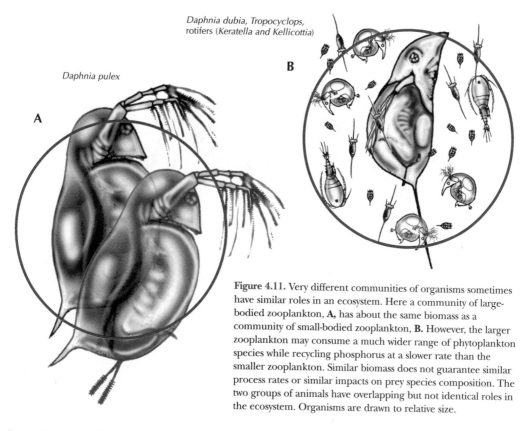

Daphnia dubia, Tropocyclops,
rotifers (*Keratella* and *Kellicottia*)

Daphnia pulex

A

B

Figure 4.11. Very different communities of organisms sometimes have similar roles in an ecosystem. Here a community of large-bodied zooplankton, **A,** has about the same biomass as a community of small-bodied zooplankton, **B.** However, the larger zooplankton may consume a much wider range of phytoplankton species while recycling phosphorus at a slower rate than the smaller zooplankton. Similar biomass does not guarantee similar process rates or similar impacts on prey species composition. The two groups of animals have overlapping but not identical roles in the ecosystem. Organisms are drawn to relative size.

olism. These are different, partially overlapping, but not completely substitutable ways of accomplishing the same process—stabilizing body temperature. Different circumstances call for different mechanisms. The temperature analogy also demonstrates why species are not expendable. Suppose that you lost one of the mechanisms for temperature regulation. You would still be able to cope with some thermal stresses, but there would be other stresses that you could not endure. If you lost your ability to perspire, you could probably cope with a polar climate or a temperate continental winter using your remaining thermoregulatory mechanisms, but a tropical climate or a temperate continental summer could be fatal. Ecosystems that have lost certain species may maintain rates of core processes—production, nutrient cycling and so forth. But they have lost the capacity to sustain process rates under certain stresses, environmental shifts, or changes in inputs. Reductions in the capacity to maintain ecosystem processes under stress are one reason for concern about declining biodiversity. How does species diversity affect the **resilience** of ecosystems? How much species loss can occur before ecosystem processes are affected? Which species are most critical, and which processes are most sensitive? Such links between species diversity loss and the **sustainability** of ecosystem processes are relevant to the future of ecosystems and people (Gunderson et al. 1995, Mooney et al. 1996).

THE ROLE OF MODELS IN ECOSYSTEM ECOLOGY

Every ecosystem study involves at least one box-and-arrow model which defines the components of the ecosystem, the critical flows among them, and potential cause-effect sequences. The box-and-arrow model may be as simple as Figure 4.8, or much more complex than Figure 4.4. The box-and-arrow model describes the ecosystem components and interactions relevant to the question under study.

Ecosystem ecologists also use a great diversity of more formal models, which may be manipulated mathematically or with the aid of computers. These models can have two general purposes. One is to describe an ecosystem or process. The other is to make predictions. A mix of these goals is also possible. Mathematical models used by ecosystem ecologists can be classified as complex, minimalist or statistical.

Complex Models: Simulation models with many compartments and many flows are often used to describe complex ecosystems. Such models have the advantage of including all components that might be important and focusing discussion among collaborating scientists. On the other hand, simulations using complex models can be difficult to interpret. Also, the parameters, their variances, and their correlations are usually not precisely known so it is impossible to calculate the uncertainties of predictions from highly complex ecological models. Consequently, ecologists use simpler models when predictions are important.

Minimalist Models: One way to simplify ecosystem models is to use the same general structure for each component of the model, while including a large number of components. Landscape models may use the same basic structure for each subunit of the landscape, and estimate parameters from detailed spatial data, to predict larger-scale landscape patterns (Chapter Three). Models of food webs may use the same equations for each ecosystem component, and estimate parameters from physiological or behavioral data, to predict flows of carbon, nutrients or contaminants (Chapter Five).

The second way to simplify simulation models is to discard unnecessary detail, by deciding which components and interactions are really crucial to the question (Scheffer and Beets 1994). These are included in the model, and other processes are ignored or greatly simplified. Often parameters can be estimated directly from field data, and predictions and their uncertainties can be calculated in straightforward ways. Fisheries management models are a good example of this approach to modeling (Hilborn and Walters 1992). This simplification strategy is similar to the approach used by population ecologists (Chapter Seven).

Statistical Models: Ecosystem ecologists often fit statistical models to describe the relationship between two or more variables. Statistical methods such as regression are used to fit curves to observations. For example, Figure 4.6 presents a statistical model that predicts chlorophyll from phosphorus inputs to lakes. Parameters of the equation that predicts chlorophyll were estimated from the data. Uncertainties, such as the variance of parameters, the scatter of observations about the fitted curve, and the variance of predictions, can be calculated. Statistical approaches are often used to fit minimalist models to data (Hilborn and Walters 1992).

These distinctions describe the range of models used by ecosystem ecologists, but are artificial because ecosystem models cannot be sorted neatly into a few categories. Ecosystem ecologists try to adapt models to diverse questions, types of ecosystems, and kinds of data. There is no single "correct" approach to modeling ecosystems. Successful models link questions with field observations in ways that lead to clear answers, new insights, and solutions to environmental problems.

How do ecosystem ecologists use models to learn about ecosystems? They compare models that imply different answers to the question under study, make different assumptions about how ecosystems work, and make discernably different predictions. Models are used to sharpen discussion, organize available data, improve the description of the ecosystem, focus the measurements that need to be made, rule out implausible explanations, corroborate plausible explanations, and help create the next set of models. Productive research projects have a rapid cycle of creation and destruction of alternative models.

TECHNIQUES SPECIFIC TO ECOSYSTEM ECOLOGY

The techniques of ecosystem ecology are drawn from a diversity of sciences and are often tailored to specific ecosystem type. Sala et al. (1998) provide a thorough overview of methods commonly used in ecosystem ecology.

For a particular research project, the appropriate methods follow from the question. Aspiring ecosystem ecologists should not structure their training around methodology. Rather, students should learn to recognize productive questions, gain a broad

background in the natural sciences and mathematics, develop abilities to collaborate with other scientists, and learn new techniques as required by the questions that arise.

Field experience is crucial in the training of ecosystem ecologists. The significant questions of ecosystem ecology arise from a deep personal knowledge of real ecosystems. Increasingly these are urban, agricultural or otherwise human-dominated ecosystems, rather than completely natural ones. In any case, they are outdoors and often large. Ecosystem ecologists process chemical samples in the laboratory and run models on computers, but observation and experimentation take place in the field. Ecosystems are not amenable to laboratory experimentation. Students can gain valuable field experience by taking courses at a biological field station, or by working on a field project with their professors.

Broad training in the natural sciences is valuable to ecosystem ecologists. Techniques and insights are likely to draw on the scientist's knowledge of mathematics, statistics, physics, chemistry, geology or biology. A basic background in these sciences will prepare students for future careers.

Aspiring ecologists should also learn to write well. Communication of ideas and career success depend heavily on writing reports, articles for journals, and proposals for projects or grants. The best way to learn to write is to read good writing, and take courses that require competent writing.

APPLICATIONS OF ECOSYSTEM ECOLOGY

Questions asked by scientists are sometimes classified as "fundamental" or "applied." Fundamental questions are motivated by issues internal to the discipline, and lead to work which is intended primarily to increase the core knowledge of the discipline. Many questions about production and nutrient cycling are fundamental to ecosystem ecology. Applied questions arise when science is relevant to a societal problem. Ecosystem ecology has been applied to such diverse issues as changes in the global carbon cycle caused by burning fossil fuels, cycling of persistent toxic chemicals, production of food and fiber, and understanding the environmental consequences of development.

The boundary between fundamental and applied science is often indistinct. In fact, important interactions occur across the spectrum from fundamental to applied research. The study of lake eutrophication (Box 4.2) was originally prompted by societal concern about the deteriorating conditions of lakes in Western Europe and North America. This application of ecosystem ecology led to new, fundamental understanding of the phosphorus and nitrogen cycles and the control of lake productivity. Fundamental ecology can also lead to new applied research. Acid rain in North America was discovered by scientists studying nutrient cycles of watershed ecosystems (Likens 1992). This work, motivated by fundamental questions of biogeochemistry, revealed excess acidity in precipitation from many parts of North America. This discovery prompted extensive research on the causes of acidity in precipitation, transport of acids through the atmosphere, and impacts of acid deposition on many types of ecosystems (Plate 4.3).

Applications of ecology bring scientists into the realm of policy (Box 4.5). Scientists can help policy makers by identifying environmental problems, contributing to new options for managing or coping with the problems, and by forecasting effects of policy choices on ecosystems.

A comprehensive summary of applied ecosystem ecology is beyond the scope of this book. Instead, we will examine four questions:

- How much of nature's services do people appropriate?
- How are toxic substances cycled in ecosystems?
- How can damaged ecosystems be restored?
- How can ecosystems be managed sustainably in a changing world?

How Much Of Nature's Services Do People Appropriate?

Ecosystems provide many services to society. These include excess harvestable production from living resources, nutrient regeneration, breakdown or sequestration of certain pollutants, pollination of crops and many other processes (Daily 1997).

Ecosystem services are often not traded in markets, have no price, and therefore do not enter into economic decision-making. Absence of a link between ecosystems and economic systems can lead to environmental problems. Consider pollution. Society relies on aquatic ecosystems to dilute or break down pollutants to acceptable levels. Ecologists can estimate the amount of a pollutant that can be added to a lake or stream without causing significant harm to the ecosystem, the ecosystem's capacity to dilute pollutants in the future, or other services that the ecosystem provides (fish production, water for drinking, irrigation, industry, and so on). But the maintenance of pollution-dilution capacity is not a part of the market system. Runoff from farms degrades aquifers, streams and lakes, but this loss of pollution-dilution capacity is not factored into food prices; industrial effluents reduce pollution-dilution capacity, but this cost is not included in the price of manufactured products. The disconnection between the ecosystem and the economic system may not be noticed when the capacity to dilute pollutants is large. But when the pollution-dilution capacity is exceeded, the aquatic ecosystems degrade and society loses of benefits from water.

BOX 4.5. ECOSYSTEM DEGRADATION, RESILIENCE AND SUSTAINABILITY

Ecosystem degradation has occurred when an ecosystem is no longer providing resources or services that people expect. *Degradation* results from human impacts such as resource use or pollutant disposal. Therefore, degradation is a property of ecosystems and society together. Degradation involves values and debate; people will differ in the conditions they find acceptable or desirable. However, there are absolute limits to the amount of impact that an ecosystem can withstand.

Resilience is the amount of a given impact that an ecosystem can withstand before it changes to another, different state (Ludwig et al. 1997). Reversal of the state change would involve a substantial intervention or change in policy. Lake eutrophication (Box 4.2) is a good example. Lakes can recover from modest inputs of phosphorus. When phosphorus inputs exceed a critical level, however, efficient mechanisms of phosphorus recycling are established which tend to stabilize the eutrophic state. Then substantial effort is required to restore water quality. Resilience depends on the ecosystem states of interest (in this case, eutrophic versus not eutrophic) and the stress (in this case, phosphorus input). Both below and above the critical level of phosphorus input, lakes are resilient to fluctuations in phosphorus input. Thus resilience can preserve either desirable or undesirable states of ecosystems.

Variability is important; phosphorus inputs vary from year to year because of weather, and the critical level of phosphorus inputs is not known precisely. Consequently, resilience may be expressed as the probability that the ecosystem can experience a given stress and not change states.

Sustainability is the maintenance of ecosystem resources and services for extended periods of time (Gunderson et al. 1995). Sustainability is a property of ecosystems and society together. Many commercial fisheries are failures of sustainability: for institutional and economic reasons, harvest almost always rises to levels that the fish populations cannot tolerate (Hilborn et al. 1995). In sustainable systems, institutions and economics act to maintain the resilience of the ecosystem. Thus scientific aspects of sustainability center on resilience. Societal debates revolve around such questions as "How much impact is tolerable?" and "How much impact is too much?" For example, how much carbon dioxide can be added to the atmosphere without producing unacceptable changes in climate, ecosystems, and sea level? How much nitrogen can people fix without intolerable changes in atmospheric chemistry and water quality? How much biodiversity can be lost without unacceptable losses of resilience and ecosystem services? Science can provide information that is very helpful in answering such questions.

To make decisions about sustainability, we need to ask "What kinds of ecosystems do we want?" and "What kinds of ecosystems can we get?" (Gunderson et al. 1995). Questions of the first type are best answered through democratic decision-making in which all interests have open access to relevant scientific information. The second type of question engages the scientific expertise of ecosystem ecologists. How will different policies affect ecosystems? Predicting consequences of policy choices is an increasingly important task for ecologists.

Causes and corrections for breakdowns between ecological and economic systems are an important research frontier. The growing field of ecological economics is attracting attention from a variety of social and natural-science disciplines. Ecosystem scientists can contribute by estimating the extent to which ecosystem services are appropriated for human use. Such estimates help gauge whether the services are overexploited. The following examples apply to the biosphere. The general approach, however, can be applied to smaller ecosystems and to a wide range of ecosystem services.

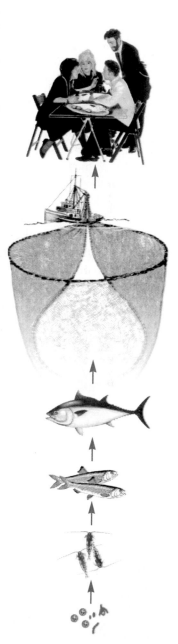

How much of Earth's terrestrial production is appropriated by people? Global primary production is about 224 x 109 met-

ric tons per year of biomass (dry weight). About 59% of the total is produced on land. Humans appropriate terrestrial production directly (as food and fiber) or indirectly (for example as food for livestock). People also choose to forego primary production, for example by replacing farmland or forest with urban land uses. The appropriation of terrestrial production was estimated by adding up the production used (directly and indirectly) and foregone through human activity. In the mid-1980s, Peter Vitousek and colleagues (1986) estimated about 35 to 40% of terrestrial primary production was appropriated by people.

How much of Earth's aquatic primary production is needed to sustain the world's fisheries? The world fisheries catch (including "bycatch"—fish killed but not harvested) was about 131 million metric tons wet mass per year in 1990. About 85% of this total comes from wild stocks, and the remainder from aquaculture. The harvest includes top predators like tuna and billfish, forage fish like herring, invertebrates like shrimp, and even algae such as nori. Daniel Pauly and colleagues (1995) assigned the various species to trophic levels, which represented the number of consumption steps between primary production and harvest. Thus algae would be trophic level 1, while tuna would be trophic level 4. Using models for the efficiency of trophic transfer, they estimated the amount of primary production needed to produce the total catch. About 8% of Earth's aquatic primary production (freshwater plus marine) is required to support the fisheries.

However, fisheries of continental shelves and upwelling areas, which account for most of the world's industrialized fishing, use 24% to 35% of the primary production. It may be difficult to increase yield from these fisheries. Most aquatic primary productivity (75%) occurs in open ocean systems which are vast but relatively unproductive per unit volume. From these areas, people harvest top predators like tunas, which roam long distances through aquatic deserts to find scattered patches of food. Presently it is not economical for people to harvest lower trophic levels from the open oceans.

How much of Earth's renewable fresh water is used by people? Only 0.77% of Earth's water occurs as fresh water in aquifers, lakes, streams, soils, organisms and the atmosphere (97.5% of global water is salty, and the rest is locked in permanent ice). Renewable fresh water is water that flows through the hydrologic cycle of evaporation and precipitation. Sandra Postel and colleagues (1996) estimated the amount of renewable fresh water appropriated by people. About 110,000 km^3 of precipitation fall annually on land. Of this precipitation, 40,000 km^3 drain into the oceans as runoff, and 70,000 km^3 return to the atmosphere through evapotranspiration. Humans use about 26% of global evapotranspiration to produce food and fiber. Since runoff occurs mostly in remote areas or in floods, of the total runoff of 40,000 km^3/year only 12,500 km^3/year is available for human use. About 54% of this available runoff is currently used for agriculture, industry, municipal needs and disposal of wastes. The growing human population will need more than 70% of the available runoff by 2025. New dam building could increase available runoff to about 13,700 km^3/year by 2025, and improved efficiency of water use in agriculture could slow the human demand for water. In the long run, water needs may prove to be more limiting than primary production to human development.

These estimates show that humans now appropriate a substantial percentage of Earth's terrestrial primary production, marine primary production, and available renewable fresh water. These findings cannot be used to calculate the number of people that Earth can support, because that "carrying capacity" depends on several other factors:

- Level of affluence of the population (wealthier people generally use more resources per capita).
- Technologies, which can either intensify or moderate resource use for a given population size and level of affluence.
- Extent of trade among regions, which can redistribute resources and environmental impacts.

Glen Canyon Dam. From a photo by Donal W. Halloran

The data do suggest that primary production and fresh water are limiting factors for the expansion of the human population. At the current annual growth rate of 1.6%, the human population will double in 43 years (Cohen 1995). Although it is unlikely that this high population growth rate can be sustained, the next few decades will surely bring a large increase in the number of humans on Earth. Given current levels of resource use, this population would appropriate well over half the terrestrial primary production, most of the aquatic primary production that flows to industrially harvestable fisheries, and most of the available renewable fresh water on Earth! This level of use by people will substantially reduce the primary production and fresh water available to wild and semi-wild ecosystems. How will this affect the functioning of ecosystems and the availability of ecosystem services and resources for future generations? Ecologists have much work to do in addressing this question.

How much of an ecosystem service can safely be appropriated by people? To some extent, the answer depends on what people deem acceptable on ethical, cultural and social grounds. Ultimately, our appropriation is limited by its impact on resilience, the capacity of ecosystems to provide services in the future. How resilient are ecosystems? How can resilience be maintained? These are timely, important questions for ecologists.

How Are Toxic Substances Cycled in Ecosystems?

Certain metals, radioactive elements, pesticides, herbicides and other compounds released into the environment reach levels that threaten human health, wildlife and ecosystem processes. Studies of the cycling and transformation of toxic contaminants in ecosystems answer such questions as:

- How fast are these substances sequestered or broken down in ecosystems?
- Where and when are concentrations most acute?
- What levels of discharge are tolerable?

The basic principles of toxic contaminant cycling are the same as those for nutrient cycling. In fact, many of the early discoveries about nutrient cycling were made during studies of radionuclide cycling in contaminated ecosystems. However, some contaminants, including metallic elements and organic compounds, can build up to very high concentrations in organisms via **biomagnification**. This process increases the concentration of the contaminant as it passes up the food chain. Primary producers obtain the contaminant through active uptake or passive chemical mechanisms. Herbivores consume producers, and retain the contaminant much more efficiently than they retain carbon or nutrients. Carnivores consume herbivores, and they also retain the contaminant much more efficiently than they retain carbon and nutrients. With each consumption step, the concentration of the contaminant in the consumer biomass is magnified.

Mercury, a biomagnified metal, is a widespread contaminant of ecosystems. Mercury occurs naturally in an oxidized state that is unavailable to organisms. Release rates to the environment are greatly increased by burning coal and some other industrial processes in which mercury is reduced to elemental form, or a reactive compound such as mercuric chloride. In anoxic environments, a methyl group is added to dissolved mercury atoms to form methyl mercury. Methyl mercury is mobile, biochemically reactive, and is biomagnified. It accumulates in the muscles of fishes, wildlife and people to levels that can be toxic.

Organochlorine contaminants are examples of organic compounds that are biomagnified. Traces of organochlorine compounds occur naturally, but most of the organochlorine contaminants on Earth were made by human industry. They include pesticides like DDT and chlordane, lubricants and insulators like PCBs, and byproducts created inadvertently when chlorinated wastes are burned or chlorine is used to bleach paper (dioxins and furans) or purify drinking water (chloromethanes). Some organochlorine compounds are biomagnified in fat deposits and can reach toxic concentrations in fish, mammals, birds and people.

Figure 4.12. Cycle of organochlorine contaminants in a lake.

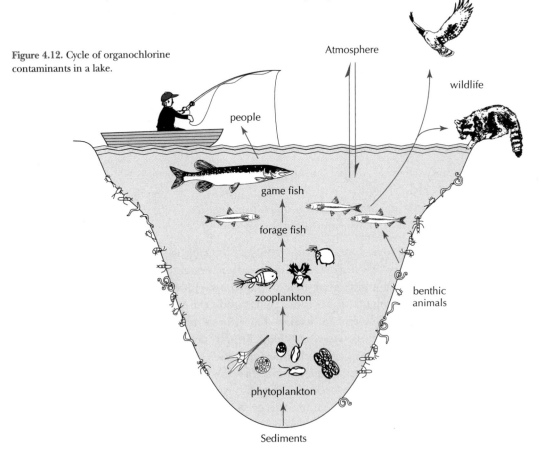

These compounds have extremely low breakdown or excretion rates in organisms, and extremely slow decomposition rates in detritus. Although hazardous organochlorine compounds such as DDT, PCBs and chlordane are no longer used in the United States, some of these compounds are used elsewhere in the world. Many organochlorine contaminants are mobile in the atmosphere, and trace amounts of them have been found even in animals from remote polar ecosystems.

The cycle of organochlorine compounds in contaminated lakes in the U.S. is driven by recycling from sediments and exchange with the atmosphere (Figure 4.12). New inputs from discharges or runoff are now very low; however, some organochlorine compounds are persistent in sediments. They are volatile, and can be lost to the atmosphere or added from the atmosphere. Organochlorine compounds are taken up by invertebrates feeding in the sediments, and by phytoplankton. They are then biomagnified through the food chain to zooplankton, forage fish, game fish, terrestrial wildlife (such as mink), fish-eating birds (such as cormorants, ospreys and eagles), and people.

PCBs in the Laurentian Great Lakes are an example of organochlorine pollution followed by a cleanup that is still under way. The Great Lakes were contaminated with PCBs in the 1950s and 1960s, and have been recovering since the manufacture of PCBs ceased in the mid-1970s. However, the rate of decline has been nearly zero since the late 1980s (Figure 4.13). There continues to be a net loss from the Great Lakes, because they are more contaminated than the rest of the environment and are losing PCBs as Earth approaches a global equilibrium distribution of the chemicals. Also, some PCBs are slowly degraded or buried in permanent sediments. Phytoplankton—microscopic, photosynthesizing aquatic organisms—concentrate PCBs about 100,000 times more than the surrounding water. This concentration step is explained by the high solubility of

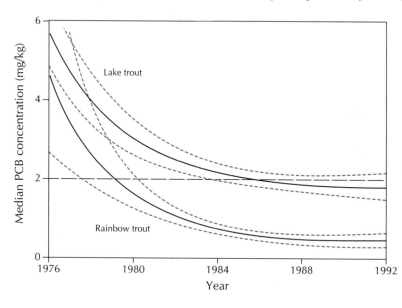

Figure 4.13. Time course of median PCB concentrations in lake trout and rainbow trout from Lake Michigan. The horizontal line is the U.S. Food and Drug Administration level at which people are advised to restrict their consumption of fish (2 mg/kg). Redrawn from Stow et al. 1995.

PCBs in the lipids of the phytoplankton. Concentrations increase further through transfers to zooplankton, forage fishes, and game fishes. Concentrations in game fish are several *million* times higher than those in the water. People are often advised to restrict their consumption of fish when total PCB concentrations in the fish exceed 2 mg/kg. Concentrations in lake trout, a slow-growing, long-lived fish with high fat content, were near 2 mg/kg in the mid-1990s. Concentrations in fast-growing rainbow trout were substantially below 2 mg/kg by the mid 1990s. This degree of recovery took about 20 years. However, the International Joint Commission that oversees the Great Lakes Water Quality Agreement has set a goal for PCBs in fishes of 0.1, not 2, milligrams per kilogram, and it will likely take decades to reach this lower level for lake trout. There is still concern that concentrations could increase again due to changes in the food web that increase biomagnification, mobilization from contaminated hot spots, or accidental release from disposal or storage sites.

Controversy about organochlorine contaminants in Lake Michigan is a typical example of debates about sustainability. How much contamination is tolerable? While high levels of PCBs are known to be harmful, we lack direct evidence of harm to human health from current concentrations of PCBs in Lake Michigan. Large amounts of PCBs still exist in storage or disposal sites, and environmental groups are concerned that restrictions on PCB release may be relaxed. Others are concerned about more toxic, but less abundant, organochlorine contaminants such as dioxins and furans in the Great Lakes. While debate continues, organochlorine contaminants in the lakes are slowly being broken down, washed downstream, or transferred to the atmosphere. Questions about organochlorine contamination of the Great Lakes and other ecosystems will likely occupy several generations of ecologists.

How Can Damaged Ecosystems Be Restored?

Can degraded ecosystems be restored? Yes, but with some significant reservations (Dobson et al. 1997). **Restoration** may be expensive and take a long time. Restoration is an experimental science. It may be necessary to try a tactic, assess its effects, and then try again. Restoration may be impossible at certain spatial scales. The ecology of small areas may be largely controlled by the surrounding landscape, and no self-sustaining restoration may be possible. For example, the University of Wisconsin's Curtis Prairie is continually invaded by plants from surrounding farms and urban areas and must therefore be continually managed to suppress the invasions (Box 4.6).

Restorations are usually motivated by decisions to re-establish or maintain some desirable ecosystem attribute, such as biodiversity, or some ecosystem service, such as erosion control, denitrification, or habitat for living resources. Ecological restoration is defined as "...the return of an ecosystem to a close approximation of its condition prior to disturbance" (National Research Council 1992). The goal is a self-sustaining system that is integrated with surrounding ecosystems.

Restoration uses information from all branches of ecology. The word "ecosystem" is often attached to restoration because the process involves the physical-chemical environment as well as the biotic community. Usually key features of the physical-chemical environment (such as water level, water flow, fire, or nutrient levels) must be changed to set the stage for biological manipulations such as reintroduction of native species. Also, restoration of ecosystem processes (such as productivity or denitrification) is usually among the goals of a restoration project. For example, wetland restoration involves the development of a characteristic plant community and associated animals. For the biotic restoration to succeed, the hydrology must be modified so that the site is flooded at the appropriate times of the year. The beneficial effects of wetlands include denitrification, retention of sediments and phosphorus, and production of humic compounds. These ecosystem processes improve water quality downstream.

Box 4.6. A PIONEERING RESTORATION PROJECT

Perhaps the first deliberate restoration of an ecosystem took place on an abandoned 60-acre field near Madison, Wisconsin (Sachse 1965). Prairie had occupied the site until the mid-19th century, when it was first plowed. The site was plowed regularly until the 1920s, thereafter heavily grazed by a herd of horses. In the 1930s, the field was purchased for the newly-formed University of Wisconsin Arboretum. In spite of the trampled, eroded soil, heavy brush and abundant thistles, botanists Norman Fassett and John Thomson were inspired by the idea of restoring a prairie on the site. Aldo Leopold, professor of Wildlife Ecology at the University, arranged for Theodore Sperry, one of the nation's foremost prairie ecologists, to direct the restoration. Sperry's crew burned the site to destroy brush and weeds, and plugged ditches to reduce erosion and restore the original drainage pattern. Volunteers combed southern Wisconsin for seeds and cuttings of the native prairie plants.

Only the steep bluffs, railroad rights-of-way, and a few roadside patches still harbored the original prairie flora of Wisconsin. Seeds were broadcast on the field, and strips of prairie sod were transplanted. Despite high mortality of the initial plantings, by 1938 the first prairie plants were blooming on the site. When a newspaper reporter asked Sperry how long it would take to complete the restoration, he replied "Roughly a thousand years." Over the 60 years since restoration began, the site has been burned every 1 to 3 years to approximate the natural fire regime of native prairie. The plant community composition now resembles native prairie. The soil is regaining the deep black color and high organic content characteristic of prairie soil. In 1962 the site was named Curtis Prairie after famed plant ecologist John T. Curtis. Curtis Prairie will always require regular burning and occasional targeted removals of invading species to sustain its restored condition (Plate 4.4, A, B, C).

The concept of alternate states is often used to describe restoration problems (Gunderson et al. 1995). Ecosystems may exist in more than one state. Each state is potentially self-sustaining if disturbances are not too large. Lake restoration provides a clear example of the alternate state concept. The problem of restoration is to shift the ecosystem from an undesirable state to a more desirable one. To accomplish this shift, the factors that maintain the undesirable state must be overcome, and the factors that sustain the desirable state must be activated.

Normally, several mechanisms buffer lake ecosystems from the effects of fluctuations in nutrient inputs. Heavy rainfalls may wash nutrients into a lake, or an accident may dump sewage or other nutrient-rich material into a lake. These inputs could potentially cause water quality problems, including outbreaks of toxic algae or protozoa, deoxygenation, and mass mortality of fishes and other organisms. However, there are many ways these impacts may be prevented. Belts of wetlands and submerged vegetation around the lake can reduce algal growth. Certain zooplankton species are effective grazers that can prevent outbreaks of algae. If excessive algal growth is prevented, then deep waters of the lake remain oxygenated. Oxygen oxidizes iron in sediments, converting it to a form that binds phosphorus , decreasing phosphorus recycling to algae. All of these processes tend to stabilize the primary production of algae at low to moderate levels. If these processes break down and excessive amounts of phosphorus are added, the lake becomes eutrophic.

The most common challenge in lake restoration is to shift a lake from a eutrophic state to the normal one (National Research Council 1992). Excessive phosphorus input from sewage, industry, and agricultural or urban runoff is the primary cause of eutrophication, and reduction of phosphorus input is therefore the first step in lake restoration. However, lakes often remain eutrophic even after phosphorus inputs have been decreased. The eutrophic state can be self-sustaining—phosphorus already present in the lake may be recycled effectively and maintain high production of algae; occasional large inputs from storms or accidents may sustain over-fertilization. Restoration requires additional steps to reduce the recycling of phosphorus, and buffer the ecosystem against occasional large inputs.

The specific choice of management actions will depend on the characteristics of the lake to be restored (National Research Council 1992). Wetland restoration goes arm-in-arm with lake restoration. Wetlands retain phosphorus pulses, denitrify nitrogen pulses, release humic compounds, and provide habitat for fishes. Restoration of submerged aquatic plants such as pondweeds and wild celery has many of the same benefits as wetland restoration. Restoration of large game fishes like northern pike can create trophic cascades that clear the water. Chemical interventions may also be needed to prevent recycling of phosphorus from sediments. Aluminum sulfate can be added to bind phosphate at the sediment surface; deep waters can be oxygenated mechanically to oxidize iron and bind phosphate. Such chemical interventions are often expensive and may be feasible only for small lakes, or as a brief shock treatment to interrupt phosphorus recycling and attempt to shift the lake to a normal state.

Restoration ecology is still a new and rapidly-changing science. What aspects of the physical-chemical environment must be changed, and by how much? What species must be introduced and what species must be eliminated? Can the restored ecosystem sustain itself? Every restoration project is an experiment, yet their numbers and diversity are increasing as society recognizes the value of restored ecosystems. As more restorations are attempted and studied, we will have the opportunity to learn a great deal more about this important application of ecology.

How Can Ecosystems Be Managed Sustainably in a Rapidly Changing World?

Ecosystem management is the process of sustaining ecosystems, their processes, and the services they provide for future generations (Christensen et al. 1996). The concept of ecosystem management is based on three premises. First, people must manage ecosystems. A hands-off approach is sure to fail in a time when people dominate most landscapes, appropriate much of Earth's production, alter most nutrient cycles, and pollute all ecosystems. Second, ecosystem processes and services can be reasonably self-sustaining under some, but not all, management regimes. Third, future generations deserve access to the same ecosystem services as we have.

Sustainability is defined in terms of values, context, and scale (Box 4.5). It depends on the size of the area to be managed. Small areas will have stronger interactions with the surrounding landscape than large areas, and will consequently require more intensive management. It depends on the rates of the processes to be sustained. Forest management plans can be projected for centuries, fish management plans for a few years. Sustainability involves the capabilities of ecosystems to persist and provide services, and the expectations of society. Both of these change through time, and they interact. Thus the study of sustainability is larger in scope than ecological science. Sustainability requires the integration of institutional systems with ecosystems.

During the 1970s, ecologists and managers invented Adaptive Ecosystem Management, or AEM (Gunderson et al. 1995). The idea of AEM was that knowledge of ecosystems is incomplete and evolving, so management actions should be viewed as experiments. Project design involved formulation of hypotheses about the effects of alternative management schemes, collection of data to evaluate the consequences of management actions, and explicit comparisons of management options. AEM could be used in an iterative fashion so that management policies improve through a sequence of trial, evaluation, comparison, and redesign leading to new trials.

How has AEM fared? Experience with a wide range of projects suggests that the interaction of management and science follows a cycle (Figure 4.14) with important differences from the original vision of AEM (Gunderson et al. 1995). During early

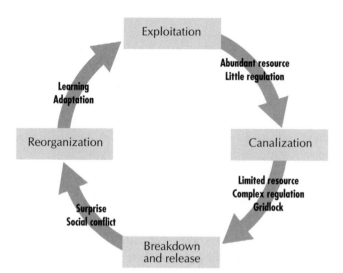

Figure 4.14. Phases in the cycle of ecosystem management Redrawn from Gunderson et al. 1995.

stages of exploitation, ecosystem resources are widely available and there is little regulation. As resources become limiting, regulations become more complex and gridlock develops. During this period, which may take years or decades, bureaucracies become more efficient but less flexible in their practices. While the bureaucracy becomes narrowly focused and rigid, the ecosystem and society are changing. Eventually the prevailing policies are clearly incompatible with the state of the ecosystem and society's expectations of the ecosystem. At this point, change can happen very quickly. The ecosystem may collapse to an undesirable state, or social conflict may develop over resource policies, or both. Eventually the process of rebuilding begins. Task forces and networks of managers, scientists and stakeholders analyze the problems with old policies, the condition of the ecosystem, and societal needs. Modeling and experimentation help discover new policies that appear to work. Creative science and innovative management thinking are likely to occur during the transition from the release phase to reorganization. During reorganization, the new knowledge is transformed into policy and a new phase of exploitation begins.

One consequence of the cycle of AEM is that progress in ecosystem management is not a continuous process but rather occurs in lurches. Long periods of apparent stasis are punctuated by brief periods of innovation, experimentation, discovery and learning.

The process of ecosystem management forms collaborations with managers and scientists from many disciplines. These collaborations are not new, hybrid sciences. The participating disciplines retain their core concepts and fundamental questions, and add new questions at the interface with other disciplines. It is simply too inefficient to reinvent the parent disciplines, and it is hard to anticipate which aspects of their core knowledge will be important to the problem-solving process.

Herein lies an important lesson for students. An interdisciplinary team is a group of specialists who need each other to act. It is not a collection of generalists all working in the same way on the same problem. Because the mix of specialties needed for interdisciplinary problem-solving is continually changing, it is impossible to predict the mix of specialties needed for future problems. As new problems arise, new teams are assembled of members with appropriate skills. Participants should be deeply knowledgeable in their core disciplines—thus a strong disciplinary base is an important goal in training students. The core skills will be recognized and valuable in future opportunities for productive interdisciplinary collaborations.

LINKS TO OTHER KINDS OF ECOLOGY

Ecosystem ecology has strong and active connections to the other kinds of ecology and to evolutionary biology. In fact, the boundaries between ecosystem ecology and other kinds of ecology are not always easy to discern. Ecosystem ecology is also linked to biogeochemistry, geology, hydrology, meteorology, oceanography and soil science, and is now building links to economics and other social sciences.

Physiological ecology (Chapter Five) has been strongly associated with ecosystem ecology for several decades. Physiologists have devised many of the methods and models used to measure ecosystem process rates, such as rates of primary production, nutrient exchange, and consumption and excretion by animals. Ecosystem ecology, in turn, studies the environment in which ecophysiological processes occur.

Landscape and ecosystem ecology have been closely associated since the origin of landscape ecology as a self-conscious discipline. Landscape ecology (Chapter Three) has influenced ecosystem ecology through conceptual approaches to problems of scale and methods for modeling spatially heterogeneous processes. Questions about large-scale spatial heterogeneity in ecosystem processes motivate a considerable amount of research that bridges ecosystem and landscape ecology. For example, how do we combine models of individual plant production and nitrogen cycling to study carbon and nitrogen flows through an entire watershed? This question requires a combination of ecosystem and landscape ecology.

Questions about the physical and chemical environment of community ecology (Chapter Eight) prompted much early work on ecosystems. For example, ecologists have studied the relationships between ecological succession and processes of production and nutrient cycling.

Population ecology (Chapter Seven) and ecosystem ecology have always been linked in the sense that population dynamics occur in ecosystems. Ecologists have also discovered many ways in which populations of a single species or a few closely related species have enormous effects on ecosystem processes (Jones and Lawton 1995). Two well-understood examples are the trophic cascade in lakes (Box 4.2) and the effects of animals on ratios of limiting nutrients (Box 4.4). Other examples are numerous. Beaver control the movement of water through landscapes, the course of plant succession near their ponds, and the flow of nutrients and trace gases from the areas they inundate. Rates of nitrification and denitrification depend on just a few genera of bacteria. In these examples, and many more, the population dynamics of one or a few species profoundly affect ecosystem process rates. Sometimes these effects hinge on behavioral properties of the organisms, thereby linking ecosystem and behavioral ecology (Chapter Six).

SUMMARY

Ecosystem ecologists ask questions about the transport of energy and matter through organisms and their environment. An ecosystem is an explicitly defined sector of the Earth, including all of the organisms and their physical-chemical environment within its boundaries. Ecosystem boundaries can be arbitrary, but ecosystems with tangible boundaries (such as lakes or watersheds) have proven especially useful for ecosystem studies. In practice, the definition of the ecosystem for a particular study depends on the question being asked.

Some of the fundamental questions of ecosystem ecology are:

- How productive are ecosystems?
- What controls their productivity? How are production of plants and animals connected?
- How are limiting nutrients, such as phosphorus and nitrogen, transformed and cycled in ecosystems?

Much work remains to be done to provide answers, and understand the processes of organic production that support life on Earth. Applied questions of environmental science also play important roles in ecosystem ecology. Some examples are:

- How much of Earth's productivity is appropriated by people?
- How are toxic substances cycled and biomagnified in ecosystems?
- How can damaged ecosystems be restored?
- How can we devise institutions that sustain ecosystem resources and services over extended periods of time?

Such questions are becoming urgent in a time when humans appropriate a significant fraction of Earth's production and available, renewable fresh water and have affected all ecosystems. We are just beginning to apply ecology to understand the interactions of people and ecosystems.

Ecosystem ecology has important connections to the other kinds of ecology, evolution, the Earth sciences, and social sciences. Cross-disciplinary interactions are a growth point for ecosystem science.

Models are used to describe ecosystems, learn about them, and predict their behavior. Learning with models involves the invention of alternative models, and tests using field observations that rule out implausible models and corroborate plausible ones. Models take on diverse forms including the definition of ecosystem boundaries, to box-and-arrow diagrams, verbal or diagrammatic hypotheses, statistical models for data, and mathematical models.

Aspiring ecosystem ecologists are advised to learn the basics of mathematics, physics, and chemistry that usually accompany bachelor's degree programs in the biological sciences. It is important to learn to write capably. Most ecosystem ecology is performed by interdisciplinary teams of specialists who need each other to accomplish a larger goal. The ideal team member has cultivated useful specialized knowledge and the ability to collaborate with other specialists. The best ecosystem ecologists are lifetime learners whose knowledge broadens throughout their careers, but they start out from a base of excellence in a core discipline.

RECOMMENDED READING

F. S. "TERRY" CHAPIN AND PAMELA A. MATSON. *Principles of Ecosystem Ecology.* Springer-Verlag, N.Y. 1998
>   *A modern, advanced textbook of ecosystem ecology.*
GRETCHEN DAILY, (ED.). *Nature's Services.* Island Press, Washington, D.C. 1997.
>   *Services provided by ecosystems in relation to economics.*
FRANK B. GOLLEY. *A History of the Ecosystem Concept in Ecology: More Than the Sum of the Parts.* Yale University Press, New Haven, Connecticut. 1993.
>   *The rise of ecosystem ecology in the U.S.*
HEINRICH D. HOLLAND AND U. PETERSEN. *Living Dangerously: The Earth, Its Resources, and the Environment.* Princeton University Press, Princeton, N.J. 1995.
>   *How humans are changing the biology and geochemistry of the planet.*
GENE E. LIKENS. *The Ecosystem Approach: Its Use and Abuse.* Ecology Institute, Oldendorf/ Luhe, Germany. 1992.
>   *Reflections on ecosystem science in relation to environmental policymaking.*
MICHAEL L. PACE AND PETER M. GROFFMAN (EDS.). *Successes, Limitations, and Frontiers in Ecosystem Ecology.* Springer-Verlag, N.Y. 1998.
>   *Critical review of the accomplishments and challenges of ecosystem science.*

Chapter 5

PHYSIOLOGICAL ECOLOGY
Tradeoffs for Individuals

James F. Kitchell

> *It would be instructive to know…what circumstances in the life-history and environment would render profitable the diversion of a greater or lesser share of the available resources towards reproduction.*
> — *R. A. Fisher*

WHAT IS PHYSIOLOGICAL ECOLOGY?

Physiology is the study of function at the scale of individual organisms, and ecology is the study of interactions between organisms and their environment. Both apply to any living thing, from bacteria to whales, and in any place they live. Synonyms for this approach include *autecology, ecophysiology, ecological physiology,* and several others, but we will stick to terms used in the chapter title and allow the reader to substitute as necessary. Physiological ecology focuses on the performance of individuals within the **constraints** of physical-chemical (**abiotic**) conditions such as temperature of the habitat, matter available as nutritional resources, or light energy that can be used for photosynthesis. These constraints define whether the individual lives or dies, with its relative success expressed as growth and reproduction. Within the boundaries created by constraints, organisms carry out the biochemical processes and express the adaptations that accomplish the balance termed **homeostasis** (literally translated as "same place" or "one state"). Homeostasis involves the maintenance of time, matter and energy budgets that allow for survival, growth and reproduction by the individual. A graphic representation of this conceptual framework is presented in Figure 5.1.

QUESTIONS ASKED BY PHYSIOLOGICAL ECOLOGISTS

Definitions are intended to be both technical and precise. What ecologists really do is better represented by the common kinds of simple questions they ask:

- Is species A present in this habitat or not? If not, why not?
- Why are members of species N much more abundant in one location than another?
- Why do members of species N grow better in one location than another?
- This organism is dead. Did some physiological stress kill it?

The limitations imposed by individual abiotic factors can be readily defined in controlled laboratory conditions. Much of the knowledge about the ecology of individual plant and animal species derives from laboratory experiments. By analogy, laboratory results express an organismal response to **Liebig's Law of the Minimum**, which comes from chemistry and states that the rate of reaction proceeds in proportion to the availability of the reactant in shortest supply. Its ecological equivalent makes much the same

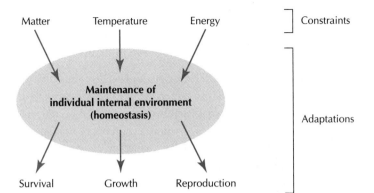

Figure 5.1. A graphic representation of the conceptual framework for the interactions of environmental constraints and adaptations that influence homeostasis as expressed in survival, growth and reproduction.

case for a critical nutrient (e.g., nitrogen, potassium, calcium, Vitamin A, molybdenum, etc.), or an environmental factor required to sustain the life of the individual, or one that constrains growth or reproduction (e.g., maximum summer temperature, minimum oxygen concentration, soil moisture, light for photosynthesis, etc.). These critical limits set the template for existence. However, organisms live in a multivariate world, and the interaction of abiotic and biotic factors becomes important in understanding trade-offs or compromises of the cost of acquiring resources or avoiding stressors, and the benefits of having done so successfully.

Key Concepts of Physiological Ecology

Three concepts are key to recognizing the role that physiology plays in an ecological context:

- *The ecological niche is the conceptual cornerstone.* This multi-dimensional framework defines all of what a species could do and all of the places where it could exist.
- *A few ecological **master factors** typically play a major role in the survival, growth, and reproduction of organisms.* Master factors are factors that vary in time and space in ways that play a major role in the survival, growth, and reproduction of organisms in the environment. For example, temperature is often the master factor in physiological studies of any kind of organism. In addition, plants and photosynthetic bacteria are strongly affected by light and inorganic nutrients such as nitrogen and phosphorus. Food is a master factor for non-photosynthetic organisms. Water balance is often an important constraint for terrestrial organisms; just as oxygen concentration often is an important constraint for aquatic organisms, especially animals.

• *Average conditions are rarely good predictors of a species' distribution in nature.* Instead, the presence or absence, persistence and abundance of a species in any habitat often traces to rare but extremely stressful events that literally push an organism to the edge of life or death.

Within these boundaries, interactions with other species play the primary roles that determine changes in abundance and distribution. Stated another way, the question that physiological ecologists ask is: What habitats are possibly occupied by this species? Typically, the suite of possibilities exceeds the realities of where members of that species are found. Therefore, other kinds of ecological approaches are often required to explain the observed abundance and distribution of the species.

Master factors are ubiquitous and powerful conditions that dictate presence or absence. For example, life is largely a set of temperature-dependent biochemical reactions taking place under **aerobic** conditions within the aqueous medium of cells. Evolutionary forces have allowed many organisms to find partial exception to that fundamental fact. These exceptions provide some of the most fascinating ways to understand how physiological processes work under physical constraints.

In the long term, the dynamics of abiotic and biotic boundaries define the biogeography of the species and its role in a community or ecosystem context. We don't have to incorporate each and every variable in our thinking. A few key ones will usually explain most of the dynamics. In fact, the evidence for the importance of temperature and water as master factors derives from the general observation that these two variables are the key descriptors of our terrestrial biomes—deciduous forest, grassland, desert, and others (see Chapter Four), and that interactions of temperature and water balance (**osmoregulation**) define the realms as arctic, temperate, and tropical plus freshwater, estuarine or marine. Oxygen qualifies as a key variable because it is a requirement for most living forms. Box 5.1 offers an important exception to that generalization.

The baobab tree pictured in Plate 5.1 offers a tangible example of the view argued above. It is one of the few trees among shrubs and grasses in the semi-arid region within Tsavo National Park in Kenya. How does a tree live in a habitat where water is scarce, and nearly all other plants present indicate that scarcity? The baobab has developed two central mechanisms for water conservation. First, the tree has a deciduous habit. Leaves are required for photosynthesis, but the open stomates that allow carbon dioxide to enter the leaf also allow water to escape as vapor. When the dry season begins, the baobab drops its leaves and forgoes photosynthesis, thereby reducing water loss. During the rainy season, the tree grows new leaves and stores water in its tissues. The tree literally swells as this water reserve builds in preparation for the next dry period. In a typical year, baobabs do not need to depend on their stored water resources. However, droughts occur in this region, and the tree's reserve is sufficient to see it through an extended dry period of up to several years. Evidence of such hard times is apparent on the trunk (bole) of the tree in Plate 5.1. Elephants have gouged tissue from this tree and chewed it to get moisture during a drought several years ago.

BOX 5.1. AN IMPORTANT EXCEPTION

Most organisms owe their life to the carbon fixation process due to *photosynthesis,* which combines carbon dioxide and water to create high-energy sugars and oxygen as a by-product. We live in that part of the world where oxygen is available. We eat the organic products passed through food webs supported by those plants, and the oxygen we breathe comes from the by-product of photosynthesis. This is the central tenet of biology textbooks. However, there is an interesting alternative. *Chemosynthesis* can take place in the absence of oxygen. Chemoautotrophic bacteria are primary producers that can use chemical energy from hydrogen sulfide (H_2S) analogously to the way light energy is used in photosynthesis, to synthesize complex carbon compounds. These bacteria form the basis for food webs in habitats that lack light and have little or no oxygen. The best-known examples are those that develop around deep-sea hydrothermal vents where hot springs seep through the sea floor. The heated water accelerates biochemical reactions, and contains an abundance of reduced chemical compounds including the methane and hydrogen sulfide that are the basic requisites of chemosynthesis by some kinds of bacteria. These bacteria become the food resource for some locally abundant, bizarre life forms such as the huge, gutless tube worm *Riftia* (so named because the hydrothermal vents develop in volcanic rifts between plates in the Earth's crust), plus an assemblage of clams, mussels and crabs found only in areas where deep sea vents have developed. Films from the submersible *Alvin* first revealed these unusual communities around hydrothermal vents near the Galapagos Islands. Since then, they've been found in many locations (Pinet 1998).

Chemosynthesis under anaerobic conditions supports a diversity of less-publicized food webs (Arp and Fisher 1995). Wherever hydrogen sulfide is abundant, some bacteria can use it as an energy-rich resource. Other organisms have evolved biochemical mechanisms for dealing with the potentially toxic sulfide compounds, which allows them to occupy these habitats and consume bacteria or incorporate them as symbionts (Childress 1995). The fauna inhabiting anaerobic muds and waters of estuaries, lakes and oceans includes many forms (e.g., protozoa, nematodes, oligochaete and polychaete worms) that consume chemosynthetic bacteria, or assimilate the organic compounds they produce as symbionts. Thus, the diversity and extent of chemosynthetic-based food webs is much greater than introductory text books would lead us to believe.

Nevertheless, photosynthesis is the primary production process that forms the basis for most ecological systems. Variability in the availability of oxygen, temperature conditions and water balance offer the first and best evidence of the role that physiological constraints play in explaining the abundance and distribution of most organisms.

The tree has since recovered from those wounds. In some areas of Tsavo, a burgeoning population of elephants destroyed many trees during an extended drought. Baobabs are not found there now, but knowledge of their physiological ecology would lead a physiological ecologist to predict otherwise. The baobabs *should* be there. The physiological mystery is solved only by knowledge of the region's ecological history. Similarly, knowing the water needs of elephants would lead us to think that these animals could not have survived such a drought—knowledge of physiological constraints must take into account elephants' ability to survive a drought by securing water from the trunks of baobab trees.

The baobab and elephant examples offer evidence of both the trade-offs made by individuals in maintaining homeostasis, and the importance of rare events. Homeostatic adaptations of the baobab are represented by the deciduous habit and the capacity to store water. Rare events such as a prolonged, severe drought and the consequent attack by elephants can cause baobabs to be absent from a habitat they might otherwise occupy very successfully.

How Do Abiotic Factors Regulate the Distribution and Abundance of Organisms?

The concept of an ecological niche is a foundation for thinking about how organisms are distributed and what controls their relative abundance. The most cogent presentation of that idea owes to G. E. Hutchinson (to whom this book is dedicated) in his description of the niche as an "n-dimensional hypervolume." What does this mean? Think of the niche as the equivalent of an ecological hologram for each kind of organism. Each environmental variable is like a band in the color spectrum that defines some part of the hologram seen in three dimensions. Hutchinson separated the niche into two parts: the **fundamental niche** and the **realized niche**. The fundamental niche is defined by the tolerance levels of an organism (i.e., the *possible),* while the realized niche is that subset of tolerable conditions actually occupied by the organism (i.e., the *observed).*

Figure 5.2 provides a visual description of this niche concept (Lampert and Sommer 1997). The largest area in Figure 5.2 is the fundamental niche displayed for the constraints of two environmental factors. For a phytoplankton cell those axes might be phosphorus concentration and light intensity, for a terrestrial plant they might be soil nitrogen content and annual rainfall, while for a fish they might be water temperature and oxygen concentration. The zone defined by the fundamental niche includes those combinations of factors that the organism might actually occupy. A smaller zone defines conditions that allow maintenance and favor growth. Successful reproduction can occur within an even smaller zone. Adding another variable (e.g., photoperiod) to the picture takes the niche into three dimensions ; adding yet another variable (e.g., food resources) takes it to four; eventually it becomes "n-dimensional," where n is defined by the number of things we've measured or can imagine might be important.

Think of the niche as a fully inflated balloon whose volume represents the potential population for a species in some habitat. That volume defines the fundamental niche—the limits of tolerance to the abiotic environment. The realized niche is represented by some deflation of the balloon. Deflation occurs because of interactions with other species which compete for available resources, prey on or parasitize the members of the species of interest, and therefore reduce the potential for growth and reproduction. Extreme abiotic or biotic stress can cause additional deflation; causing local extirpation of the population—the species becomes rare or even disappears from a place where it once occurred. Predators or competitors may prevent the species from effec-

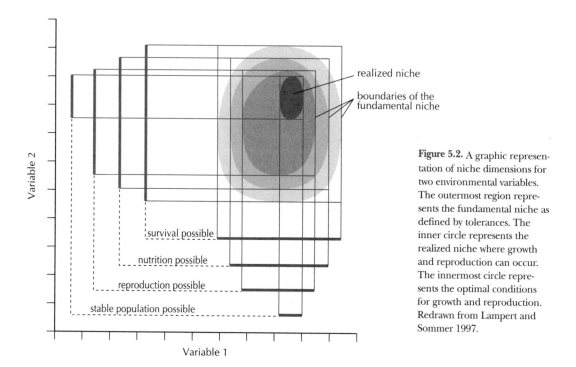

Variable 2

survival possible

nutrition possible

reproduction possible

stable population possible

Variable 1

realized niche

boundaries of the
fundamental niche

Figure 5.2. A graphic representation of niche dimensions for two environmental variables. The outermost region represents the fundamental niche as defined by tolerances. The inner circle represents the realized niche where growth and reproduction can occur. The innermost circle represents the optimal conditions for growth and reproduction. Redrawn from Lampert and Sommer 1997.

tively colonizing a habitat where abiotic conditions might otherwise allow it to survive and flourish. Each species has its own "balloon"—a set of tolerances, preferences, predators and competitors that define its n-dimensional hypervolume.

Some ecologists bicker about the definition of niche because they have trouble with the concept of an *empty niche*. Can the niche only be defined by what an organism does? Should it be defined as where an organism lives? Both definitions seem insufficient, since neither can account for the continuing parade of invasions by non-native species. A more holistic and functional view may be found in the larger question: Species N is found over there, so why isn't it found here? Three answers follow:

- Species N hasn't dispersed into this place.
- Species N could live here under typical conditions, but occasional extreme events prevent it from establishing a sustainable population.
- Some set of biotic interactions prevent Species N from persisting here.

The order of possible answers leads to greater insight into the importance of abiotic and biotic constraints. The boundaries of possibility are defined first by tolerances, then by minimum growth and finally by successful reproduction. In combination, this hierarchy of constraints defines the possible limits of biogeography, relative abundance, and habitat use by members of a species. Finally, biotic interactions can reduce or expand the boundaries of the realized niche, creating the actual distributions observed in nature.

More baobab trees might survive the droughts at Tsavo if elephants were less abundant. Once again, the key idea is the framework for that first question: Why isn't species N found in (or more abundant in) this place?

A Walk Along the Rio Grande

The discussion presented above is based on "evidence of absence" and can be illustrated by an example drawn from a current controversy over water regulation in rivers (Carpenter et al. 1992). If we were to walk along the banks of the Rio Grande River south of Albuquerque, New Mexico, we would be among a stand of magnificent cottonwood trees growing in the floodplain. All of the cottonwoods are large, old, and produce copious amounts of seed each spring. Yet there are no young cottonwoods. How will this population be sustained if there is no successful reproduction?

The physiological perspective starts with asking whether young trees are able to grow within the deep shade of the existing adults. In other words, is light availability the limiting factor? If we keep walking, we'll discover that some of the oldest trees have died, leaving large gaps in the forest canopy. The absence of seedlings in gaps tells us that light limitation is not the answer. Could competition for water or nutrients limit seedling growth? We note that much of the vegetation beneath the tree canopy is dominated by an invading species, the salt cedar or tamarisk (*Tamarix* spp.), which came from Asia in the last century and is spreading across the water courses of southwestern United States. When gaps develop, salt cedar can invade and dominate. Nevertheless, we find recent gaps where salt cedar is not yet dominant, cottonwood seeds were plentiful last year, and those gaps still have no young cottonwoods. This tells us that neither light nor competition for water or nutrients are limiting the development of cottonwood seedlings.

Examining the terrain, we see that the forest is in a floodplain, but floods are rare these days. Water reservoirs built on mountain tributary streams capture snowmelt and store it for gradual release to the downstream cities and summer irrigation needs of agriculture in the Rio Grande valley. Is a critical combination of timing and water availability required for successful germination by cottonwoods? Are the natural spring floods required? Dikes protect the agricultural areas, but the floodplain between the dikes and the river still experiences floods at regular intervals. However, the floodplain also lacks cottonwood saplings in recent gaps.

What's left? Mature forests of the Great Lakes region have very little natural reproduction by the dominant trees because deer are now more abundant than at any other time in this century. Deer eat virtually all the young trees. Taking this idea into context in New Mexico, perhaps the area that still has floods can support reproduction, but the herbivores eat the seedlings as they sprout? This question can be answered with an experiment—fence an area to exclude the herbivores from a gap within the remaining flood plain. If that doesn't produce the answer, our empirical data still leaves two possibilities.

The salt cedar somehow alters the environment (probably the soil), or some combination of timing and magnitude of flooding has changed to prevent successful sprouting and growth of cottonwood seedlings. However, salt cedar effects have been tested by clearing and plowing a gap within the flood plain, and this still doesn't produce cottonwood saplings. Only the critical-timing question remains. In fact, dams and water storage in the headwaters not only reduce the magnitude of the natural spring floods, but also delay these floods by several weeks. That change may de-couple the environmental conditions under which cottonwoods evolved their seed production pattern, because flowering and seed set are cued by seasonal changes in photoperiod and temperature. The flood now comes in the summer instead of during late spring. In dry years, most of the river flow is retained by upstream reservoirs, and no flood occurs. If the presence and timing of a flood event are the key to cottonwood regeneration, an experimental flood is the next step, and it should be timed to simulate historical floods before humans began modifying river flows.

This simple (and real) example illustrates the "deductive engine" of ecological science (Walters 1985). It starts with physiological constraints represented by the abiotic limits of the fundamental niche (light and water), leads to key biotic components of the realized niche for this species (competition and herbivory), then back to the abiotic coupling of organism with rare events in the environment (floods and their timing). While the original question—why aren't young cottonwoods here?—can't yet be answered, the logic above leads to a reasonable way to test for the likely explanation.

What about the possibility of an interaction effect? Cottonwoods may regenerate only under the conditions of a natural flood cycle *and* low herbivory. Although deer are native herbivores, cattle and sheep are not. Exploring the question of interaction effects is important because a large assemblage of plant and animal species are strongly associated with the presence of mature cottonwoods. Sustaining this riparian ecosystem will require an understanding of critical events in the early life history of one species. Water is the key, and competition for its allocation is fierce in this arid region. Cottonwood regeneration may need a flood, and the cities or agricultural interests in the Rio Grande Valley may need to compromise their water demands if maintaining the cottonwood forests is a desired outcome. In this case, as in many, the role of the ecologist is to help determine the possible alternatives, but the choices must be made in a public forum.

How Do Abiotic Factors Interact to Control Growth and Reproduction?

The approach most commonly used in adopting an organismal view starts with asking for evidence of the limiting conditions for abiotic factors—temperature, humidity, pH, critical ions, essential nutrients, etc.—as depicted in Figure 5.1. Limiting conditions are first expressed as *tolerances;* e.g., what are the upper and lower tolerance levels for temperature? After defining what's possible (the fundamental niche), ecologists then refine those limits to determine actual performance (the realized niche). Again, the approach is based on estimating the possible growth, or **scope for growth,** which in turn is based on a balanced energy budget.

The scope for growth

Questions about relative performance start with an estimate of the possible: What temperature conditions are required for growth and for reproduction? How do resources constrain feeding and growth? We then seek measures of preferred or optimal conditions—those that provide for highest efficiencies and maximum rates of growth or reproduction. Collectively these define the scope for growth, which includes the potential for gamete production. The scope for growth is defined by the balance of an energy budget.

Energy budgets

Energy budgets offer a universal way to compare organisms of all kinds because all organisms must acquire energy to maintain themselves, grow and reproduce. Energy is the common currency of life. Budgets are helpful because organisms must obey the Laws of Thermodynamics, i.e., the budgets have to balance. Growth occurs when all other terms of the energy budget are satisfied, and energy can be stored in the form of biomass. **Autotrophs** (organisms that photosynthesize, but see Box 5.1) use light energy to convert the chemical energy in water and carbon dioxide into biomass. **Heterotrophs** (e.g., *Daphnia*, sharks, bacteria, elephants, etc.) convert consumed nutrients and various organic compounds into their own biomass. A universal energy budget states that outputs must equal inputs, which means that natural selection will operate to favor the ability to capture energy and use it efficiently, thus most likely leading to greater reproductive output. Stated as an equation,

$$\text{Inputs} = \text{Outputs}$$
$$\text{or}$$
$$\text{Inputs} = \text{Metabolism} + \text{Losses (wastes)} + \text{Storage (growth and reproduction)}$$

There are conditions and constraints to achieving a budget that allows some allocation of energy to storage. In fact, there is an incredible array of solutions to the energy budgeting problem. Some of those are developed in the example described in Box 5.2, skipjack tuna vs largemouth bass.

Efficiency and the optimization problem

Efficiency in energy use is defined as the ratio of the rate of energy storage to the rate of energy acquisition. As an equation, that is,

$$\text{Efficiency} = \text{Storage} / \text{Inputs}$$

Energy budgets may be altered in two ways to improve the efficiency of energy use. One way is to maximize inputs (benefits), the other is to minimize outputs (costs). This optimization approach was borrowed from economics and serves as a very valuable framework for evaluating how organisms solve the budgeting problem (Townsend and Calow 1981).

Think of optimization of energy use in terms of the problems a typical student faces in dealing with her financial budget. First, the student must solve the income problem; that may mean finding a job, but there's a constraint of time—the student can't be in class and at her job at the same time. Other organisms face the same problem; they must gather "metabolic income" but have a finite amount of time. Spending time acquiring resources precludes spending time on other activities (avoiding predators, sleeping, mating behaviors, etc.). For humans, money and time are important currencies; for other organisms, the currency is some form of energy acquisition tallied as photosynthetic rate or prey eaten per hour. Returning to our student's dilemmas, her next decision deals with prioritizing costs. She needs housing; therefore rent must be paid. Other organisms "pay rent" in metabolic units, and that cost increases with their size. Our student must pay taxes that are roughly proportional to income. Members of other species "pay taxes" in the form of waste losses (indigestible parts of prey, disposal of toxic metabolic by-products, etc.), and these expenses increase in proportion to resource intake. Finally, our student may have some money left to "store" in her checking account. Similarly, other organisms can allocate extra energy to storage as somatic (body) growth, or invest it in reproductive tissues. Both students and other organisms can tolerate debt for a time, but eventually it must be repaid. Excessive, long-term debt causes the economic or ecologic unit to lose its place in the habitat.

The idea of a hierarchy or a sequential order of allocation is important because it emphasizes the many ways that natural selection may act to increase efficiency by reducing costs or by increasing inputs. Both methods work to increase the positive components of the budget, reflected in its last allocation—storage for future uses. In the student example, success is measured by completion of a degree program; in the organismic equivalent, by reproductive output. Our simple analogy was developed in familiar currencies. In fact, the optimization problem is a central theme of work in behavioral ecology and is treated in much greater detail by Chapter Six.

How Do Organisms Alter Their Environment?

As a consequence of the organism's activities, the local environment changes and homeostatic mechanisms adjust accordingly. Shade develops as a plant grows. This can enhance or reduce the potential of the immediate environment for an adjacent plant. While shade reduces light availability, proximity to another plant can have advantages where desiccation is a constraint. The microenvironment downwind from that plant is slightly moister,

BOX 5.2. ALTERNATIVE ENERGY BUDGETING STRATEGIES: SKIPJACK TUNA AND LARGEMOUTH BASS

This example offers a comparison for two fishes that eat similar prey (invertebrates and small fishes), are of the same size (1 kg) and carry out their energetics under comparable temperatures (24°C) (Kitchell 1983). The largemouth bass is abundant in lakes and rivers throughout the temperate zone. Native to warmwater habitats in North America, it has been introduced to lakes in Central America, Cuba, South America, Europe and Africa, because anglers hold it in high regard. A behavioral ecologist would view this fish as a "sit-and-wait" predator. It forages by ambushing prey or by slowly searching in its preferred habitat, the highly productive littoral zone of lakes and shallow waters of rivers. In the course of a day's searching activity, a bass may swim very little or as much as a few kilometers. The largemouth bass has a swim bladder for buoyancy regulation, spends very little energy to counteract its density relative to that of water, and is therefore confined to very slow changes in depth. If it were to rapidly swim through 10 meters of depth change, its swim bladder would double in size. A depth change of 30 meters would cause the swim bladder to burst.

The skipjack tuna is abundant in subtropical and tropical seas throughout the world. It has an enormous habitat but occupies an environment that has very low levels of primary productivity (Chapter Four). A behavioral ecologist would describe it as an "energy speculator"—it forages very actively and finds widely dispersed prey by maximizing encounter rates through intense, continuous searching. Its deep muscle mass contains a vascular network that functions as a counter-current heat exchanger. Blood circulating in the capillaries of active muscles is heated by the biochemical reactions that cause muscle contraction. Blood returning to the heart flows through a venous network adjacent to an analogous network of arteries carrying oxygenated blood from the heart. Arterial blood has been cooled by exposure to sea water temperatures surrounding the gills. The result of this special vascular anatomy is that arterial blood is warmed as it moves to the muscles, warmed again as a consequence of muscle activity, collected in veins and routed through the heat exchanger, where it again warms arterial blood before returning to the heart. This mechanism for conserving heat maintains an internal body temperature that is consistently several degrees above water temperature, a difference that increases in proportion to activity. The efficiency of converting biochemical energy to mechanical energy (work) is also increased by the higher temperature. If a skipjack is forced to swim against resistance and thereby prevented from adequately irrigating its gills, its body temperature will continue to increase to above 40°C, and it literally denatures its own proteins. This fish works like an air-cooled engine. It does fine if it's moving, but sitting still at high rpms will cause it to overheat.

Estimated energy budgets (J • kg^{-1} • day^{-1})
for 1-kg largemouth bass and skip jack tuna at 24°C.

Fish	Ration	Energy budgets (J • kg^{-1} • day^{-1})						
		C	R_{St}	R_{Act}	SDA	U	F	ΔB
Largemouth Bass	Maintenance	8.1	4.9	0.5	1.1	0.6	0.8	0.0
Skipjack Tuna	Maintenance	44.0	13.5	15.5	7.0	2.0	7.0	0.0
Largemouth Bass	Maximum	31.0	4.9	0.5	4.4	2.5	3.2	15.6
Skipjack Tuna	Maximum	250.0	13.5	80.0–142.0	38.0	12.0	38.0	7.0–69.0

Skipjack have no swim bladder for hydrostatic regulation. This allows them to exploit the entirety of the open water habitat without regard to pressure changes. They can swim from depths of 400 meters to the surface in a few seconds, making them extremely effective at pursuing prey. Of course, there is a price for this adaptation. Skipjack cannot ventilate their gills and oxygenate their blood sufficiently unless they swim continuously. Just meeting its minimum oxygen requirements demands that the skipjack swim at least 60 kilometers in a day. A skipjack must swim at about one body length per second for all of its life.

The contrast presented in the table of feeding and output rates (and the pie chart that presents the outputs as % of total) demonstrates that a skipjack tuna conducts energy budgeting at about five times the rate observed for a largemouth bass. Respiratory costs for standard or maintenance metabolism (Rst or "rent") are almost three times greater for the skipjack. Its cost of swimming (Ract) in order to encounter prey at a maintenance ration level is 30 times greater. The metabolic cost of digestion and assimilation of those prey (SDA or specific dynamic action), of losses due to excretion

(U) of wastes (mostly toxic nitrogen as ammonia), and the losses due to fecal wastes (F) are proportional to the amount of food that must be eaten in order to sustain the current mass for both species, i.e., growth (ΔB) is zero. A skipjack tuna must eat more than five times the prey requirements for a largemouth bass.

A 1 kg skipjack tuna is about one year old and has inhabited warm tropical seas (23°–25°C) all of that year. A bass weighing 1 kg would be 4–5 years of age in most temperate lakes. The bass spends an extended period of each year at the low water temperatures of winter conditions. As demonstrated in Figure 5.5, its scope for growth is very low for much of each year. Under summer temperature conditions (24°C) and high food availability, it can grow rapidly and very efficiently because it pays lower costs to foraging activity than the skipjack.

The point of this example is that strikingly different animals must still obey the Laws of Thermodynamics. Adaptations that minimize foraging costs (largemouth bass) or maximize encounter rates with prey (skipjack tuna) can produce a similar result in the arithmetic evidenced of growth.

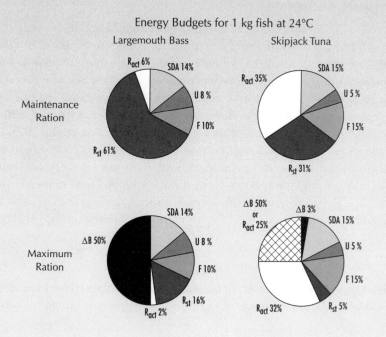

Energy Budgets for 1 kg fish at 24°C

and the reduction of one benefit (light) may be more that compensated by the decrease in cost (evaporative water loss). This effect is most apparent in stressful habitats such as those near the tree line, the upper limit of tree distribution on mountains. Light and nutrients may be available in abundant supply at the tree line in high alpine habitats, but water loss constrains growth in two ways. During winter, water is simply not available because it is frozen, and the frequent high winds cause desiccation of plant tissues. Trees at the tree line are typically stunted (very small for their age) and aggregated in groups because the microenvironment provided by adjacent trees reduces wind effects.

The trees didn't aggregate through behavioral mechanisms in the way that animals can. Instead, the seedlings that grew close to an existing tree survived because of the protection of the local microclimate. Seeds that sprouted some distance from existing trees may have grown rapidly during the first season because competition for light, water and nutrients was minimal, but their microclimate was not protective when the winter winds came, and they did not survive. Similar effects of *wind pruning* are apparent in many coastal areas where the persistent onshore winds also carry salt spray, which exacerbates the water balance problem by creating additional osmotic stress for the plants. Functionally, these coastal habitats are like deserts, and similar effects can be seen in the vegetation that grows along water courses in arid regions. In this case, the evidence of niche limits is very obvious. A single tree could not tolerate the conditions at such a site, but the microenvironment provided by proximity to other trees creates new physical boundaries for the fundamental niche and expands the realized niche.

Ecological stoichiometry

Chemists define **stoichiometry** as the ratio of abundance of different atoms in a compound. In an ecological sense, stoichiometry applies to measures of the relative availability of different nutrients. Animals excrete wastes that can serve as nutrients to the primary producers supporting their ecosystems. Thus, the negative effects of a consumer can be partially offset by its role as a recycler. The ideas expressed in a new field—ecological stoichiometry—argue that each organism serves as a selective filter in altering the biogeochemical cycles of its environment (Hassett et al. 1997). For example, oceanographers and limnologists know that the chemical composition of phytoplankton grown in the presence of abundant nutrient supplies has characteristic ratios of carbon, nitrogen, and phosphorus, termed *Redfield ratios*. Those ratios are usually C:N:P = 106:16:1. Deviations from Redfield ratios indicate the degree of nutrient limitation and the differences in physiological requirements for different groups of algae.

The paradox of the plankton

The well-mixed surface waters of lakes and oceans contain scores of algal species in what would otherwise appear to be relatively homogeneous environment, where nutrients are equally available to all. How do these species co-exist? Why doesn't competition reduce the species diversity? This phenomenon was termed the "paradox of the plankton" by G. E. Hutchinson. He explained the paradox by reasoning that slight differences in rates of nutrient uptake may favor one species over another in the intensely competitive environment. If nutrient concentrations are variable at the scale of very small patches, for example the nitrogen in ammonia excreted by and immediately adjacent to a copepod, then algae with a slightly better uptake rate for ammonia would be favored. Another alga may be more efficient in removing the phosphorus excreted by that copepod, and another might be larger or grow spines and be less susceptible to grazing by the copepod. The various combinations of nutrient uptake rates and susceptibility to grazing contribute to the exceptional diversity observed in most plankton communities, where a single milliliter of water may contain 100 algal species!

Some phytoplankton such as diatoms also require high levels of silica (Si) to build their cell walls (frustules). When phytoplankton are eaten by herbivorous zooplankton such as *Daphnia*, the ratios are altered by two processes. *Daphnia* excrete the C, N, and P in excess of their growth demand as CO_2, NH_4 and PO_4. These chemicals are immediately available as nutrients for algae. When diatoms are grazed by *Daphnia*, most of the silica is egested as silicate crystals in feces and rapidly sinks from the water column. The selective removal of silica changes the supply of this nutrient for diatoms, and alters the ratio of available nutrients to favor other algae that do not require much silica. The consequence is a change in the competitive interactions within the algal community. Other species (e.g., green algae and cyanobacteria) come to dominance in proportion to the intensity of grazing by *Daphnia*. This process continues up the food web. Fish require approximately four times the phosphorus present in the tissues of their prey because they build phosphate-rich bone tissue as they grow. As a result, they retain phosphorus and excrete much more nitrogen. This, too, alters the competitive environment for algae and favors those species that grow more efficiently when phosphorus is less available.

During the summer period when surface waters of many lakes and much of the ocean are warm, biological processes occur at rapid rates, and nutrients are in short supply. Excreted and recycled nutrients are the primary source for sustaining primary production by phytoplankton. As outlined in Chapter Four, phosphorus is typically the limiting nutrient in lakes, and nitrogen is usually the limiting nutrient in marine waters. In some cases, as much as 80% of the phosphorus uptake by algae will have passed through a fish on its way back to the level of primary producers (Schindler et al. 1993).

When the consumer is feeding at some maintenance level but not growing, it simply excretes the nutrients back to the environment in proportion to their concentration in foods. In the case of silica, the consumer *(Daphnia)* depletes the supply. In the case of nitrogen and phosphorus, the consumer enhances the nutrient supply for those primary producers that have not been eaten. Rapid changes in algal community composition can occur when some species are immune to grazing and inherit the nutrient capital of their consumed competitors. Similar effects develop in heavily grazed pastures. Plants such as thistle and mullen are avoided by cows and flourish in an environment where the competition has been eaten or trampled, and a significant share of the nutrients are deposited on the soil surface where leaching can make them readily available. Ecosystems are the appropriate context for the nutrient cycles described above, but they actually derive from the unique attributes of individual organisms.

The paradox of enrichment

Hutchinson reasoned that variability in local nutrient supply and differences in competitive ability could explain the high diversity of planktonic algae. Logically, adding nutrients would alleviate that competition and perhaps allow for an even greater diversity. In fact, the opposite can occur. The "paradox of enrichment" occurs when abundant nutrients produce conditions in which diversity declines and only a small number of algal species come to dominate the community. For example, lakes that are enriched by high levels of nutrient input produce blooms where cyanobacteria (bluegreen algae) can account for 90% or more of plankton biomass in the water column. As expected, increasing the nutrient supply results in higher productivity, but why does diversity decline? In this case, the cyanobacteria form large colonies which are not susceptible to grazing by zooplankton. These colonies become so abundant that light penetration is reduced, and their competitors are reduced to low abundances.

In some cases, ecosystem productivity can be directly traced to unique physiological capacities of a small group of organisms. For example, as discussed in Chapter Four, tropical rain forests and wetlands are among the most productive ecosystems. Tropical rain forests and freshwater wetlands support high plant species diversity, but the very productive wetlands along our marine coasts have remarkably low diversity (Mitsch and Gosselink 1986). Salt marshes and mangrove swamps occur in areas where each tidal cycle brings new nutrients and removes accumulated wastes and organic detritus. Primary production rates are exceptional. Why, then, have so few species managed to occupy this potentially rich and productive environment?

Mangroves occupy soft sediment areas of the intertidal extending from the tropics north and south to a boundary delineated by hard winter frost. Their distribution in the United States is limited to coastal areas of the Gulf of Mexico to southern Florida, where they dominate the intertidal regions of the Everglades. Salt marshes occupy the intertidal wetlands from the subtropics on to higher latitudes. They are interspersed with mangroves in the Gulf of Mexico wetlands and extend from northern Florida's Atlantic coast to the Arctic. Enormous salt marshes lie behind the barrier island system from

Georgia through Chesapeake Bay and to Cape Cod. They are also found in some estuarine bays of the West Coast (e.g. San Francisco Bay).

Salt marsh ecosystems are generally regarded as important energy sources to the adjacent estuaries because their detrital output fuels highly productive estuarine and coastal food webs. They serve as storage sites for nutrients derived from the rivers upstream and slow the loss of those nutrients to the ocean. Salt marshes are also an essential nursery habitat for many valuable shellfish and fish populations. Salt marshes are typically dominated by a few grass species, which can comprise 95% or more of total plant biomass. Their primary production rates approach those of the heavily fertilized and pesticide-protected cornfields that grow on rich Midwest prairie soils.

A small group of grass species (*Spartina* spp.) dominate this intertidal habitat along the Atlantic coast because they can deal with the stressors of water balance and oxygen shortage through unique adaptations. Water stress develops because evaporative water loss during low tide in the summer can concentrate salts in the soil interstitial water to levels as high as 60 parts per thousand (sea water is 35 ppt), and these levels would be lethal for most plants. Since a heavy rainfall can dilute salt concentrations to very low levels, *Spartina* must tolerate a wide range of salt concentrations. In addition, intertidal soils are typically saturated; oxygen concentrations a few millimeters below the sediment surface are exhausted due to intensive microbial decomposition of organics. Based on experience with potted plants at home, we've learned that constantly saturated, and therefore anoxic, soils are lethal to most plants. Intertidal soils are ostensibly lethal to plants for three general reasons:

- Oxygen and dissolved nutrients diffuse very slowly in interstitial water.
- Soils develop toxic levels of decomposition by-products such as hydrogen sulfide, ammonia and methane.
- Nutrient uptake by roots requires energy (ATP) and occurs most efficiently in the presence of oxygen.

Efficiency of nutrient uptake by plant roots is very low in saturated soils because the biochemical cost (energy required to produce ATP) of anaerobic processes is approximately 15–17 times greater than that of aerobic processes.

Figure 5.3 depicts the strong chemical gradients that develop in the saturated soils typical of most wetlands. An oxidizing environment at the mud-water interface converts the key nutrient, nitrogen, to nitrate, which makes it available to root uptake. Chemical gradients are very strong over the few millimeters below the soil surface. Oxygen is rapidly depleted with soil depth due to the intense rate of decomposition of organic matter by bacteria. This produces a reducing environment that converts nitrogen to ammonia (which is difficult to absorb), sulfur to hydrogen sulfide, and carbon to methane. These chemicals are toxic to many organisms.

Although nutrients are abundant in salt marsh soils, getting them is very costly for most plants. *Spartina* has developed special mechanisms for dealing with this stressful

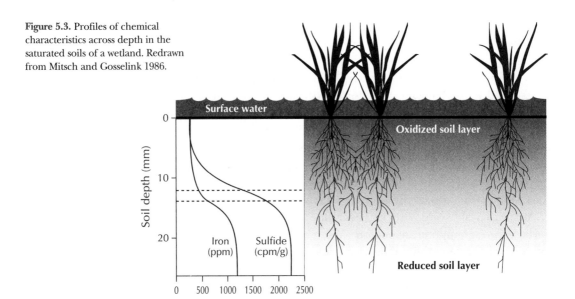

Figure 5.3. Profiles of chemical characteristics across depth in the saturated soils of a wetland. Redrawn from Mitsch and Gosselink 1986.

yet potentially advantageous environment. First, they have special tissues that concentrate and excrete excess salt as crystalline deposits on their leaves and stems. This allows them to regulate their salt concentration in root, stem and leaf cells much more effectively than other plants. When salinity levels rise to levels that repress most other plant species, *Spartina* can flourish. Second, they have air chambers in their stems and roots that allow oxygen to diffuse deep into their root systems and even a millimeter or two into the adjacent soils. This provides a microenvironment where toxic compounds are oxidized to non-toxic forms (e.g., sulfate, carbon dioxide and nitrate), and allows greater efficiency in nutrient uptake (nitrate) in a place where nitrogen is the most commonly limiting nutrient.

Why haven't a variety of plant species developed these adaptations? One could ask that question of many habitats where evolution seems to have produced only a handful of successful inhabitants. Specialization such as that demonstrated by *Spartina* species allows them to dominate and be very productive in an otherwise harsh environment where many other plant species cannot flourish. That specialization has a cost, however, as *Spartina* is a poor competitor in habitats other than those of the intertidal marshes.

Ectotherms, Endotherms and Behavioral Thermoregulation

Most plants and animals have internal temperatures approximately the same as those of their external environment. They are termed **ectotherms,** literally meaning "outside heat." Their metabolic rates follow the basic temperature dependence of rates of reaction for their enzyme systems (Figure 5.4). Feeding rates tend to parallel the enzyme

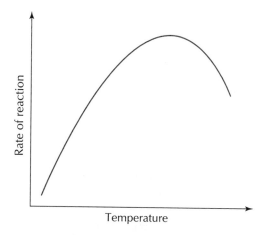

Figure 5.4. Effects of temperature on the rate of reaction for an enzyme-mediated biochemical process.

curve, but metabolic rates do not show a gradual decline at higher temperatures. Instead, metabolic rates keep increasing until temperatures approach the upper lethal tolerance level and then drop quickly as enzymes are denatured, and the organism succumbs and dies. Growth and reproduction are also temperature-dependent, but depend upon the net balance of inputs and outputs because of the hierarchy of energy allocation described above. Under conditions of high resource availability, net energy storage can occur over a broad range of environmental temperatures. When resources are more limited, growth can occur over a smaller range of temperatures (Figure 5.5). The difference between inputs (energy eaten) and the total of energy losses (wastes, metabolism, locomotion, etc.) can be represented as the scope for growth, which will have a maximum at some temperature intermediate between that of upper and lower tolerance limits.

Temperature tolerances

Although living things have similar biochemical processes and occur over a broad range of temperatures, the molecular structure of enzymes limits their activity to a lesser range of temperatures. Some residents of arctic or antarctic seas may spend their entire life in a water temperature range of $-2°C$ (high salinity prevents freezing) to $+2°C$ in the heat of polar summer. Bluegreen algae growing in the thermal springs of Yellowstone National Park may never experience temperatures below $50°C$. Bacteria found in our water heaters at home grow best at $70°C$. However, most familiar organisms would perish under any of these conditions.

Members of every species seem restricted to a limited range of temperatures—i.e., no species is able to function across all possible conditions. The ectotherms include broad classes of organisms such as fishes that include members of several thermal guilds, i.e., species that seem to do best in a relatively restricted range of temperatures. For some, such as the rainbow trout, that range is about $1–20°C$. Their upper temperature

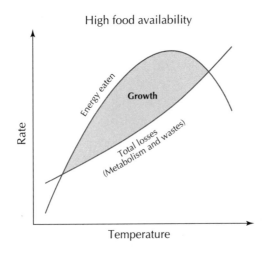

High food availability

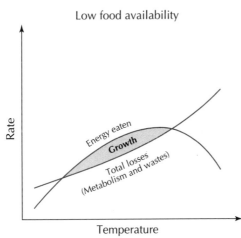

Low food availability

Figure 5.5. Effects of temperature on energy budgeting processes for an ectotherm when food availability is high or low. The upper line describes the rate of feeding. The lower line is the sum of energy outputs to metabolic activities and energy lost as wastes. The shaded area defines the scope for growth.

tolerance is about 25°C. Other species such as a yellow perch may tolerate low temperatures (1°C), but grow only at temperatures above 5°C, cease growing at temperatures above about 25°C. and die at about 28°C. Others such as largemouth bass can tolerate temperatures of 1°C, grow at temperatures between 10°C and 30°C, and survive up to about 37°C. These conditions define the scope for growth for each species as presented in Figure 5.6, and their corresponding habitat preference during summer months in a thermally stratified lake where oxygen and food are sufficiently available.

High temperature-tolerance and growth rates are strongly dependent on energetic status. If we put a fish in a temperature gradient and provide abundant food, it will select a temperature where its growth rate is maximum. For rainbow trout, that would be about 15°C, for yellow perch, about 23°C, and for largemouth bass, the preferred temperature would be about 32°C. If we stop feeding, each species will gradually reduce its preferred temperature. This behavior is indicative of the trade-off between costs and benefits. When food is available, animals seek temperatures that maximize the net benefit of growth. When it is limited, they seek to minimize metabolic costs.

Endotherms in a heterogeneous environment

Another group of organisms has escaped the most common constraints of environmental temperature conditions. **Endotherms** use the heat generated by metabolic processes to establish and maintain a relatively constant body temperature, which is usually in the range of 35–38°C. This high-cost physiology confers major benefits. Most mammals and some birds can be active under cool to cold conditions, thereby having access to habitats and activity times not available to ectotherms. For example, at temperate latitudes cool night temperatures reduce or prevent activity by many plants, insects, reptiles and amphibians, except during the summer periods. By contrast, fur or feathers retain the internally generated heat of endothermy and insulate against both the hot sun and the cold wind. Color of the pelage or plumage can influence activity by absorbing or reflecting radiant energy from the sun, and the choice of microhabitat during extreme conditions can extend a species' geographic range (Houseal and Olson 1995). Compared to ectotherms, mammals and birds can forage and mate under a much wider range of environmental conditions.

In effect, endothermy expands the temperature boundaries of the fundamental niche that correspond with space and time. There are limits, of course, as dictated by the rate of heat loss, which is a function of body size and temperature extremes. Cold induces shivering as an involuntary muscle contraction that produces additional heat. Anatomical juxtaposition of the arteries carrying warm blood and the veins returning from cooled limbs creates an average tissue temperature that allows bare feet and legs to be fully functional in cold air or water.

Other limits are imposed at high temperatures where cooling is required to maintain body temperatures below levels that would denature enzymes. Cooling is energetically expensive, requiring about five times as much metabolic cost as heating. Again, behavioral regulation becomes important under these conditions. Horses and humans perspire to cool by evaporation, dogs and ostriches can't perspire so they pant, cows seek shade, and hippos and pigs seek cool water or mud.

Hibernation such as that practiced by bears, and seasonal migrations such as those by birds, offer longer-term solutions to the problem of extreme cold and low food availability, but these activities require stored reserves sufficient to last through an extended period of energy debt. These requirements dictate the size of survivors, and the genetic feedback expressed in the next generation. Endotherms gain the benefit of increased access to resources, but they pay a price. The metabolic rates required by endothermy

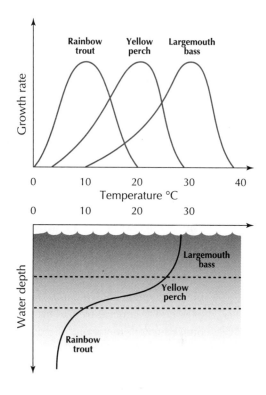

Figure 5.6. Upper graph, the scope for growth for each of three fishes that represent thermal guilds; a rainbow trout (coldwater guild), a yellow perch (coolwater guild) and a largemouth bass (warmwater guild). Lower graph, the distribution of the same thermal guild fishes when allowed to select their habitat in a thermally stratified lake where oxygen and food are available in all habitats.

produce a net cost that is approximately five times that of the metabolic rate for an ectotherm of comparable mass and occupying the same habitat.

Behavioral thermoregulation: an example in the ecological context

As described above, behavioral strategies allow organisms to select their position in heterogeneous habitats, and therefore provide a means for minimizing stressors while maximizing gains. The landscape scene in Plate 1.1 provides some examples of how physiology, ecology and behavior are expressed for a range of animals over the course of a diel cycle. Imagine yourself standing on the shore of a deep pond that receives the stream seen in the foreground of Plate 1.1. Not seen in this midday photo is the nocturnal raccoon whose body temperature is high and relatively constant. Its prey at the pond's edge would be a crayfish whose body temperature is lower and approximately that of the water. Last evening above the pond a little brown bat foraged for emerging insects to sustain the high body temperature typical of most mammals, but at dawn it retreated to the cool (10°C) of a nearby cave, and entered into a temporarily altered metabolic state equivalent to very deep sleep. By noon, when this picture was taken, the bat's body temperature had dropped by about 15°C. This allows it to continue digestion but reduces its metabolic demand and allows a greater share of last night's energy acquisition to be stored as fat for its autumnal migration south. At sunset, the hormonal system that regulates its internal clock will activate the metabolic processes which make the bat fully active again.

A turtle on the pond bank has been basking in the sun until its body temperature rose about 15°C above that of the pond water. Earlier in the season, it had migrated to an exposed, sandy site to lay its eggs. That site accelerates incubation time due to the sun's warmth, and sustains a soil moisture level that will keep the eggs from desiccating before they hatch. The turtle will alternate between diving to feed and basking to maintain the high temperature that allows it to be active and more rapidly digest its food. Like the bat, the turtle must lay down a reserve of fat tissue in preparation for the long Wisconsin winter, which it will spend hibernating in the pond's bottom muds. The turtle cannot surface to breathe through the ice, nor does the mud offer much oxygen, but the very low temperature allows the turtle to respire through its skin and cloacal tissues at a rate sufficient to maintain its extremely low metabolic rate.

Like the turtle, dragonflies on the cattails bask in the sun. Their flight muscles and agility in pursuing prey are most efficient at temperatures substantially above that of the ambient air. At night, a dragonfly roosts in cold torpor. Adult dragonflies will die before winter arrives, but their nymphs share the pond mud with the turtle. Deer come to this pond to drink, but they do so at night when the biting flies populating the near-shore vegetation are inactive with lowered body temperatures that preclude active flight. On the other hand, mosquitoes are most active during the evening period when temperatures drop and humidity increases to levels that does not stress their capacity to maintain water balance. Each mosquito species has a special combination of temperature and relative humidity that allows it to sustain activity for about an hour, before conditions change sufficiently that another mosquito species' special adaptations allow it to occupy the habitat. Most mammals may be aware of mosquito abundance, but don't necessarily appreciate the elegant physiological intricacies of their behavioral succession.

A minnow forages actively in the pond water, its body temperature equivalent to the surface water temperature, except during the hottest periods of July and August, when it descends to the cooler waters of greater pond depth where it can maintain a positive energy budget by reducing high metabolic costs. It may lose weight during that time because the volume of habitat available for foraging is much restricted. At this time of day, *Daphnia* in this pond will have body temperatures equivalent to that of the bottom waters, which are very cold relative to the surface waters. As night falls, the *Daphnia* will migrate upward into the algae-rich surface layers to feed; thereby gaining two advantages— elevating their body temperatures so that their feeding rate is very rapid, and avoiding predation by the visually-dependent minnow. At sunrise, *Daphnia* will migrate back into the cold, deep waters where metabolic rates are much lower. This behavior has some cost—those waters have very low oxygen concentration at this time of year. If we were to examine a *Daphnia,* it would be very dark in color instead of translucent like the *Daphnia* we may have seen in an introductory biology laboratory. This dark color comes from the invertebrate hemoglobin the *Daphnia* produces. This hemoglobin is synthesized at some biochemical cost, but it allows the tiny crustacean to extract oxygen and sustain aerobic respiration at very low oxygen concentrations in a habitat where minnows cannot forage.

The point of this vignette is that behavioral regulation of body temperatures operates in every aspect of the picture in Plate 1.1, and the energetic consequences play out over longer periods than those represented by a snapshot. A physiological view of this ecological context would envision dynamics different from those apparent in a picture taken in the middle of a nice summer day.

THE ROLE OF EVOLUTION IN PHYSIOLOGICAL ECOLOGY

Tolerances, efficiencies and preferences can differ among individuals and are generally heritable. These are the materials for natural selection and for the array of adaptations expressed by organisms that occupy an enormous diversity of habitats (Hutchinson

1965). Organisms succeed in habitats ranging from the ocean benthos with high pressure, no light and very low temperatures, to mountain tops, to hot, dry deserts. While a number of bacteria and algae species exist at $+110°C$ and at $-70°C$; none of those species exist at both. Abiotic gradients are typically partitioned among an array of species because the basic constraints of enzyme biochemistry dictate a finite range of environmental conditions for each species.

The physiological characteristics of an organism may differ slightly from those of other members of its species in the same habitat. When a stress such as drought or extreme cold develop, many individuals may succumb, but a few may survive. Changes in local populations get fixed in genetic form because the survivors reproduce. Immigration or emigration can mix those genotypes with others derived under a different set of conditions. The point is simply that neither the environment nor the organism's responses are static. Big effects can derive from rare events.

A sense of time and place

Chapter Two gives excellent background for, and examples of, the importance of recognizing the historical context for ecological questions. Such lessons apply in the physiological view, too. For example, we are currently in the middle of the fourth interglacial period of the last million years. The current habitat and the adaptations of organisms to it, are, in fact, transient features in time and space. The list of species present on the North American continent 12,000 years ago was probably little different from that for our local ecosystems today; they were simply found someplace else. As the glaciers retreated toward the poles and up the mountains, organisms followed them north, south, or upslope, invading newly-available habitats including ones where these species occur today. This recolonization was based on the principles embodied in a physiological ecologist's view of what regulates distribution and abundance.

Individuals of a species now found in the southern limit of its range have an environmental history very different from that of individuals found at the species' northern limits. Members of a species differ in their tolerances and preferences, and those differences have a genetic component. Population geneticists demonstrate that there are genetic differences between populations of the same species now found in different places. What else would we expect? They have been exposed to environmental gradients that changed over time and across space. Habitats are changing rapidly again, but in this case the cause is different. Powerful ecological forces are brought about by the advent of human populations and their technological advances, evident in the effects of the wheel, the plow, the axe, the fish net and, at the largest scale, the use of fossil fuels.

THE ROLE OF MODELS IN PHYSIOLOGICAL ECOLOGY

The principles of homeostasis allow a direct, empirical approach to understanding the construction and use of models for physiological systems. The idea behind energy-budget models is simply that budgets must balance and that outputs cannot exceed inputs. The balance of gains and losses is expressed in the living individual. This principle applies to all materials utilized in the biochemistry of living organisms. While defining the budget is easy, characterizing its dynamics is more challenging. For example, the body temperature of a basking lizard is a function of direct solar radiation, radiation reflected from adjacent rock surfaces, and radiation reflected from the soil surface. Wind speed governs the rate at which acquired heat is lost. The lizard can pant to decrease body temperature through evaporative cooling, so its water balance becomes directly tied to its behavioral thermoregulation. An equation defines each of these functions. The lizard's preferred body temperature, and those temperatures bounding upper and lower limits for its ability to forage, grow and reproduce, define the times and places where this lizard will succeed. Within the general field of physiological ecology, approaches based on bioenergetics and biophysical ecology have developed sophisticated approaches to these complex problems. The bass and tuna example developed in Box 5.2 is representative of the energetics approach.

Energy-budget models usually describe the dynamics for an average organism. Another approach involves recognition of the variability among individuals and in their day-to-day ecological experience. For example, a tree may produce thousands of viable seeds each year, but the large majority of those will not survive to the next year, and very few grow to maturity. In simple terms, the average seed is dead after less that one year. Success through time of the survivors may be due to some mix of good luck and good genes. Models constructed to include environmental and/or genetic variability are termed *individual-based*. Individual-based models allow comparative descriptions of how one plant seed germinates and grows in the shade of a tree while another does so in an open patch in the forest. Those initial conditions strongly determine each plant's relative success over time. Their location may have been determined by the random act of where an individual seed fell, but, due to differences in local context, these two seeds produce very different results from those of the average for all members of that year's group of seeds. While individual-based models typically include some rules for growth, such as those in an energy-budget model, individual-based models allow for differences in the conditions under which energy budgeting occurs, and produce a statistical measure of the variability that develops in time and space. This approach offers important insights about the causes of mortality, the constraints to growth, and the level of confidence that can be ascribed to the ecological descriptions of the average individual—that individual being one of the few that survived (DeAngelis and Gross 1992).

TECHNIQUES SPECIFIC TO PHYSIOLOGICAL ECOLOGY

Among the novel techniques employed at the individual scale are such uses of telemetry as those that record heart rate, respiration rate or body temperature in animals and evapotranspiration responses by plants. These technologies can be employed in the continuous and distant measurement of responses by free-living organisms to changing environmental conditions. For example, satellite systems allow physiological ecologists to monitor the heart rates, feeding rates and body temperatures of seals diving beneath the Antarctic ice sheet, or of grizzly bears in the mountains of Alaska. Similarly, instruments attached to sharks, tuna or marlin can accumulate a continuous record of water temperatures, depth and geographic position. When released by time-activated devices, these recorders float to the surface and broadcast their data to satellites. This provides a detailed and long-term record of the environmental conditions in habitats occupied by wide-ranging pelagic species. Data sets of this kind could not be economically acquired through any other means, and they are vital to our ability to define habitat in a way that corresponds with the organism's perception and behavior.

Direct measures of physiological processes are commonly pursued by controlled laboratory studies of metabolic rates through measurement of oxygen consumption rates or rates of feeding and growth under specific temperature conditions. In fact, all terms of a mass balance for water or energy can be determined from these approaches. Results find important applications in evaluating the performance of an organism in response to manipulation of one or more variables. However, those conclusions may be questioned because the organisms are separated from their ecological context due to the isolation required for controlled laboratory techniques. Several novel techniques provide measures of integrated physiological processes and their rates. These include the use of **radioactive isotopes** that provide a means for measuring rates of carbon fixation by primary producers (^{14}C) or rates of nutrient cycling (^{32}P) between trophic levels. These studies are limited, of course, because one cannot release large quantities of radioisotopes in ways that would remove the constraint of learning from the larger context.

A relatively new approach involves following the dynamics of naturally occurring **stable isotopes** such as those of oxygen, carbon or nitrogen. These isotopes are rare in nature and difficult to measure. Recent advances in technology allow us to determine their abundance with high precision. This provides an alternative to radioisotope use and a safe means for measuring the basic biochemical processes involved in maintenance and growth. Stable isotopes accumulate in living tissue at rates that can be determined from controlled studies, and they are ubiquitous. In other words, they can be used for deductive purposes wherever ecological processes are of interest.

The higher atomic weight of the stable isotope of nitrogen (^{15}N) compared to the common isotope (^{14}N) makes the former less likely to cross a cell membrane as biochemical processes proceed. This means that the heavier isotope accumulates slowly in tissues of all organisms as a function of age, and is effectively *bioaccumulated* as prey organisms are eaten by their consumers (Chapter Four). The relative proportion of this stable isotope in a consumer's tissue is, therefore, an integrated record of what the consumer has eaten over time. Ecologists who study food webs use these ratios to deduce the kinds of foods eaten by consumers that chew or shred their prey (e.g., crabs, mink, eagles) and as a means for estimating the relative importance of prey from each of many potential categories. For example, predatory fish that prey on invertebrates will have lower nitrogen isotope ratios than those that eat other fishes. This can be particularly helpful in understanding how food webs are structured.

Another potential benefit of stable isotope methods derives from the fact that changes in isotope ratios appear to be highly correlated (Figure 5.7) with the documented bioaccumulation of organochlorine contaminants (e.g., toxaphene, DDT, PCBs) and heavy metals such as mercury (Kidd et al. 1995). This has powerful potential because the mechanisms of trophic accumulation are known, and analyses on many samples from many places can be done for a much lower cost than analyses required for contaminants that have not been bioaccumulated. Perhaps more importantly, stable isotopes can trace the sources of contaminants. The technological ability to analyze low levels of contaminants—those found in water or in the algae that form the base for the bioaccumulation processes—is limited. However, the stable isotope composition of components at the base of the trophic system can readily be determined and followed up through food webs.

APPLICATIONS OF PHYSIOLOGICAL ECOLOGY

In its most basic applications, physiological ecology provides an understanding of dimensions for the environmental template of the fundamental niche where biotic interactions occur. In other words, it provides the first principles to guide expectations for

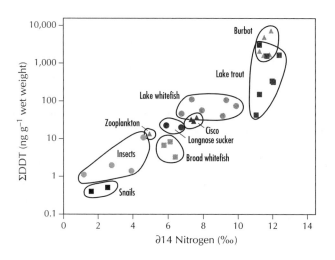

Figure 5.7. Correspondence between concentrations of DDT and changes in stable isotope ratios for nitrogen in the tissues of invertebrates and fishes in a Canadian lake. Redrawn from Kidd et al. 1995.

understanding the abundance and distribution of organisms, which have applications in all scales of ecological endeavor. The understanding of niche dimensions apply equally to all species, abundant or rare, ubiquitous or endangered, native or exotic (introduced). The examples that follow show how physiological principles are used to deal with such diverse and difficult problems as the "invention" of new species through genetic engineering, the ecological effects of toxic compounds, and the profound challenges caused by invading species.

Genetic engineering in an ecological context

Genetic changes in physiological function due to directional and artificial selection are prominent in animal and plant husbandry, which have a long history, and in more recent work using **genetic engineering**. Results of selective breeding are expressed in the variety of domestic plant and animal species employed as primary food sources. Such results derive from the variability in physiological properties (e.g., growth efficiencies and tolerances) of organisms, and the canalization of artificial selection, practiced by selecting parents for the next generation. Many varieties now widely employed in horticulture, agriculture and animal husbandry bear only distant resemblance to their ancestral stocks (Brody 1945). The principles of physiological ecology are directly employed in pursuing these practices, and in understanding the importance of trade-offs (e.g., growth rate vs. disease resistance) in developing these strains. Genetic engineering now offers tools to accelerate this process.

One anecdote demonstrates the potential and the problems of artificial selection. A genetic engineering firm in California developed and established genes that promoted rapid growth, disease resistance and drought resistance in a strain of tomatoes. Much fanfare and publicity accompanied the announcement of this success. When the new variety was tested in Florida, it performed as advertised but with some limitations. The new fruit grew rapidly, but in the humid climate of Florida (unlike that in dry, irrigated California) it grew so rapidly that it burst its skin. Tomatoes that burst have relatively modest market value. The genetic manipulations had altered the water balance of this new form at one edge of its tolerance, but not at the other. Moreover, public resistance to genetically-engineered food products continues. This resistance may not be rational, but it is a real market constraint.

The potential of genetic engineering is vast, but the warning about ecological context must be heeded. Evolution of today's organisms resulted from a history of many kinds of selective forces and their interactions. Little is known about that history. On the other hand, genetic engineering will proceed before we have the full list of possibilities and despite the fact that we have a small list of answers. There is important opportunity for much scientific investigation here.

Ecotoxicology—The physiological ecology of toxic compounds

Physiological processes are amenable to laboratory studies and have therefore provided an array of methods now widely practiced in field applications. Widely developed tests used to estimate tolerance levels now form the basis for many of the environmental standards employed by federal and state regulatory agencies. In these cases, the criteria for the standards usually derive from controlled laboratory studies across a gradient of treatments to define the equivalent of LD_{50} levels (the concentration of a toxin or the critical temperature at which 50% of a test population dies or shows evidence of chronic stress). These testing procedures have also been employed in the larger ecological context of effects on the performance of organismic processes, including photosynthesis, evapotranspiration, nutrient uptake rates, osmoregulation, respiration, feeding rates, excretion, locomotion, growth and gamete production, etc. Time and mass-balance budgets serve as the basis for synthesis.

The field of **ecotoxicology** arose from the need to estimate physiological effects of anthropogenic (man-made) chemicals on organisms. The United States Environmental Protection Agency maintains a Red Book of standards for the tolerable levels of a long and growing list of chemical compounds. Standards are set based on bioassay results such as the effects of compound X on reproduction by a zooplankter *(Ceriodaphnia)* and survival of the fathead minnow *(Pimephales promelas)*. On the other hand, naturally produced toxins, synthesized analogues, and their physiological action are the basis for much of our pesticide industry. The real challenge is to find some balance of wise use and the minimum of unwanted environmental effects.

Toxins are a major component of the continuing contest between plants and their herbivores, between prey and their predators. In one well-known example, the toxicity of milkweed plants prohibits most herbivory, but the larvae of the monarch butterfly has developed the capacity to thrive on milkweed, which in turn imparts a powerful toxin to the adult butterfly. Birds that attack monarch adults quickly sicken and vomit (explosively)—but recover to remember that the brightly colored orange-and-black monarch is a poor prey choice. Other butterflies (e.g., the viceroy) mimic the adult monarch's color pattern, and although they lack the toxin, can deceive some of their potential predators.

The field of chemical ecology is enormously complex and full of fascinating examples; adaptation has produced a diversity of toxins as part of many natural processes. So why haven't all plants or animals developed a suite of toxins that dissuade their predators and parasites? In a biochemical sense, most toxic compounds are difficult to synthesize and cost much precious energy that might otherwise be channeled to growth and/or reproduction. Imagine this as the analogue of providing for a large defense budget. Resources devoted to chemical or physical armaments may provide for longer survival, but they occur at the cost of better maintenance and possible investments in the future (e.g., schools for humans as the equivalent of seed production for plants). In addition, natural selection will occur among potential consumers of an otherwise protected resource. Some organism may, by random mutation, develop an antidote or a behavior that reduces the effect of the defensive chemical. That animal and its progeny now have exclusive access to a new resource, and their populations flourish. Selection now favors those plants with greater or alternative chemical protection.

The chemical industry has copied some of the naturally occurring compounds in developing many of the pesticides currently in use (e.g., compounds that inhibit chitin synthesis in insects, or overdose the hormonal control of growth in plants). Some pesticides are synthesized based on knowledge that analogous chemical structure can produce analogous biochemical actions and equivalent ecological results.

Some elements classed as toxins rarely occur deleteriously in nature (e.g., mercury and lead), but their concentrations in natural systems have been substantially increased due to human activities such as mining and industrial effluents. Some ecotoxic effects are obvious, others occur indirectly. For example, the effects of acid rain on vegetation are immediately apparent in leaf damage. Acidification effects on soil tend to increase the availability of aluminum in soluble form, which is toxic to plants and inhibits their growth. These indirect effects are obvious only in the fact that the plants do not grow as well. Acidification effects on fishes are less obvious because they can act through the food web by reducing zooplankton populations. Like most crustaceans, zooplankton molting and reproduction demand calcium carbonate, which becomes less available in acid lakes. Aluminum concentrations increase with reduced pH, creating additional physiological stress because aluminum ions inhibit the sodium pump required for effective function of nerves. Because zooplankton are key food sources for small fishes, their numbers also decrease. Reducing prey resources lessens the energy available for reproduction by adult fishes, and those fry that do hatch in turn find the cupboard bare. Eventually, the effects of acidification produce the local extinction of fish populations as seen in lakes of the Adirondacks, central Ontario and Scandinavia. These lakes are crystal clear and beautiful to look at, but they hold no fish. The mechanism of toxicity is not direct, but very effective.

Many of the worst problem toxic compounds were invented by chemists and are manufactured (e.g., DDT and PCBs). They have no history in the evolution of physiological tolerances. Some compounds such as DDT can mimic and block the biochemical action of hormones in vertebrates, and, in the case of birds, disrupt the hormonal cycles that mobilize the calcium stored in bones, from which birds create egg shells. The effects of egg-shell thinning caused reproductive declines in many bird populations, and declining bird abundance was the basis for Rachel Carson's book, *Silent Spring*. Historians of science recognize Carson's book as a cornerstone of the environmental movement that swelled through the 1960s and 1970s. That concern continues as we deal with the consequences of human population increase (Cohen 1995).

The Zebra Mussel Invasion—An example of applied physiological ecology

Exotic species are among the most important challenges to resource managers and ecological researchers. Many invaders do not succeed, or simply show up as modest additions to a local species list. Some succeed and flourish at the expense of native species. North Americans spend billions of dollars per year battling exotic species like the sea lamprey, common carp, purple loosestrife, Eurasian water milfoil, Brazilian pepper and Asian salt cedar. The following example describes how the principles of physiological ecology can help deal with the problems from a recent invader.

The zebra mussel *(Dreissena polymorpha)* is native to eastern Europe but new to North America. After the last glaciation, it spread slowly out of a refugium from rivers at the north end of the Caspian Sea and Black Sea, then into the lakes and rivers of Russia. The zebra mussel is an ecological analogue of the familiar blue mussels so common on rocks, piers and piling along U.S. marine coasts. Zebra mussels attach to solid substrates using strong byssal threads and feed by filtering the overlying water column. They produce tremendous numbers of gametes that become the planktonic veliger form and disperse to new habitats. As canals and shipping developed throughout Europe during the nineteenth century, they provided avenues and the means (attachment to barges) for zebra mussel colonization. This accelerated the spread of zebra mussels into the rivers and lakes of western Europe. Zebra mussels now occur from Scandinavia to Italy and west to the waters of Britain.

In 1988, zebra mussels were discovered in Lake St. Clair near Detroit and along the western shores of Lake Erie (Figure 5.8). They are presumed to have arrived in the ballast water of some ship from a port in Europe. By 1996, zebra mussels had spread throughout the Great Lakes, up the St. Lawrence River, down the Mississippi, and up into its major tributaries (the Ohio, Tennessee and Arkansas Rivers). Via the Erie Canal, they invaded the finger lakes of New York, Lake Champlain and the Hudson River. They have also invaded 79 inland lakes.

The main mechanism for dispersal is the planktonic veliger stage, which can live in bait buckets, ballast tanks, etc. Dispersal associated with humans also occurs when adult mussels attach to vessels that move upstream and to trailored boats that move overland. Zebra mussels can remain alive and well for more that 48 hours of continued exposure to air. In a cool and wet place they can last longer. Between 1993 and 1996, agricultural inspectors at the California state border intercepted eight trailored boats unknowingly carrying zebra mussels. Some of those boats carried live zebra mussels. Given their rate of dispersal to date, zebra mussels are expected to continue their march of invasion across the North American continent.

Zebra mussels quickly become very abundant and create major economic and ecological problems (Plate 5.2). By attaching to intake pipes, lock gates, wharfs, pilings, boat

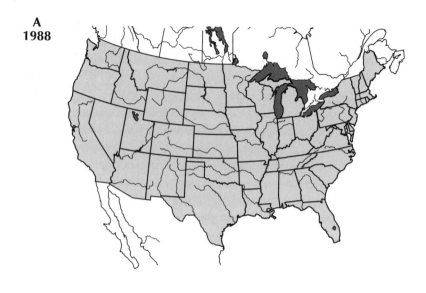

Figure 5.8. Distribution of the invading zebra mussel *Dreissena polymorpha* in North America. **A.** 1988 sites. **B.** 1996 sites. Solid areas indicate heavy infestations. The most recent distribution of zebra mussels can be found on the WorldWideWeb at http://www.nfrcg.gov/zebra.mussel/.

bottoms, etc., they create a fouling problem that annually costs hundreds of millions of dollars. They also colonize the shells of native mussels, overgrow and can eventually kill them. Native mussels of the Mississippi, Ohio and Tennessee River systems constitute nearly one-third of the endangered species list for freshwater organisms in the United States. Zebra mussels are a serious threat to the native mussels.

The filtering activity of zebra mussels removes planktonic material at prodigious rates. Water transparency in Lake Erie nearly doubled as zebra mussels came to occupy virtually all available substrates, and comparable effects occurred elsewhere. Increased water clarity allows greater growth by aquatic macrophytes. Large shallow bays such as Green Bay of Lake Michigan and Saginaw Bay of Lake Huron now have a major macrophyte "problem" as perceived by boaters. As zebra mussels become more abundant, they change the habitat for nearly all local organisms. Many fishes and birds eat them, but none can actually control their population growth. Zebra mussels grow rapidly to sexual maturity and have very high fecundities; their densities increase until space or food resources limit them. Where newly settled juveniles over-grow older colonies, massive mortalities can occur which cause taste and odor problems for drinking water supplies to many cities. The recreational beaches of many lakes are now deeply covered with the shells of dead zebra mussels.

The long-term effects of zebra mussels on lake and river ecosystems are unknown. Zebra mussels reduce algal densities and therefore cause declines in the zooplankton populations that serve as prey for some fishes. On the other hand, their feces and pseudofeces concentrate organic matter on the benthic surfaces, thereby providing greater densities of food for detritivores such as amphipods, a preferred prey of many fishes. Increased water clarity allows greater growth of macrophytes that enhance the littoral habitat of many juvenile fishes and protect them from predation. Thus, zebra mussels provide possible benefits in the ecological processes that produce fishes. Do those benefits outweigh the detrimental effects of zebra mussels? We will learn some answers to those questions in the near future.

The above information is background for the role that physiological ecology can play in this continuing issue. Zebra mussels do not flourish in all lakes or rivers of their ancestral eastern Europe. Reviews of the European literature reveal that zebra mussels have distinct abiotic constraints (Ramcharan et al. 1992). The ability to forecast where zebra mussels are most likely to become a problem in North America would allow focus of preventative efforts through enforcement and education, would reduce their rate of spread, and delay or prevent their negative effects.

Similarly, the abiotic parameters of the fundamental and realized niche can identify waters that will not be suitable for zebra mussels. Some places need not be defended and efforts can be focused on those that are vulnerable. For example, the upper thermal tolerance temperature of zebra mussels is 26–32° C; they will not survive in warmer waters. Optimum growth rates and veliger settlement occurs between 15–17° C. Many of the lakes in our southern states are not vulnerable because they are too warm in summer. A calcium concentration of 8–12 mg/liter is required for survival, but growth and reproduction do not occur at calcium concentrations less than 15–20 mg/liter. Adult mussels can survive at pH levels of 7.0, but the larval stage (veligers) require a minimum pH of 7.4–7.5. Softwater lakes of the Canadian Shield (which holds 90% of North America's lakes), and mountain reservoirs (vital water sources to western cities) are not vulnerable because they have low calcium concentrations. Many native mussels exist in many places with calcium and pH levels below those required by zebra mussels; these native species are safe.

Zebra mussel population growth requires an adequate supply of phytoplankton and sufficiently high temperatures to provide energy for somatic growth and reproduction. A bioenergetics model of zebra mussel growth forecasts that cold, unproductive lakes such as Lake Superior may support populations in the harbors and river mouths, but the main body of the lake is not vulnerable because neither food supplies nor temperatures are sufficient to sustain zebra mussels (Schneider 1992). Thus, the world's largest freshwater lake is safe from this invasion.

Evolutionary effects play a part in this drama. Zebra mussels in the United States have temperature tolerances at the higher end of the range reported from European studies. It appears that this invasion originated from the southerly regions of the mussel's native range, where local adaptation had selected for higher temperature tolerance. This genetic strain of the zebra mussel does not grow at temperatures below 10–12° C. That offers some cold comfort in the prospect that this group of mussels may be even less tolerant of the long cold periods in northerly latitudes.

Unfortunately, a second exotic mussel species recently appeared in the deep, cold waters of Lake Ontario and Lake Erie. This one is called the quagga mussel *(Dreissena bugensis)* and occupies a habitat that suggests less constraint by low temperatures. At present, the quagga mussel's fundamental and realized niche dimensions are not known (McMahon 1996).

A total of 146 exotic species are recorded for the Great Lakes at this point in time (Mills et al. 1995); it is not possible to prepare for all these species and much less so for others that may soon follow. The exotic species problem is one of the most important challenges for ecology. It requires a fundamental understanding of physiological ecology and the capacity to forecast biotic interactions that may be expressed at all levels of ecological organization.

LINKS TO OTHER KINDS OF ECOLOGY

This chapter offers an overview of ways that physiological ecologists view the interactions between individual plants or animals and their abiotic environment. Like bricks in the wall of a building, the process of assembly begins with the individual units. The success of individuals is dictated in part by their behavioral responses to a heterogeneous environment. Individuals are assembled in a collective unit—the population—which is set in the larger context of communities where species interact, then in the context of ecosystems where species interactions are set in a larger abiotic context, and finally in the context of the assemblage of ecosystems seen as landscapes. Still, the basic unit of physiological activities and ecological interactions is the individual.

The basic unit of evolution is the integrated, overall measure of success by an individual in providing for the next generation, i.e., its *fitness*. Individuals that grow and reproduce most effectively will be most likely to populate the next generation. Their offspring become a part of the population component of the community within an ecosystem that is set in a landscape. Perhaps the best explanation of the relationship between a physiological perspective and that of other kinds of ecology will derive from examples developed in this book. Chapter One's description of the Wisconsin landscape includes that underlying fact that this state is in the temperate zone—Wisconsin enjoys the genuine change of seasons. The combination of frigid winter and hot summer limits the number and kinds of species, both native and introduced, that can occur. Chapter Two, Historical Ecology, develops the evidence that fire tolerance (a combination of morphological and physiological properties) of yellow pine was the key to understanding why the Blue Mountain forest ecosystems changed when humans began to manage them. Chapter Three, Landscape Ecology, demonstrates that the nutritional requirements of individual elk—a focus of physiological ecologists—are a major component of ecological dynamics at the landscape scale in Yellowstone National Park. Ecosystem Ecology, Chapter Four, includes the strong, ubiquitous effect of temperature on production processes at all trophic levels and recognizes that we are altering the most general of physiological conditions— ambient temperature and carbon supply. Chapter Six's review of Behavioral Ecology demonstrates how habitat and prey choices are set in the context of physiologically important environmental gradients. Population Ecology, Chapter Seven, looks to the role of density-independent agents (e.g., the weather and habitat quality) as a first principle in explaining dynamics; and Chapter Eight, Community Ecology, starts the process of assembling groups of species based on physiological tolerances to the abiotic conditions of different environments. Again, the brick analogy applies as we try to envision how the relative success of an individual organism contributes to the progressively larger and more complex whole.

Defining the physiological conditions for maintenance of an individual offers only the broadest ecological view. Be cautioned—that view may be necessary and helpful, but it is probably insufficient. A self-sustaining population of any species occurs only in the context of historical constraints resulting from geographic accidents, evolutionary processes, and interactions with other species. This book was organized and written as a guide to using a full suite of ecological perspectives.

SUMMARY

Physiological ecology provides a general framework for seeing the role that abiotic conditions play in regulating the performance of individual organisms in an ecological context. Using the niche as the general framework emphasizes the importance of ecological master factors: temperature effects, water balance constraints and oxygen availability. The examples discussed show how evolutionary principles play a role and how applications to the real world may derive from an ecological perspective based on tolerances, preferences or constraints, and the ways that behavior can modify those parameters.

Like all other chapters in this book, this one offers a specific view of ecological processes and a modest selection of the fascinating variety in ecological systems. Understanding the complexity of ecology requires a polytypic perspective, and is an increasingly important obligation for members of our society. Humans are the dominant ecological force of today's world (Vitousek et al. 1997), literally changing the globe and doing so at an accelerating pace. Human decisions will determine the structure and function of many ecological systems. While only a small fraction of the readers of this book will become professional ecologists, all of them will be given the chance to vote. An informed view of ecological issues will help ensure wise use of that power.

RECOMMENDED READING

The titles listed below represent a selection of "classics"—they are among the intellectual cornerstones of physiological ecology.

GEORGE EVELYN HUTCHINSON. 1965. *The Ecological Theater and the Evolutionary Play.* Yale University Press, New Haven, Connecticut.
> *This short book describes the main ideas behind the concepts of fundamental and realized niche.*

WARREN P. PORTER AND DAVID M. GATES. 1969. "Thermodynamic equilibria of animals with environment." Ecological Monographs 39:227–244.
> *An excellent synthesis that defines the approach now known as "biophysical ecology."*

COLIN R. TOWNSEND AND PETER CALOW, EDS. 1981. *Physiological Ecology: An Evolutionary Approach to Resource Use.* Blackwell Scientific Publications, Oxford, England.
> *This edited volume emphasizes an evolutionary or "optimization" approach in evaluating how animals and plants adapt to their environment.*

Chapter 6

BEHAVIORAL ECOLOGY
Investigating the Adaptive Value of Behavior

Robert L. Jeanne

It seems to me that no man need be ashamed of being curious about nature. It could even be argued that this is what he got his brains for and that no greater insult to nature and to oneself is possible than to be indifferent to nature.
—*Niko Tinbergen, Nobel Laureate*

INTRODUCTION

Let's return to the stream bank shown in Plate 1.1 and learn how to think like a behavioral ecologist. The first step is to ask questions about what we see. Since behavioral ecologists study behavior, this means asking questions about what the animals we see are doing. Our best bet is to find a spot along the stream, sit down, and keep quiet so we don't disturb our subjects. Asking questions is easy; all we have to do is be alert for movements and sounds near us, watch closely, and turn the dial on our "curiosity amplifier" to the max. All "why" questions, questions that ask why the animal behaves as it does, are fair game.

The first behavior we notice is the raucous calling of the redwing blackbirds nesting in nearby cattails. The males—the ones with the red and yellow patches on the wings—seem to be defending territories. Why do they do that? Why don't the females defend territories, too? Females are brownish, without any red or yellow; why are the males strikingly colored and conspicuous while females are drab and more cryptic? Next we see a bumblebee visiting the flower spikes just in front of us and wonder why she starts at the bottom of the inflorescences and works her way up (Figure 6.1). What makes her decide to leave one plant and fly to the next, or to return to the nest? We turn over a flat rock nearby and discover ants nesting underneath. Why are they under the rock? When we peer closer and watch the workers scramble to carry their brood to safety below ground, we detect the strong smell of citronella. What is the function of this odor? Next we see two damselflies linked together and flying in tandem (Plate 6.2). Why do they do this? Now we peer over the edge of the streambank. On the water are whirligig beetles skimming about on the surface. Why do they stay in such

Figure 6.1. Bumblebees foraging for nectar on vertical inflorescences start with the older flowers at the bottom and work their way up. Redrawn from Krebs & Davies 1981, after Pike 1979.

tight groups? A little farther out we spot some water striders skating over the surface. Why don't they stay in tight groups like the whirligigs? Now we notice some movement under the surface—small fish. Surely the striders and the whirligigs ought to make easy pickings for the fish. Why don't they become fish food?

Given the luxury of time to linger along the stream, we continue to let our curious eyes—not to mention ears and noses—focus on first this and then that detail of the myriad lives being played out around us, becoming so absorbed that time passes without notice. In no time, it seems, we've come up with dozens of questions about why the animals we see behave the way they do, all within a few square meters of stream-edge habitat. We've taken the first step toward becoming a behavioral ecologist.

WHAT IS BEHAVIORAL ECOLOGY?

Notice that most of the questions we asked above begin with *why. Why* questions ask about the biological function of a behavior pattern. By *function* in this context, we mean the role that the behavior pattern or structure plays in the survival and reproduction of the animal. In other words, we mean that the behavior is **adaptive.** By adaptive, we mean that the trait has been favored by **natural selection** because in past generations that specific behavior allowed its ancestors to produce more offspring compared to others that had a slightly different behavior. In scientific terms, successful ancestors had greater **fitness,** which is defined as the number of surviving fertile offspring produced by an individual, relative to other individuals in the population. Another way of looking at *why* questions is to think of them as questions about *ultimate causation.* The ultimate cause of a particular behavior pattern is differential reproduction over the history of the species. The behavior we see today is the effect of this ultimate cause (Box 6.1).

Behavioral ecology is the subfield of ecology that seeks to understand the functions, or fitness consequences, of behavior. It is called behavioral ecology because the adaptiveness of an individual's behavior has meaning only in the context of the physical and biotic environment in which it lives. Therefore, as in other branches of ecology, most of the work of behavioral ecologists is done in the field, studying animals in their natural environments. Unlike most other branches of ecology that address interactions from the perspective of the population, community, or ecosystem, behavioral ecology, like physiological ecology, focuses on the individual and its interactions with its environment. Many of the phenomena that ecology deals with, including niche selection, predator-prey interactions, and competition, are ultimately behavioral phenomena. In this sense, much of modern ecological research is concerned with behavioral problems (Mayr 1982).

Niko Tinbergen, who together with Konrad Lorenz and Karl von Frisch won the Nobel Prize in 1973 for contributions to the study of animal behavior, divided the kinds of questions we can ask about behavior into four categories (Tinbergen 1963; Sherman 1988):

- Questions about physiological mechanism, including cognitive processes: *What is the stimulus that causes a particular behavior to be elicited? What goes on in the nervous system to mediate the response?*
- Questions about ontogenetic mechanism: *How does the behavior develop in the individual? What are the genetic and environmental inputs? What is the role of learning in the development of the behavior?*
- Questions about phylogenetic history: *What has been the evolutionary history of the behavior?*
- Questions about function: *What are the functional consequences of the behavior? Why does the animal behave in that way? What is its effect on fitness? How does the behavior contribute to the survival and reproduction of the animal?*

Behavioral ecology focuses primarily on the last of these questions. It seeks to understand the adaptive significance of behavior, that is, how a given behavior helps the animal survive and reproduce in its natural environment. Study of functional consequences, however, often gives rise to questions about underlying mechanisms, and behavioral ecologists commonly pursue these leads. For example, in studying why distasteful insects are often so brightly colored, we are led to ask whether their potential predators learn to associate distastefulness with a bright color more readily than a dull color. A complete understanding of function requires linking up with the other three of Tinbergen's four questions (Davies 1991).

The behavioral ecological approach to studying adaptations is based on several principles (Krebs and Davies 1993):

BOX 6.1. WHAT IS "BEHAVIOR?"

In this book we've already defined what ecology is, and in the paragraphs above we've explored what behavioral ecology is. But what do we mean by *behavior?* There are many definitions in the literature, but perhaps the simplest and most satisfactory is that behavior is "anything an animal does." This is generally taken to mean motor responses of all kinds, from movements as simple as an eye blink, to patterns as complex as the courtship ritual of the woodcock. By defining behavior in this way, behavioral ecology effectively excludes plants. With the exception of the small group of carnivorous plants that trap their prey with rapid movements, plants don't "behave"—but perhaps the field

shouldn't be so exclusive of plants. Behaviorists are quite willing to include among the things animals "do" the emission of exocrine glandular secretions—pheromones and allomones—that modify the behavior of other individuals; and many plants certainly produce such secretions— floral odors and nectar, for example. Plants evolve life history strategies, resource allocation patterns, and strategies for attracting pollinators, all topics that could be included in a broadly defined field of behavioral ecology. However, because the field of behavioral ecology has grown up with a focus on the motor activities of animals, animal behavioral ecology will be the focus of this chapter.

- Natural selection favors individuals whose life history strategies maximize their contribution of genes to the next generation, that is, who pass on more copies of their genes than do their competitors. **Life history strategy** refers to how the animal balances the trade-offs between traits affecting survival and traits relating to reproduction.
- The optimal life history strategy for an animal depends on its ecological context—the animal's physical environment, food supply, predators, competitors, and mates.
- Because animals interact with their environment primarily through their behavior, natural selection will therefore act on behavior to make animals efficient at foraging, competing, defending, and reproducing.

Underlying these principles of behavioral ecology are two fundamental assumptions. The first is that behavior evolves by natural selection. In order for natural selection to cause behavior to evolve, the behavior in question must first show *variability* among individuals in the population. All traits show at least some variation, but the range of vari-

---

Box 6.2. Behavior is adaptive only in the context in which it evolved

The importance of the role of ecological context in shaping and maintaining behavior in a population readily becomes apparent when the environment changes. Remote islands are excellent natural laboratories in which to see this. Because of their distance from source populations, the faunas on such islands are a non-random subset of nearby mainland faunas, heavily weighted in favor of groups that are better dispersers, but seasoned with an abundance of chance. Thus birds and bats and flying insects are more likely to be represented than are freshwater fishes, amphibians, and large mammalian predators. Because large predators are typically missing from island faunas, descendants of successful colonists often lose the defensive behavior that enabled them to coexist with these predators on the mainland. Common adaptive responses among birds on remote islands, for example, include the evolution of flightlessness and tameness.

The island of Mauritius lies some 850 km east of Madagascar. In the late sixteenth century, Dutch sailors began using it as a stopover on voyages across the Indian Ocean. They found it inhabited by quantities of wild game, which they used to restock their ships' larders. Among the game was the dodo, a large flightless descendant of a species of pigeon that managed to colonize the island in the distant past. Because they faced no predators on Mauritius, dodos had evolved to be flightless and extremely tame. They were well adapted to their environment.

The arrival of humans suddenly changed the dodo's ecological context: now it faced a predator. Because wariness was not a trait that had been adaptive in generations past on Mauritius, the dodos were so unafraid they could be approached by the sailors and clubbed to death. By the end of the seventeenth century, the last dodo had been eaten, earning the species the distinction of becoming the most famous human-caused extinction in history. Behavior that had been adaptive in the environment in which it evolved suddenly became maladaptive when that environment changed. Selection against tameness was so strong over the next 100 years that the population did not have enough time to evolve more adaptive behavior: fleeing at the sight of men.

ability may differ widely, depending on the trait. There is probably little variation in the chemical composition of the sex attractant (or pheromone) emitted by female silk moths, for example, because it is important that this key signal be unambiguous. On the other hand, the menu of food plants that rabbits include in their diets may vary greatly across a population of rabbits or through the history of a population.

The second requirement for behavior to evolve is that the behavioral variability in a population must have **heritability,** or a genetic basis. A genetic basis has been shown for all behavioral traits examined to date. Several techniques are used to study the genetic input into behavior. A population of fruit flies can readily be selected over a few generations into two lines, one positively phototactic (moves toward light) and the other negatively phototactic (moves away from light). When individuals in a lineage are bred with one another, their offspring inherit the behavior of their parents. Even learning ability comes under the influence of natural selection. While the learned behavior itself is not directly genetically encoded, the context in which learning can take place, what kinds of behavior can be learned, the speed of learning, and the duration of memory, are all shaped by natural selection. As long as some of the behavioral variability in a population is heritable, natural selection acting on the behavior of individuals will cause the variants that are better adapted in that environment to increase their representation in the population.

The second assumption critical to behavioral ecology is that the behavior studied is adaptive, that is, that natural selection maximizes fitness within the constraints that may be acting on the animal (Box 6.2). Some, notably Gould and Lewontin (1979), have attacked this assumption, arguing that theories about adaptations are little more than untestable stories, much like Kipling's imaginative tale about how the leopard got his spots. While Gould and Lewontin were right to call attention to the possibility that some behavior may in fact not be adaptive, to take this as a working assumption in all cases would have the effect of cutting off inquiry. The kinds of behaviors that interest behavioral ecologists tend to be complex and/or conspicuous behavior patterns that are costly to maintain. It is reasonable to assume that those patterns are maintained in a population by natural selection because their benefits outweigh their costs, that is, because they are adaptive. In fact, it is hard to imagine that a complex behavior pattern with no function could continue to exist in a population. Variants lacking the pattern would have the advantage of allocating time and energy saved from having to perform it to behavioral traits that are adaptive, and their descendants would increase in frequency in the population.

To illustrate how natural selection can move a population toward adaptive behavior, consider clutch size, the number of eggs laid in a brood. How many eggs should a female bird lay in a clutch? At first glance, it might seem the more the better, since natural selection will favor individuals that leave the most offspring over their lifetimes, as long as some of the variance in clutch size is heritable. But what counts when tallying relative fitness is not the number of eggs laid, but the number of offspring that survive to reproduce themselves—and a lot can happen between the egg stage and the reproductive-age adult.

Why doesn't natural selection simply select for ever-larger clutches? Let's take the example of the great tit *(Parus major)* in England, a bird related to the North American chickadee (Plate 6.1). A decades-long study of a population near Oxford (Perrins 1965, 1979) showed that at small clutch sizes, each offspring's chance of survival is high because the parents have no trouble collecting enough food to rear healthy, well-nourished chicks. However, as clutch size increases, it becomes increasingly difficult for the parents to provide enough food for all the young. Consequently, as clutch size increases, mean nestling weight declines (Figure 6.2A). Undernourished, underweight chicks have less chance of surviving after fledging (Figure 6.2B). This means that very large clutches will actually produce fewer surviving adult offspring than clutches intermediate in size. Perrins & Moss (1975) experimentally manipulated brood sizes to obtain clutches over a wider range of sizes than occurs naturally, then measured the survival of the fledged offspring. The results are shown in Figure 6.2C. The peak of the curve predicts an optimal clutch size of 9–12, the size that will yield the greatest fitness for the parents per brood. As long as some of the variance in clutch size is heritable, selection will act most strongly against individuals at the two extremes, favoring clutch sizes toward the middle value; a process known as **stabilizing selection.**

Surprisingly, actual mean clutch size in the same population is about nine eggs (Perrins 1979), somewhat smaller than what Perrins & Moss's results predict. The reason is thought to be that the manipulated nests showed what parents could raise *per brood* and does not take into account the full costs to the parent of raising

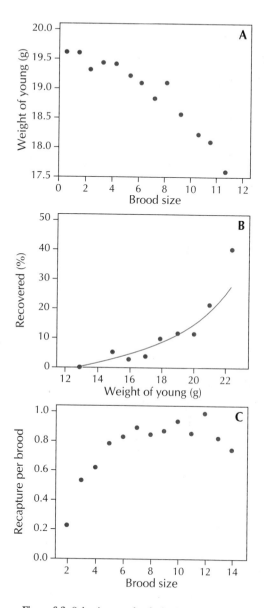

Figure 6.2. Selection on clutch size in great tits.
A. As clutch size increases, the average weight of the young at age 15 days declines (Perrins 1979).
B. Heavier chicks have a greater chance of being alive 3 months after fledging (Perrins 1965).
C. Standardized number of recaptures per brood, for experimentally modified clutch sizes (Perrins 1979). The peak value predicts the optimal clutch size, the size at which maximal fitness will be realized.

these larger broods. For example, the female with a larger brood must spend more time foraging, exposing herself to a greater probability of predation. Natural selection acts to maximize reproductive output over the entire lifetime of the female, and since tits survive and reproduce for several years, females that produce clutches of around nine may be more likely to survive to reproduce another year than those that raise larger clutches. Thus, there is a trade-off between a large clutch in a single year and the number and size of clutches over the lifetime of the parent.

HOW BEHAVIORAL ECOLOGISTS TEST HYPOTHESES ON THE FUNCTIONS OF BEHAVIOR

Suppose you've identified a behavior pattern to investigate. The next thing to do is to develop tentative explanations of the adaptive value of the behavior. If you observed and described the behavior in some detail, several ideas may already be jostling around in your head, competing for your attention. The next step is to formulate these into distinct alternative hypotheses, the more the better. To test each hypothesis, use it to make predictions about the effects of the behavior. A hypothesis can make a variety of predictions, and here ingenuity is required to come up with testable predictions.

Behavioral ecologists have two techniques available to them for testing hypotheses. One is *experimental manipulation* and the other is the *comparative method*. With the experimental approach, an ecologist tests the prediction of a hypothesis by holding certain variables constant while varying the one in question to see what effects it has. Often, however, this is not feasible. Sometimes the variation one needs to test a prediction does not exist in the population, and cannot be created artificially. If an experimental approach is not workable, then ecologists often resort to the comparative approach. In effect, the comparative method takes advantage of the natural differences among species and among environments to test hypotheses on the survival value of the behavior under study.

To illustrate these two approaches, let's look at Niko Tinbergen's classic studies on eggshell-removal behavior in gulls. Black-headed gulls *(Larus ridibundus)* nest in colonies on dunes along the coast of Britain. Tinbergen noticed that within an hour or so after a chick hatches, the parent picks up the eggshells in its beak and flies off to drop them at a distance from the nest (Figure 6.3). Tinbergen felt sure that the behavior was adaptive because it has a cost: when a parent is off the nest, the eggs and chicks, even though camouflaged, are vulnerable to predation by carrion crows, foxes, and even gulls on neighboring nests. Tinbergen and his coworkers speculated about what the function of the behavior could be. One of several hypotheses they considered was that the behavior nevertheless reduced the overall risk of predation by removing the eggshells, whose bright white interior could serve to alert predators that eggs or chicks are nearby (egg exteriors and chicks are cryptically colored). One alternative hypothesis was that the sharp edges of the shell might injure the chicks.

A prediction made by the predation hypothesis is that intact eggs placed near eggshells would be more likely to be found by predators than eggs more distant from shells. Tinbergen tested his prediction with an experiment in which he placed intact eggs out in the sand dunes, each paired with an eggshell at a distance of 15, 100, or 200 cm. Crows found 42% of the eggs close to shells, 32% at 100 cm from shells, and only 21% of those 200 cm from shells, supporting the hypothesis that the conspicuous eggshells tip off predators that food is nearby (Tinbergen et al. 1963).

Figure 6.3. Eggshell removal by a black-headed gull.

Another prediction the predation-reduction hypothesis makes is that a population that had never been exposed to predation on chicks in the nest would lack eggshell removal behavior. On the other hand, the alternative hypothesis, that the shells might injure the chicks, would predict that shell removal would still occur in such a predator-free population. Unfortunately, no black-headed gull populations are known that are not exposed to at least some predators, and in fact all known black-headed gull populations perform eggshell removal. Here Tinbergen and his students resorted to the second approach, the comparative method. The kittiwake *(Rissa tridactyla)*, a relative of gulls, does provide a test of this prediction. Because they nest on narrow ledges on cliffs, predation on kittiwakes chicks in the nest is extremely low. As predicted, kittiwakes do not remove eggshells from their nests (Cullen 1957), supporting the hypothesis that eggshell removal is an adaptation that reduces predation on the chicks by flying predators, but not the hypothesis that the shells are removed because they might injure the chicks.

THEORY IN BEHAVIORAL ECOLOGY

Tinbergen showed that if gulls did not remove the eggshells, they would be likely to suffer higher predation on the chicks and remaining eggs in the nest. His hypothesis on the survival value of eggshell removal made mostly *qualitative* predictions about the effects of the behavior. One of the most significant developments in behavioral ecology in recent decades has been the adoption of a more *quantitative*, cost-benefit approach to the measurement of fitness. This method of analysis, co-opted from the field of economics, has been instrumental in putting behavioral ecological theory on a solid quantitative footing. The units of cost and benefit are not dollars, of course, but units of fitness. Recall that individual fitness is defined as the number of surviving fertile offspring produced by an individual, relative to other individuals in the population. How do the activities that an animal engages in contribute to the animal's overall fitness? Here the behavioral ecologist is faced with two problems. One is that it is often extremely difficult to measure the relative fitness of individuals in a population. This involves tracking the rate of successful production of offspring by a set of individuals under natural conditions. The other problem is that the fitness of an individual is a consequence of its entire **phenotype** (all the characteristics of an individual) and not just the behavior we are interested in. The behavior under study is not the only factor affecting survival and reproductive success; many variables interact, often in subtle ways, making it difficult to isolate the ultimate fitness effects of a given behavior, even if clever experimental ways are devised to get at them. Faced with these difficulties, behavioral ecologists have taken the approach of assuming that natural selection not only optimizes or maximizes overall fitness, but will favor individuals that maximize the *benefit:cost ratio* for each activity, within the limits of existing constraints.

It is important to recognize that organisms face constraints of many kinds. The overarching one is the physiological limit to the amounts of energy and materials an individual can commandeer for its own use. Every organism allocates its fixed energy budget among activities related to growth, maintenance, and reproduction. Different individuals or species may allocate energy among these differently; how an animal chooses to do so constitutes its life history decisions. If an animal allocates a large proportion of its energy to growth, for example, it will have less to allocate to reproduction. At a finer level, environmental constraints are imposed on particular activities by the animal. A foraging bird may be constrained from foraging for food at its maximum rate by the need to avoid predators. Alternatively, its foraging rate may be constrained by the rate at which it can obtain information about the location of food sources. Because of such constraints, the animal often cannot maximize the benefit:cost ratio of an activity, but can only optimize, i.e., attain the highest net return possible within the limits imposed by its internal and external environment. It does the best it can in the environment it faces and with the equipment it has.

The assumption that natural selection favors individuals that maximize the benefit:cost ratio for a given activity raises the question of what units a scientist should use to measure cost and benefit for that activity. Although the ultimate measure of fitness is the number of copies of one's genes passed on to the next generation relative to others' genes, that wouldn't be very useful when we're looking at the adaptive value of specific behavior, for the reasons noted above. We need some measure of the costs and payoffs relevant to the behavior being studied. For example, if we're interested in food-finding behavior, we may measure its benefit in terms of the rate of energy gain that results from the behavior, and measure cost as energy expended while foraging. (Energy is measured either in calories or joules.) On the other hand, if we're studying male mating behavior, the appropriate units may be the number of matings achieved. In each case, we assume that the units chosen translate into fitness units over the life span of the individual. In other words, we are taking calories and matings as valid "currencies" of fitness in nature's economy, just as money is the currency for goods and services in the human economy.

In summary, the aim of behavioral ecology is to gain insight into the adaptiveness of behavior, or why animals behave as they do. The working assumption of behavioral ecology is that natural selection favors individuals who perform necessary tasks of finding food, competing for resources, avoiding predation, and reproducing with the highest benefit-cost ratio. Understanding adaptiveness calls for gaining insights into the constraints (historical, physiological, ecological, informational) acting on a particular behavior, and to identify the currency of fitness being maximized in particular cases.

QUESTIONS ASKED BY BEHAVIORAL ECOLOGISTS

The questions posed at the head of this chapter are questions a curious but naive observer might ask about behavior seen in a local environment. The questions taken up in the rest of this chapter all had similar beginnings, but are more mature in the sense that they are framed within the context of modern behavioral ecological theory. That is, they focus on some of the numerous fronts along which behavioral ecological research is active.

The first question we'll deal with below is food finding, to illustrate the application of optimality theory to behavioral ecological questions. Although optimality theory has been applied to reproductive behavior, most of the studies involving optimality theory have involved food finding because costs and benefits can be measured in terms of energy units or time. The second question adds the ingredient of intraspecific competition. When competitors are involved, optimizing behavior becomes more complex. In addition to having to take account of the distribution and properties of resources, the animal must factor in what its competitors are doing. In these circumstances, there may not be a single, best response, but two or more alternative strategies; game theory is the appropriate technique to use. The third question deals with issues centered on sexual reproduction and how conflicts of interest between the sexes give rise to differences in reproductive strategies between male and female. We examine the consequences of intrasexual (between members of the same sex) and intersexual (between the sexes) selection. Finally, in the last section we take up the question of how cooperation can evolve.

What Influences the Decisions Animals Make as They Forage for Food?

An animal searching for food—foraging—makes a host of decisions, such as what habitat to search in, when to search, how long to spend looking in each patch of habitat, and whether to eat all potential food items encountered or to select only those of certain types. (The term "decision" here is not meant to imply that the animal makes a conscious decision, but is used as shorthand for the exercising of an appropriate behavioral option.) If the animal is a parent foraging for its offspring, or a social insect (such as a bee or ant) gathering food for its colony, it must make additional decisions about how far from the nest to travel to hunt for food and how much of a load to gather before returning to the nest. As a working hypothesis, we assume that an animal (or a colony) making these decisions in ways that incur less cost while obtaining a given amount of food ought do better than one that incurs more cost. How best to do this is essentially an economic decision, and behavioral ecologists, in searching for ways to analyze how animals make these decisions, have adopted an analytical approach from economics called **optimality theory.** Optimality theory says that natural selection will favor the variant of a behavior with the greatest net benefit; i.e., for which the value of the benefit of the behavior minus its cost is maximized. Because both benefits and costs of obtaining food can be measured directly in terms of energy units or their equivalents, the optimality approach has been widely applied to foraging behavior, where it is called **optimal foraging theory.**

Optimality Modeling

The foods utilized by animals often occur in patches, or islands of habitat, that are separated from one another. The foraging bumblebee mentioned in the introduction to this chapter is a good example. Each spiked inflorescence visited by the bee can be thought of as a patch, consisting of a number of individual, nectar-yielding flowers. We saw that the bee worked the flowers of each inflorescence for only a brief time before flying on to the next. As the foraging bee uses up the nectar in an inflorescence, it becomes less and less profitable for it to stay there continuing to search. But how should it decide when to go the next inflorescence? If it stays too long, it will sip more nectar, but its average rate of nectar intake will drop because it will be working flowers with ever-lower nectar amounts. However, if it leaves too soon, it will also experience a low rate of nectar intake because it will be spending a large proportion of time flying between inflorescences rather than gathering food. The optimality approach assumes there is an optimal time for the bee to spend at each inflorescence if she is going to maximize her rate of energy intake. Optimal foraging theory provides a way of determining what information the bee is using to make her decisions, what constraints she is acting under, and what currency units she is using in her accounting system to tally costs and benefits.

The first step in developing and testing an optimal foraging model for a species is to decide how to measure costs and benefits. Given the reasonable hypothesis that animals gathering energy at the highest rate with the least cost will have the highest fitness, a plausible working hypothesis might be that the forager behaves so as to maximize its net rate of energy gain. *Rate* is the amount of energy collected per unit of time in the form of food items; *net rate* is the rate of energy collected minus the rate of energy spent in the search. Because it is seldom practical to determine the energy content of each food item, we make the simplifying assumption that the quantity of energy gained in each unit of food is proportional to the size of the unit. In other words, we assume that food amount—volume of nectar or number of prey items of a given type, for example— is a reliable surrogate "currency" for energy. Likewise, in lieu of trying to determine how many calories the forager actually burns in each patch, we make the simplifying assumption that the time spent per patch is a reasonable currency for the energy cost. These shortcuts are justifiable as long as we acknowledge that, if our data fail to support our hypothesis, one reason might be that our assumptions about these currencies are wrong.

To collect data on the rate of food uptake in a patch, we observe the animal actually foraging in a patch, and score whenever it finds a food item. A forager first arriving at a patch likely finds abundant food, and rate of food gain is high. As the forager depletes the food in the patch, however, finding each new unit of food becomes increasingly harder. As a result, a plot of cumulative food uptake (benefit, or gain) as a function of time spent in an average patch (cost) should level off asymptotically, as shown in Figure 6.4A.

As the forager continues to work the patch, there comes a point on this curve of diminishing returns when the rate of collection of new food items is so low, the forager benefits by leaving for the next patch. But flying to the next patch is an activity that has a net cost, because the forager is spending energy in movement without taking energy in as consumed food. We show this graphically by adding travel time to the x-axis, which is our graphical measure of cost in terms of time spent. This is done by extending the x-axis to the left of the base of the gain curve, to represent the average amount of time it takes the forager to get to a patch. Suppose we measure this a number of times and find the average time to be T, as shown in Figure 6.4B. Now we know how long it takes to fly from one patch to another, on average. But we still don't know how long the forager should stay in the patch once it gets there. How do we solve for that?

Our hypothesis says that the forager maximizes the net rate of energy gain, or benefit:cost ratio. In our graphical representation, energy gain (benefit) is given on the y-axis and time spent (cost) is given on the x-axis, so the net rate of energy uptake will be a line starting at T and rising with a slope of y/x. The maximal net rate of energy gain will therefore be given by the line y/x that has the steepest slope. We know the line must begin at point T at the left end and that it cannot rise higher than the gain curve at the right end, so the steepest line is the one that lies tangent to the gain curve, i.e., just touches it at its highest point, as shown in Figure 6.4C. The value of x at this tangent point, F, will be the optimal time that our model predicts the forager should spend feeding in each patch to maximize its overall rate of energy intake. If it spends less or more time than this, then the slope of the energy rate curve will be less than this optimal value.

Note that the model also predicts that the shorter the travel time between patches, the less time it pays to stay in each patch. This prediction of optimal foraging theory is shown graphically in Figure 6.4D.

Figure 6.4. Optimal foraging modeling. **A.** The curve showing rate of energy gain as a function of time spent in a patch rises more and more slowly as the food is depleted. **B.** Travel time, T, to the average patch is represented by extending the x-axis to the left of the time, A, the forager arrives in the patch. **C.** The optimal foraging time, F, in the patch is solved graphically by finding the point at which line TG falls tangent to the gain curve. This gives the maximum slope, y/x, for the net rate of energy gain. **D.** Optimal foraging theory predicts that optimal residence time in a patch should decrease (increase) as travel time between patches decreases (increases). A shorter travel time, T_s, predicts a shorter optimal residence time, F_s, in the patch.

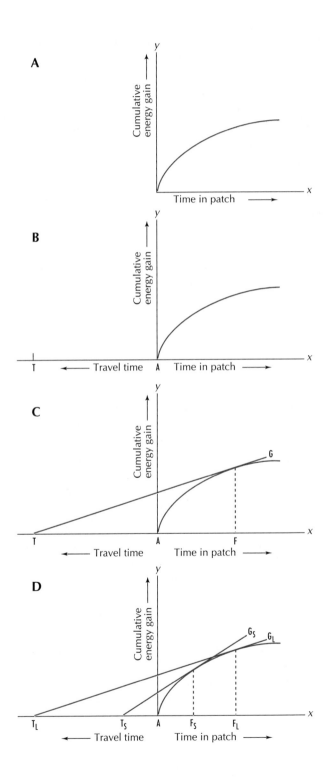

A test of a simple optimal foraging model is Cowie's study of great tits in England (Cowie 1977). Working with birds in an aviary, Cowie trained six of them to search for food in patches. Each "patch" consisted of a cup containing six bits of mealworm mixed with sawdust. As a foraging bird discovered and removed each mealworm bit, its time to discovery of the next one increased, because fewer prey were left. The curve of energy gain therefore had the general familiar shape shown in Figure 6.4A, above. To vary the time it took to reach each new patch, Cowie made the birds remove lids from the containers. One type of lid was easy to remove and the other was more difficult, simulating short and long travel time between patches, respectively. The results, shown in Figure 6.5, revealed a good fit to the predicted values for net rate of food intake. In other words, the tits appeared to be maximizing their net energy gain rate, or gain in energy (food) minus the cost of collecting it per unit time.

Sometimes empirical results don't fit such a simple model. When this happens, it suggests that the model needs to be modified. One possibility is that the model includes costs and benefits that are incorrect. For example, perhaps different types of prey have different nutritional concentrations, so that if we simply count prey items we will get an inaccurate estimate of the rate of energy gain. Or it may be that the energy cost of flying to and from a patch is higher than the cost of searching at that site, so that simply equating cost of both activities to time spent gives an inaccurate representation of the true energy cost of performing the activity.

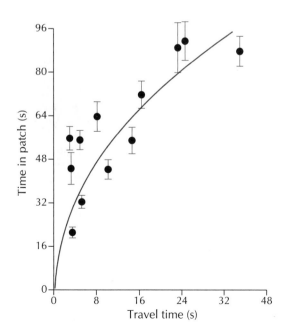

Figure 6.5. Results of Cowie's test of an optimal foraging model for great tits. The data do not differ significantly from the result predicted by the model, shown by the solid line. Each point gives the mean time (with standard deviation) spent by a bird in a patch as a function of how long it took it to get to the patch. Individual birds varied in how long it took them to remove the lids. Since each bird was tested on "short travel time" and "long travel time" patches, there are 12 points, each representing six trials. Redrawn from Cowie 1977.

A second possibility is that our assumption about the currency is wrong and the animal is not maximizing net rate of energy gain, as in the case of the great tit. Honeybee foragers, for example, maximize efficiency, measured as net energy unit collected per energy unit expended (Schmid-Hempel et al. 1985).

A third possibility is that our tacit assumption that the animal has access to all the information it needs is wrong. Animals are not omniscient; they don't have instant access to everything they need to know to make optimal decisions.

Still a fourth possibility is that the animal may be acting under constraints we haven't recognized, such as the risk of predation, or the need to obtain nutrients from a variety of sources, and that these are compromising what would otherwise be a purely optimal strategy. An example of this is food plant selection by moose (Belovsky 1978, 1981). During the summer, the moose on Isle Royale in Lake Superior feed on deciduous leaves, herbaceous plants, and aquatic vegetation. Deciduous leaves and forbs (herbaceous plants other than grasses) have the highest energy content, and a model based on maximization of caloric intake predicts that they should specialize on these food plants alone. But the moose don't behave according to this prediction. Instead, they spend a considerable amount of time feeding on aquatic plants. Sodium is an essential element in the diet of moose, and aquatic plants are much higher in sodium than are terrestrial plants, so by including aquatic plants in their diet, the moose are maximizing their energy intake within the constraints of meeting their daily minimum requirement for sodium, all within the upper limit of food bulk imposed by the size of the rumen (Figure 6.6).

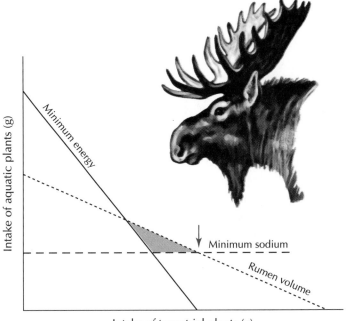

Figure 6.6. Diet optimization in the moose. The diet (arrow) maximizes energy intake within the bounds of three constraints. To obtain the minimum amount of energy required for survival, the diet must lie above the line labeled *minimum energy* and to meet the minimum sodium requirement, it must lie above the *minimum sodium* line. Imposing an upper limit on energy intake is the volume of the stomach, or rumen, indicated by the *rumen volume* line. Because aquatic plants are bulkier and contain less energy per gram than terrestrial plants, the rumen volume line has a shallower slope than the minimum energy line. Redrawn from Belovsky 1978.

Intake of aquatic plants (g)

Minimum energy

Minimum sodium

Rumen volume

Intake of terrestrial plants (g)

A common misconception is that the aim of optimality modeling is to test whether natural selection optimizes. Optimality is not what's being tested; it is assumed. The goal of optimality theory applied to behavior is to test hypotheses on the adaptiveness of behavior. What is being tested are the assumptions about cost-benefit measures, currency, and constraints built into the model. When the empirical data don't fit the predictions of a model, we revise certain of these parameters in the model, make new predictions, and retest. Because many of the predictions can be made in quantitative terms, we can often reach quite a detailed and precise understanding of what goes into the decisions animals make in foraging for food.

Thus, optimality modeling analyzes how animals make decisions in carrying out their activities, and why—in terms of their fitness consequences—they make the decisions they do. This knowledge can be applied in other fields of ecology. Understanding how predators make decisions about hunting for prey, for example, can be used by community ecologists as a basis for building detail into their models of predator-prey interactions to make them more realistic and accurate.

Why Do Some Species Have Alternative Strategies for Competing for Resources?

In the discussion of foraging behavior in the preceding section, there was a single best option for a forager to adopt to maximize energy input. Analyses of foraging behavior typically assume that the forager's behavior is unaffected by the activity of other members of the same species in the area. In other words, it is assumed that these *conspecifics* do not affect the abundance of food in the habitat. The forager must merely match its searching behavior to the current conditions of food distribution and abundance, and to the curves of handling and travel costs. Under such conditions, there is typically a single best feeding response to adopt—a single optimum.

However, in many cases a species may exhibit two or more alternative routes to gaining a resource. These are especially common among males seeking matings with females. A widespread pattern among both invertebrates and vertebrates is one in which some males defend females with whom they mate, while others simply try to sneak copulations with females. In other species, especially a number of insects, some males defend mating territories, while others search for females at large. In still other species, including some shrimp, crabs, and fish, some males behave like females, and, if they can fool the dominant male, quickly attempt to copulate with the female that the dominant male has attracted. In many cases, males exhibiting the different strategies are of different sizes or even different morphologies, and evidence indicates that in some cases the development of different morphs is genetically controlled (Lank et al. 1995).

What could be going on? If there is a single, optimal behavior, why don't all males opt for that? The answer is that there is not a single optimum.

Scorpionfly

Game theory

In situations where competition for the resource is intense, the behavior of some of the competitors modifies the competitive environment, creating constraints on options available to others, or providing opportunities for others to pursue a novel behavioral **strategy.** A strategy is a behavior pattern or morphology used in competing for a scarce resource. Applying simple optimality models when competition is a significant factor does not work, because the "best" strategy may depend on what strategy others in the population are adopting. To use an example from everyday experience, if you are driving to the mall at two in the afternoon, the fastest way may be the direct route via main roads; but if it is rush hour, a longer route using secondary roads may be the quickest. When competitors are engaged in contests over a resource, the appropriate analytical technique is **game theory.** Evolutionary game theory is the quantitative analysis of the evolution of two or more alternative strategies when the fitness of each depends on its frequency in the population. First developed as a tool in economics, game theory has been effectively applied to behavioral ecology, thanks largely to the pioneering work of John Maynard Smith (1982).

Much of the work on alternative strategies has focused on reproductive behavior, especially on how males compete with one another for access to females. The reason is that competition among males for mates often is intense (see below) and so plays an important role in decision-making. For some species, it has long been known that not all males behave alike in their search for females; this alternative behavior was previously assumed to be abnormal. Now we recognize that males, by exercising one of two or more different strategies, are probably doing the best they can to maximize their reproductive payoff, given the competitive context in which they find themselves.

The most common form in which alternative behavioral options are maintained in a population is the **conditional strategy.** Individuals are capable of following any of the potential strategies shown by a species, but they perform only one, according to their abilities relative to those of others. Male scorpionflies *(Panorpa latipennis),* for example, offer nuptial gifts of food to females to feed on during copulation (Thornhill 1981). Not only is a female more likely to choose to mate with a male who offers a larger nuptial gift, but the larger the gift, the longer she will mate with him. Male scorpionflies adopt one of three strategies, depending on their ability to produce a nuptial gift. Thornhill showed experimentally that males that could find and defend a dead insect such as a cricket as a nuptial gift were on average 89% successful in obtaining copulations, and that success rate was correlated with prey size. Males that were unsuccessful in obtaining a dead insect still managed to gain some reproductive success, however, by producing masses of nutritive salivary secretion as nuptial gifts. In Thornhill's experiment, 56% of the males with salivary masses were able to obtain copulations. Males that could neither find a prey nor had the nutrient reserves to produce a salivary mass adopted a third strategy—trying to mate forcibly with females. These realized only a 7.4% success rate. Thus the payoff to a male, in the form of the likelihood of inseminating a female, depended in large part on the male's ability to offer and defend a nutrient source.

In many species, males establish and defend territories containing resources required for the successful rearing of offspring. Male crickets and bullfrogs advertise their ownership of a territory, and attract females to it, by calling (Cade 1979; Howard 1978). Often, however, suitable territories are in short supply and not all males obtain one. In bullfrogs, generally the larger, older males are successful in gaining and defending territories. For smaller males, the strategy of trying to establish a territory too close to a larger, calling male doesn't work, because the big territorial male will attack and drive away rivals calling nearby. Consequently smaller males adopt the alternative of lurking around the margins of defended territories and remaining silent. These *satellite males* are sometimes successful in sneaking copulations as females make their way toward a calling male, but their reproductive success is not as high as it is for territorial males. That the satellite-male strategy is a facultative one, adopted when population density precludes them from holding territories, can be demonstrated by removing territorial males. Satellite males move in at once, adopt the vacated territories, and begin calling—showing they are clearly capable of exercising the more successful strategy, but when prevented by competition from doing so, they make the best of a bad situation and salvage at least some fitness by adopting alternative behavior.

Males of the cricket *Gryllus integer* show similar alternative strategies, but here satellite males may be gaining more than just a marginal success rate. A parasitic fly *(Ormia ochracea)* locates its cricket hosts by homing in on the male's territorial call. Satellite males, by remaining silent, suffer a lower rate of parasitization than their calling rivals (Cade 1979); thus the presence of the flies can increase the survival payoff to satellite males relative to calling males.

Conditional strategies such as these depend on differences in size or ability among males. However, alternative strategies can occur even in a population without such differences. The adoption of a strategy to obtain a resource by the majority of the population may create an opportunity to compete for the same resource via an alternative strategy. A good example is nesting behavior in the great golden digger, *Sphex ichneumoneus*, a large, solitary wasp. A female *Sphex* excavates a burrow ending in one or more enlarged cells (Figure 6.7). She provisions each cell with paralyzed katydids, lays an egg on one of the prey, then seals the cell. The developing larva feeds on the prey, overwinters, and emerges as an adult the following year (Brockmann et al. 1979).

The most common option for a female *Sphex* is to dig her own burrow ('digging' strategy). However, there is always a low percentage of burrows that are, for whatever reason (e.g., predation on the paralyzed prey; female killed), abandoned before being fully provisioned and sealed. Under these circumstances, entering an existing burrow is a viable alternative strategy to digging, because considerable time and energy are saved

Figure 6.7. A digger wasp female inspecting a burrow.

by doing so. But a female who chooses to enter an existing burrow has no way of knowing if it is truly abandoned, or if the owner is merely away on a foraging trip. In fact, both the enterer and the owner of the nest may provision such a nest for some time without encountering each other. Sooner or later, however, they do meet, and when they do, they fight. The winner goes on to complete the stocking of the cell, lay the egg, and seal the nest, while the loser goes off to dig her own burrow or find another one to enter.

In the great golden digger wasp, the strategy of digging creates an opportunity for entering. If digging is the predominant strategy in a population, then there may be lots of abandoned nests. Given this situation, an enterer will do well, because she is likely to find an abandoned burrow, and this will give her a considerable saving of time and energy over digging her own. Thus, in a population of pure diggers, the frequency of the entering strategy will be likely to increase over several generations (assuming that digging and entering behaviors are inherited and that burrow abandonment rates are high). However, if entering is the predominant strategy, then there will be few abandoned nests and enterers will frequently encounter occupied nests, from which they are likely to be ejected. Because of the time lost entering nests, enterers will not do well in terms of eggs laid per unit of time, and it will pay them to exercise the digging strategy. The success of each strategy is dependent on the frequency with which it is played in the population; each does better when it is rare, so the stable situation should be a mix of the two. In fact, game theory predicts that natural selection should act to move the mix of the two strategies in the population to the point where their payoffs are equal. This point is sometimes called a mixed **evolutionarily stable strategy,** or **ESS.** The important, and counter-intuitive, point about an ESS is that, while the payoffs are expected to be equal, they are less than the payoff of either strategy when played by 100% of the population. In other words, the stable mix is not the optimal strategy, but it is the evolutionarily stable one. The calculus of how this works is shown by the Hawk-Dove game, explained in Box 6.3.

In a New Hampshire population of the wasp *S. ichneumoneus* studied by Jane Brockmann, each female could play either the "digger" or the "enterer" strategy. Females dug their own nests 59% percent of the time, while 41% of the time they entered an existing burrow. By calculating the time spent and number of eggs laid by females pursuing each option, Brockmann et al. (1979) showed that the payoff was 0.96 eggs/100 hours for digging vs. 0.84 eggs/100 hours for entering. While not identical, these two values do not differ statistically, leading to the conclusion that the best explanation for

BOX 6.3. THE HAWK-DOVE GAME

John Maynard Smith (1982) was instrumental in applying game theory to the modeling of animal contests over resources. The classic version of the theory is the Hawk-Dove game."Hawk" and "Dove" refer to two strategies played in a population. The game is designed to show how natural selection may drive a population to a mix of two strategies in which the average payoff is less than the payoff to either strategy when pure, i.e., when played by the entire population.

The game is played as follows. The Hawk strategy is always to fight over the resource; the Dove strategy is simply to display and never fight. Suppose the winner in the contest gains a resource worth 50 fitness "points," while the loser gains nothing, or 0 points. In addition to these payoffs, there is a cost to each strategy. Hawks risk serious injury at a cost of –100 fitness points. Because Doves don't fight, they never risk injury, but the displays they perform in each contest take some energy and therefore cost 10 points.

The rules of the game are that when a Hawk attacks another Hawk, it wins half the time (for a payoff of +50) and suffers injury the other half (–100). Likewise, when a Dove attacks another Dove, it wins half its contests (+50 for the resources gained and –10 for the cost of displaying) and loses half (–10). When Hawk attacks Dove, Hawk always fights and wins; when Dove attacks Hawk, Dove always retreats without displaying and loses.

From these rules, we can construct a 2 × 2 matrix of the four possible types of encounters, and calculate the average payoff to each contestant (table).

To see how evolution would proceed, we begin by asking what would happen in a pure population of either all Hawk or all Dove. From the payoff matrix, if 100% of the population played the Hawk strategy, the average payoff would be –25. But this situation is not evolutionarily stable, because it can be invaded by the Dove strategy. A Dove in a population of all

Hawks would do very well, because its average payoff would be 0, much higher than the -25 the Hawks were suffering among themselves. The Dove strategy would spread in such a pop-

Attacker	Victim	
	Hawk	Dove
Hawk	0.5(50) + 0.5(-100) = -25	+50
Dove	0	0.5(50-10) + 0.5(-10) = +15

ulation. But Dove would not go to fixation by completely supplanting Hawk. In a population of pure Doves the average payoff is +15, but an invading Hawk would enjoy a payoff of +50, because every contest would be against a Dove. Natural selection would drive the frequencies of the two strategies to the mix at which the payoffs to both strategies were equal. The figure below shows the payoffs of each strategy as a function of its frequency in the population. If one of the strategies increases in the population, the fitness payoff to the other increases; selection will favor an increase in the frequency of the other strategy, bringing it back to the equilibrium point again (dashed line).

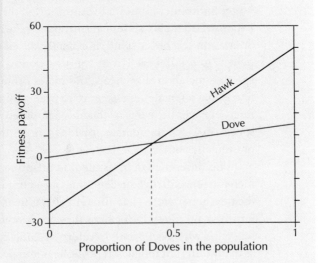

the existence of digging and entering in this population of *Sphex ichneumoneus* is that they form a mixed ESS.

While behavioral ecologists tend to apply game theory to cases of competition within species, as in the *Sphex* example, the same principles apply to competition between species. Thus, the game theory approach could be applied by community ecologists to understand the details of interspecific competition. Many cases of interspecific competition involve more than two species, however, making them *n*-player games, which are more complicated to analyze than the 2-player game illustrated above.

How Has Sexual Selection Shaped the Evolution of Reproductive Behavior?

Courtship was previously considered an entirely cooperative enterprise between the sexes. After all, what could be more in the common interest of a courting pair than to collaborate to produce and rear their mutual offspring? We now recognize, however, that there are considerable conflicts of interest between the sexes, and that male and female are far from being in complete agreement over the details of reproduction. The sexes may disagree, for example, over how much each should invest in rearing the offspring. Some conflict can be expected simply because mates are not genetically identical, but this is not enough to explain why the sexes behave so differently during the various stages of reproduction. In many species, for example, males tend to be ardent suitors and to compete for access to females—any females. Females, for their part, tend to be choosy about which male they mate with. Why do differences such as these occur? Are they adaptive, and if so, how?

The causes of the conflict arise out of the most fundamental difference between male and female—the size of the gametes produced by each (Trivers 1972). Eggs are large, immobile, crammed with food for the developing embryo and are therefore expensive to make, so females produce relatively few in their reproductive lifetimes. Sperm, on the other hand, are small and motile—little more than "DNA with a propeller," to use Robert Trivers' phrase. Compared to the cost of an egg, manufacturing sperm is metabolically cheap, and males produce huge numbers of them. The bottom line is that a female invests more in each gamete than does a male. This greater investment in each offspring by females compared with males is even more accentuated in species in which the female provides care for the developing embryo during pregnancy or even after birth.

This difference in investment in offspring by the two sexes means that selection to increase reproductive success—measured by how many copies of one's genome get into the next generation—has different effects in males and females. Since he can potentially fertilize eggs much faster than they are produced by females, a male can increase his reproductive success, and therefore his individual fitness, by finding and inseminating as many different females as possible. A female, however, cannot do the same by mating with many males. The only way she can raise her reproductive success is by turning food into eggs at a faster rate, a process limited by the high physiological cost of each egg. This difference means that it pays males to seek out more matings. For a female, by contrast, a single copulation is enough to achieve more or less maximum reproductive success. Consequently, females are a *de facto* scarce resource for which males will compete. Because the potential payoff to a male is very high, there is strong selection for male ability to acquire matings. This kind of selection is called **sexual selection,** and it can work in two ways. First, as noted above, it can favor the ability of males to compete directly with one another for inseminations. Darwin called this **intrasexual selection,** because it is imposed by males on males. Second, it can favor traits in males that attract females. Darwin called this **intersexual selection,** because it acts via females selecting on males.

This competition among males, combined with their variability, results in some males doing very well in fathering offspring and others doing poorly (Figure 6.8). Intrasexual selection has led to a number of forms of "beating the competition." In some species, males have evolved structures that are used in fighting with other males. In deer, sheep, and some scarab beetles, for example, males have evolved horns that are used in male-male combat over access to females (Figure 6.9). In songbirds and dragonflies, males establish territories from which they exclude other males and to which they attempt to attract females (Plate 6.1). In many insect species the female is promiscuous, but it is the sperm from her last mating that inseminates most of her eggs. In some of these species, the male guards the female after mating to prevent rival males from usurping his paternity. Males of some damselflies, for example, remain "in tandem" with the female, holding her by the thorax with his abdominal claspers, effectively preventing another male from taking this position to court her (Plate 6.2). In some

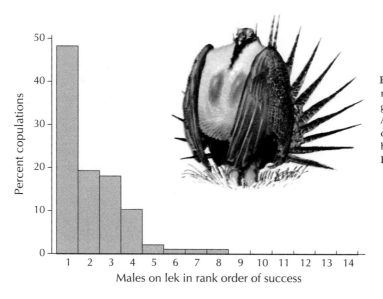

Figure 6.8. Distribution of matings among male sage grouse at a lek in Wyoming. Almost all of the 105 copulations were performed by a few top-ranked males. Redrawn from Wiley 1973.

Figure 6.9. Two male scarab beetles fighting for access to the female, right. The horns are used to pick up and throw the opponent on his back. The female stands on a fifty-pence coin (about the size of a half dollar).

damselfly species, the male's penis even has a specialized appendage designed to scoop sperm from a previous mating from the sperm storage receptacle *(spermatheca)* of his mate, to be replaced with his own, so his sperm will father her offspring (Waage 1973). In other insects the male inserts a mating plug in the genital opening of the female, effectively blocking copulations by rival males. Still another effect is the evolution of low thresholds for mating attempts. Male *Aedes aegypti* mosquitoes, for example, will attempt to mate with a tuning fork tuned to the frequency of the female's wingbeat, and males of an Australian buprestid beetle will try to copulate with beer bottles whose orange color and mammilated surface texture resemble females (Plate 6.3). Sperm is cheap, after all.

Our understanding of how intersexual selection works is not so straightforward as it is for intrasexual selection. Because each egg represents a relatively large proportion of a female's lifetime productivity compared to sperm for a male, the female in most species invests more than the male in each mating, and therefore stands to lose more if anything goes wrong. This is thought to be why females are choosier about their mates during courtship than are males.

What are females choosy about? One criterion is that her mate be of her own species. If she mates with a male of the wrong species, no viable offspring will be produced and her fitness will be zero. (It is also in the best interest of the male to mate with a female of his own species, but for the reasons given above, he is likely to be less careful than the female.) Therefore, mate location or courtship communication typically include species-specifying signals, and these come early in the courtship sequence. They include female sex pheromones that serve to attract conspecific males and filter out non-conspecifics.

The second thing females should be choosy about is the relative fitness of the male. Females do discriminate among conspecific males. Do courtship signals provide information to the female about the male's genetic quality? How can female preferences evolve?

One way in which female preferences can evolve is if they directly benefit the female's survival or fecundity. If males provide some resource for females, such as a feeding territory, nutritive spermatophore, or nuptial gift, females will evolve a preference for mating with males that provide the largest resource. For example, in the scorpionfly introduced earlier, the female chooses a male on the basis of the size of the prey item he offers (Thornhill 1976). The female benefits not only in acquiring a large food resource, but her sons will inherit their father's ability to catch large prey, enhancing their reproductive success.

In most species, however, males don't provide any contribution to the fitness of the female during courtship, nonetheless females still appear to be discriminating among males. For example, in birds such as sage grouse and manakins, males form *leks*—groups of a dozen or so individuals in which they compete for positions of dominance, and to which females come to select a mate. The males in a lek are highly variable in their mating success. One male typically inseminates most of the females, with a few others inseminating the rest (Bradbury et al. 1986) (Figure 6.8). It is thought that female preference depends on the evolutionary exaggeration of one or more characters in the male that are genetically correlated with the preference behavior shown by the female. What can those characters be? Two models have received most of the attention in recent years.

The older of the two is the *runaway hypothesis*. Ronald Fisher (1930), who first proposed this as an explanation for the peacock's tail (Plate 6.4), argued as follows: If females prefer males with unusually long and showy tails, such males will obtain more matings. If difference in tail size has some heritability, the female's sons will inherit longer tails and thereby enjoy greater reproductive success. Selection will also enhance any tendency in females to prefer unusually long tails, because these females will select males who will pass on the trait to their sons. Once started, positive feedback should lead to a runaway process, with males evolving ever more elaborate tails and females preferring to mate with males having the most elaborate tails. This continues to the point where sexual selection favoring larger tails is just balanced by the cost such a tail imposes on males' ability to feed and to escape predators.

How can such a runaway process get started? One possibility is that originally females select males on the basis of some morphological or behavioral trait that indicates something about male fitness. Perhaps tail length is an indicator of a male's ability to fly, which correlates with his ability to forage for food or escape predators. Once female preference for the trait is established, the runaway process can exaggerate the trait far beyond its optimal size in its original context. In other words, according to Fisher, the peacock's tail is an ornament that has no meaning other than to appeal to the female's preference for large and showy tails.

An alternative model is the *handicap hypothesis.* First proposed by Zahavi (1975), this says that elaborate structures such as the peacock's tail do contain useful information for the female: they are indicators of the male's vigor. The male's tail in effect demonstrates his ability to survive despite the handicap a long tail imposes on his daily life. Thus the tail indicates underlying genetic quality, and the females are using it to recognize and mate with males manifesting these "good genes." Hamilton & Zuk (1982) proposed a variant on the handicap hypothesis that says male sexual displays convey information about the male's ability to resist diseases and parasites. Only males with large tails in good condition will be disease-free. A female who mates with such a male gains an evolutionary advantage by passing on those traits to her offspring.

Empirical tests have so far failed to distinguish between these two hypotheses. While a number of studies have shown that females have a preference for males with exaggerated displays or ornaments, it is difficult to demonstrate whether the displays communicate information about the male's health and vigor (handicap hypothesis), or are merely ornaments for ornament's sake (runaway hypothesis).

A few species reverse the usual roles of sexes, with the male investing more in the offspring than does the female. As sexual selection theory predicts, in these cases females compete for access to males. What can the male invest? We've already seen that his gametes are a trivial contribution. But he can provide nutrients for the female, or he can provide care for the offspring later. The male mormon cricket, for example, produces a spermatophore—sperm-containing packets rich in protein—that can cost him as much as 27% of his body weight. After the sperm enter the female's spermatheca, the female eats the remaining spermatophore, shunting the nutrients it contains into the elaboration of her eggs (Gwynne 1984). This investment by the male exceeds the female's investment in the brood, and so it is the females that compete for access to singing males (Gwynne 1981).

Perhaps the most well-known example of role-reversal is the giant water bug. The female lays her eggs on the back of the male, where he broods them for the 27 days they take to hatch. Males invest a considerable amount in brooding, and male back space is a limited resource, so we see females aggressively competing with one another for access to males. Likewise, female dendrobatid frogs compete for males that provide the best care for their eggs (Summers 1992). There is even an Australian katydid in which the relative investment ratio shifts between male and female, depending on the environment (Gwynne 1990). Females elaborate eggs from nutrients obtained both from pollen of certain plants on which they feed, and from the male spermatophore. In the spring, when only flowers producing small amounts of pollen are available, the male's investment in the spermatophore exceeds the female's investment in eggs, and females compete for males. Later, however, when the females can feed on more productive flowers, the female's investment in the eggs exceeds that of males, and males compete for females.

If Individuals Behave In Their Own Best Interests, How Can Altruism Evolve?

In many situations animals compete for access to resources; in other situations we see cooperation instead. In the Florida scrub jay *(Aphelocoma coerulescens)* and a number of other birds, some young individuals help at the nest of a relative rather than reproducing on their own; female ground squirrels work together to defend each other's young; lionesses cooperate in bringing down large prey; vampire bats that foraged successfully share their blood meals with less successful roostmates; honey bee workers give up their own reproductivity to contribute to the effort of their mother, the queen, to rear more of her own offspring. Given that natural selection favors individuals that behave so as to maximize their own fitness, i.e, to act in their own selfish best interests, how can such seemingly altruistic behavior arise and be maintained in a population? The social insects are extreme examples of altruism in the animal world, posing a dilemma that Charles Darwin recognized:

> *With the working ant we have an insect differing greatly from*
> *its parents, yet absolutely sterile; so that it could never have*
> *transmitted successively acquired modifications of structure or*
> *instinct to its progeny.*

In other words, how can sterile workers, who make the seemingly supreme sacrifice of giving up their personal reproductivity to toil selflessly for the good of their colony, pass these traits on to the next generation? Darwin himself hit upon the essence of the answer when he recognized that it is the "family," that is, the colony, rather than the individual, that natural selection acts on.

William D. Hamilton (1964) reinterpreted Darwin's explanation and provided a mathematical formulation in terms of selection acting on the individual. Hamilton's explanation recognizes that direct production of offspring is not the only route to fitness for an individual. What counts is getting copies of one's own *genes* into the next generation, and reproduction is only one way. Because relatives other than sons and daughters share one's genes, an alternative route to gaining fitness is to help a relative produce offspring. We share genes with our siblings, for example, so we can get copies of our genes passed on to the next generation by helping our parents produce more siblings than they would have without our help. This process is called **kin selection.**

To calculate the fitness equivalents of each extra relative so produced, Hamilton discounted each offspring by the degree of relatedness, r, that the helper shares with the relative helped. For example, because we share on average half of our alleles with a brother $r = 0.5$), who in turn shares half his alleles with his daughter, we can expect to share one fourth (0.5×0.5) of our alleles with our niece. So if we can help a brother rear two additional nieces whom he would not have been able to produce without our help ($r = 2[0.5 \times 0.5]$), it would give us the same fitness gain as if we had reared one of our own sons or daughters ($r = 1[0.5]$). Or as J. B. S. Haldane put it, he would be willing to give up his life to save that of two brothers or eight first cousins. Hamilton coined the term **inclusive fitness** to signify fitness gained via assisting all relatives, whether one's own descendants or not.

Thus we can see how *altruism*—helping a relative produce offspring instead of producing our own—could spread in a population. If the ratio of fitness benefit, *B*, to fitness cost, *C*, exceeds the reciprocal of *r*, the coefficient of genetic relatedness of the helper to the aided relative, the helping trait will spread. *B* can be thought of as the number of additional offspring the relative produces because of the helper's aid, and *C* as the number of the helper's own offspring given up by virtue of helping. This condition can be expressed in the following concise mathematical form, known as Hamilton's rule: Thus "altruism" as used here is not the selfless contribution to the welfare of others that its everyday meaning connotes.

$$\frac{B}{C} > \frac{1}{r}$$

In most situations, the inequality will not be satisfied. If the aided relative is a parent or a sibling, $r = 0.5$, which means that *B* must be $> 2C$. In other words, because of the relatedness discount, an altruist's aid must increase a sibling's output by at least two offspring for every offspring of her own the altruist gives up. There are, however, two ways the balance can be tipped in favor of the altruistic option. One is by skewing the left side of the inequality, the ecological side, upward. The other is to skew the right side, the genetic side, downward.

Certain ecological conditions can sufficiently skew the benefit, *B*, of helping above the cost of the personal fitness given up, *C*, to satisfy the inequality and favor the evolution of altruism. For example, for some nesting species the risk of predation on the offspring while the parent is off foraging may be so high that few offspring survive. If two siblings share a nest, however, survival may be much higher because one can guard the young while the other is away. Under these conditions, it can pay even if one sibling dominates the other into giving up her reproduction entirely. The helper still enjoys greater fitness than she would by fruitlessly nesting on her own.

Helping at the nest occurs in over 2% of bird species. In the white-fronted bee-eater (*Merops bullockoides*) studied in Kenya by Stephen Emlen (1990), helpers tend to be the older offspring of the pair they are helping. Why do they opt to help their parents rather than forming pairs and producing their own offspring? In the first place, they are gaining some indirect fitness by virtue of their relatedness to the beneficiaries of their aid. In most cases they are helping to rear siblings, to which they are as closely related as they would be to their own offspring ($r = 0.5$), so the genetic benefits of helping to produce one sib are equivalent to producing their own offspring. The addition of each helper increases the output of a breeding pair by half a fledgling (Figure 6.10). Opting for the

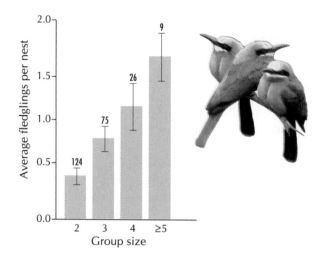

Figure 6.10. Reproductive success of pairs of white-fronted bee-eaters as a function of the number of extra-parental helpers. Redrawn from Emlen & Wrege 1989.

direct reproduction route may yield lower payoffs than this. For one thing, because bee-eaters are colonial nesters, it is difficult for pairs to find enough food near the colony without help. Second, young pairs lack experience that may be crucial to successful nesting; helping at the parent's nest may enhance their success when they do reproduce on their own. Third, helping may put a bird in an advantageous position to take over as a breeding individual should the parents die. If ecological conditions are such that the indirect fitness gain through helping is enough greater than the gain achieved if the helper were to nest on her own, Hamilton's inequality would be satisfied and helping behavior should evolve.

The Florida scrub jay is another cooperatively breeding bird. Breeding pairs are accompanied by 0–6 adult nonbreeders, usually their offspring, that act as helpers by feeding the young of the breeding pair. In an experimental test of the effects of helpers at the nest, Mumme (1992) experimentally removed helpers from some pairs while leaving other pairs and their helpers as controls. Offspring of breeding pairs from which helpers were removed suffered higher rates of predation as nestlings and lower rates of survival as fledglings than did offspring of control pairs. Experimental groups produced an average of only 0.56 independent juveniles, compared to 1.62 for the controls. Mumme concludes that the nonbreeding helpers increase their inclusive fitness by helping to rear the young of relatives, and therefore the behavior is favored by natural selection.

The other way the balance can be tipped to favor altruism is if r is unusually high. For most animals, which are *diplodiploid* (both sexes are diploid), even closest relatives cannot enjoy a coefficient of relatedness of more than 0.5. In the Hymenoptera (bees, ants, wasps), however, r is skewed upwards to 0.75 for sisters. This comes about because the Hymenoptera are *haplodiploid:* males are haploid and females are diploid. Males are haploid because they develop from unfertilized eggs; the sperm a male produces are all genetically identical. The daughters of a male-female pairing will all have their father's genetical contribution identical by common descent, while they share the usual 0.5 genetic contribution of their diploid mother. Thus, among sisters, the average relatedness is $r = 0.5(0.5 + 1.0) = 0.75$. This skew in relatedness relaxes the conditions on the ecological side of the inequality such that each extra female sib a daughter helps her mother rear is worth 0.75 fitness units, *more* than the 0.5 units she would gain by rearing her own daughters. This skew in favor of daughters helping their mother rear more daughters is thought to have favored the evolution of social behavior in the Hymenoptera, where it has evolved independently in at least 12 lineages of bees, ants, and wasps.

This genetic skew in the Hymenoptera predicts there should be a conflict between the mother (queen) and her worker daughters over the sex ratio produced by the colony. Male haploidy means that while sisters are related to each other by 0.75, sisters are related to their brothers by only 0.25 (as long as their mother mated with only one male). So, from a female worker's point of view, her sisters are three times as valuable as her brothers as vehicles for getting copies of her genes into the next generation. She should therefore favor a 3:1 female:male ratio of investment among the reproductives produced by the colony. The queen, on the other hand, is related equally, at r = 0.5, to both her daughters and her sons, so she should favor a 1:1 sex investment ratio. Note that if one sex is larger than the other, that sex requires a greater investment to rear, so the ratios of actual individuals must be weighted by the relative investment in each sex.

Trivers and Hare (1976) attempted to test whether the queen or the workers win the conflict, by measuring the ratio of investment in the two sexes for 21 species of ants. Using the comparative approach, they measured the colony investment in each sex. The results, shown in Figure 6.11, are consistent with the conclusion that, in most species, the workers are more in control of the sex ratio than is the queen. More recent studies

Figure 6.11. Ratio of colony investment in males and females in 21 species of ants. The left-hand line represents a ratio of 1:1, the right-hand line a ratio of 3:1 females:males. As predicted if workers control the sex ratio, most species lie closer to the 3:1 ratio. Redrawn from Trivers & Hare 1976.

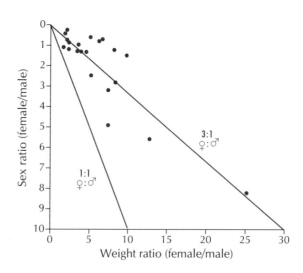

on ants as well as other social insects also support the prediction that workers can control sex ratios in the face of opposing queen interests (Nonacs 1986; Boomsma & Grafen 1990; Queller et al. 1993a).

In most Hymenoptera, workers have functional ovaries that can develop and produce eggs. Because they cannot mate, however, any eggs they lay will be uninseminated and will therefore develop into males. Nevertheless, worker offspring of a singly-mated queen should gain greater fitness by rearing their own sons, to whom they are related by $r = 0.5$, and even the sons of their siblings (their nephews), $r = 0.375$, than the queen's sons (their brothers), $r = 0.25$. In fact, worker-derived males are quite common in some species, but not in others. In the honey bee, for example, they occur at the rate of only about 1 in 1000. Why is this so? The answer lies in the fact that the queen mates with 10 to 20 males and uses the sperm of all of them to produce her worker offspring. This multiple paternity among the workers reduces the mean relatedness among them to $r < 0.5$. In turn this means that a worker's relatedness to her nephews, the sons of her sisters, is now $r < 0.187$, less than the $r = 0.25$ relatedness she shares with her brothers. A worker should therefore prefer to rear male offspring of the queen over those of their sister workers. Even though they still prefer their own sons ($r = 0.5$) over those of the queen, each worker is alone in this preference, and any eggs she lays are eaten by other workers. In other words, the workers "police" each other to prevent the rearing of worker-derived males in favor of queen-produced males (Ratnieks 1988). Multiple mating by the queen brings the genetic interests of all the workers into alignment with hers, and thus enhances cooperation among colony members (Ratnieks & Visscher 1989).

Tests of more detailed predictions of kin-selection theory have been facilitated in recent years by new developments in molecular genetic technology. Microsatellite markers in particular have considerably sharpened the resolution of measures of genetic relatedness (Queller at al. 1993b). Microsatellites are (usually) noncoding loci in nuclear DNA containing tandem repeats of 1–6 nucleotide base pairs. For example, a locus containing the trinucleotide sequence …thymine-adenine-guanine… might contain 8 repeats: TAGTAGTAGTAGTAGTAGTAGTAG, denoted as $(TAG)_8$. There are several advantages to using microsatellites to measure relatedness. First, these loci are often highly variable among individuals, usually in the number of repeats in the sequence. These differences in length can readily be detected by running the DNA segments on polyacrylamide gels. Furthermore, the loci can be amplified to many copies by the polymerase chain reaction (PCR), which means that only minute amounts of DNA are needed to determine the length of the repeat. Usable genetic information has been obtained from saliva, semen, hair and feather follicles, fingernails, and even sloughed-off gut cells egested with feces (Queller et al. 1993b). Moreover, loci containing such tandem repeats occur in large numbers in eukaryote organisms, providing enough variation to enable the unique characterization of individuals.

Microsatellites provide increased resolution for determination of paternity, for example to measure the extent of extra-pair matings in birds and other animals. The

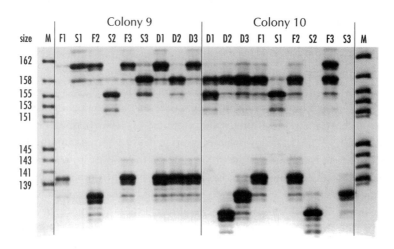

Figure 6.12. Microsatellite genotypes. Shown are amplified genotypes for locus PBE492AAT from members of two colonies of the social wasp Polistes annularis. Each of the numbered columns, or lanes, represents an individual. Amplified microsatellites from each individual are places at the top of the gel and migrate downward in response to an electric current. the larger the microsatellite fragment, the shorter the distance it migrates. The left- and right-hand lanes, labeled M, are calibration markers, indicating the positions reached by fragments of known sizes, shown by the numbers at the left. The genotype of each foundress (F) is shown to the left of the genotype of the sperm stored in her spermatheca (S). In each case, stored sperm showed a single allele at each locus, indicating that each foundress mated with a single male. The female foundresses are diploid and therefore show up to two alleles. The genotype of each daughter (D) is consistent with only one foundress-sperm combination. Data from Queller et al 1997.

enhanced resolution has been particularly useful in recent studies of social Hymenoptera. Because members of hymenopteran colonies are often closely related, earlier techniques such as allozymes, DNA fingerprints, RFLPs (restriction fragment length polymorphisms), and RAPDs (randomly amplified polymorphic DNA) did not allow enough resolution to genotype them. Furthermore, male parents are typically long-dead by the time the colony begins its development, which means their genetic input into the colony could not be assessed using less precise methods. Microsatellites, combined with PCR amplification, are sensitive enough to genotype the males a queen mated with, using only the minute amount of sperm stored in the queen's spermatheca after each copulation—finally making it possible to determine how many males inseminated a queen without having to genotype the progeny. Figure 6.12 illustrates the power of this method by showing how microsatellites can identify the parentage of the workers in a colony of the paper wasp, *Polistes annularis*.

Another prediction of kin selection theory is that helpers should be able to recognize kin and to direct their aid toward them, rather than to non-relatives. This has stimulated a great deal of research on kin recognition in recent years, resulting in considerable evidence that animals can distinguish kin from non-kin. This is an example of how hypotheses on the adaptive value of behavior, or *ultimate* causes, can lead directly to questions about mechanism, or *proximate* causes.

Not all cases of cooperation need involve kin, however. In some circumstances, cooperation may be advantageous even among non-relatives. One form this can take is for two or more individuals to engage in a mutualistic interaction, whereby each gains a net benefit. A possible example of **mutualism** in the defense of winter feeding territories by pied wagtails *(Motacilla alba)*. When food renewal rate in the territory is high, or when intrusion rate by neighboring birds is high, it pays a territory holder to share his territory with another individual, because two can more effectively keep out intruders than one, yet each still enjoys a greater net feeding rate on the territory than either would if defending it alone (Davies & Houston 1981).

A second form of cooperation among non-kin is *reciprocal altruism* (Trivers 1971). If an altruistic act provides more benefit to the recipient than it costs the altruist, and if the recipient reciprocates at a later time, both benefit and the behavior should spread in the population. A possible example occurs among vampire bats in Costa Rica. Wilkinson (1984) found that bats that failed to obtain a blood meal during a night's foraging sometimes successfully begged for blood from successful roostmates, even though they were not always relatives.

Although mutualism and reciprocal altruism represent cooperative behavior, they are not based on the kind of "altruism" seen among kin. In kin-based altruism, the altruist suffers a loss of personal fitness, i.e., fitness gained through personal reproduction. True altruism of this kind cannot evolve except through kin selection. Mutualism and reciprocal altruism are quid pro quo systems in which neither party in the relationship gives up personal fitness, although it may be temporarily deferred.

APPLICATIONS OF BEHAVIORAL ECOLOGY

Although behavioral ecology is still largely a basic science, the results of the field are beginning to be applied in a number of other areas. Knowledge about how social dominance, reproductive competition, dispersion, and mortality influence population dynamics can lead to improved population models used to develop conservation strategies for endangered species (Strier 1997; Smith et al. 1991; Caro & Durant 1995; Arcese et al. 1997). Similarly, an understanding of the ultimate causes of such behavior as territoriality and social hierarchies in domestic animals can facilitate pet training and breeding programs. Behavioral ecology can even provide insights into human behavior. Stephen Emlen, for example, has called attention to the similarity between avian family systems, such as found in the white-fronted bee-eater, and those of humans (Emlen

1995). He argues that experimentally unraveling the decision rules for family systems in birds can help sociologists understand the evolutionary history and adaptiveness behind cooperation and conflict in human families.

LINKS TO OTHER KINDS OF ECOLOGY

Behavioral ecology overlaps most extensively with physiological ecology, because both focus primarily on the individual. Physiological ecology takes the adaptations of individual organisms as givens, and investigates how extrinsic factors affect those organisms' survival and reproduction. Thus, physiological ecology can provide useful information to behavioral ecologists seeking to understand how a particular behavior is adaptive in a given environment. Likewise, behavioral ecology can aid the investigations of the physiological ecologist. Go back to the moose example for a moment (Figure 6.6). Belovsky (1978) showed how the sodium requirement shapes the optimal diet of the moose. This information is useful to a physiological ecologist attempting to explain the distribution and abundance of moose across a habitat with varying amounts of sodium-rich aquatic vegetation. Through such connections to physiological ecology, behavioral ecology links on upwards to landscape ecology. If it becomes possible to use satellite imagery to map the density of sodium-containing vegetation in lakes, for example, one could use knowledge obtained by behavioral ecologists to begin to predict the distribution of moose using GIS (geographic information systems). Similarly, carrying such an analysis back in time would establish a link to ecological history.

Behavioral ecology also has strong ties to population ecology. An understanding of the adaptiveness of behavior dovetails with the population ecologist's desire to understand the population effects of that behavior. For example, behavioral ecologists might discover that a particular predator species adopts one foraging strategy at a low prey density and another at a high density, with the result that the population of the predator is a non-linear function of the density of the prey species. The population ecologist would be interested in the consequences of the relationship for the populations of the two species at different densities.

The selective pressures studied by the behavioral ecologist occur, of course, in the context of the environment the animal lives in. Thus, behavioral ecology links with

community and ecosystem ecology. The ptarmigan is a bird that spends its life in the arctic, where species diversity is low and ecological communities relatively simple. Much of the ptarmigan's behavior is adapted to abiotic factors, such as extremely low temperatures and short day lengths during winter, while its interactions with other species are relatively few. On the other hand, the behavior of a toucan living in the Amazon, a much more diverse and complex ecosystem, will be shaped to a greater extent by interactions with other species than by the relatively benign abiotic conditions.

Sometimes even complex communities can be dominated by one or a few species. For example, most of us probably assume that ungulates dominate the primary consumer niches in African savannas; but in some areas, at least, the biomass of termites per hectare may equal or exceed that of grazing mammals (Wood & Sands 1978). The behavioral ecologist can elucidate how the social organization of termites enables them to gather enough food to maintain dense populations. Understanding the behavior of such dominant species can help community and ecosystem ecologists to understand the energy and materials flow in those systems.

SUMMARY

We have seen that behavioral ecology focuses on animal behavior, the part of the phenotype through which animals most intimately interact with their environment. By seeking to explain how behavioral traits are adaptive, behavioral ecologists attempt to identify the selective forces that have molded animal behavior. By now it is apparent that evolution by natural selection is the concept most central to this science. It is where the field of ecology meets Darwin.

Two primary methodological approaches are available to behavioral ecology. Experimental tests of hypotheses on function are commonly used when it is feasible to manipulate the variables under investigation. If not, the comparative approach may be used, whereby we take advantage of the natural variation among species and environments to test the survival value of the behavior in question.

The field of behavioral ecology has matured greatly over the past three decades, from a largely qualitative science with only a rudimentary theoretical base, to one in which quantitative theory has been developed in many areas of inquiry and is actively being tested in empirical studies on a wide range of animal species. Despite the criticisms leveled at the use of optimality theory, this approach has proven valuable and has produced detailed and rigorous explanations for the adaptiveness of behavior in many instances. Cost-benefit models have proven of great value in revealing the subtleties of how constraints and hidden costs shaped adaptive strategies in a wide range of animal species. Game theory has rapidly expanded our understanding of why animals behave as they do in competitive situations, and particularly why alternative strategies may be adaptive and evolutionarily stable, rather than being pathological abnormalities as was once thought. The evolutionary approach has made significant progress in understanding why the two sexes behave so differently during courtship and reproduction, yet many questions remain. The runaway process and the handicap principle, two unresolved hypotheses on how females discriminate among males, exemplify just one of the many areas of active research in behavioral ecology. Another phenomenon attracting a good deal of attention by theorists and empiricists alike is the evolution of cooperation. Again, we saw how seeking explanations in terms of fitness costs and benefits has led to considerable progress.

Despite this remarkable progress made in recent decades, however, many interesting problems remain unsolved. The basis of female choice is still puzzling, many of the details of social behavior remain to be worked out, and there is still the largely unexplored realm of the mechanisms that are assumed by so many of the models, to name just a few areas of active research. The future is bright and there is much to be done.

RECOMMENDED READING

JOHN ALCOCK. *Animal Behavior: An Evolutionary Approach,* Sixth Edition. Sinauer Associates. 1997.

Widely used textbook that focuses on functional questions.

RICHARD DAWKINS. *The Selfish Gene.* Oxford University Press, Oxford. 1976.

In this thought-provoking book, Dawkins makes the case for viewing the individual organism as a mere throwaway "survival machine," blindly programmed to preserve its immortal genes.

JOHN R. KREBS AND NICHOLAS B. DAVIES. *An Introduction to Behavioural Ecology,* Third Edition. Blackwell, London. 1993.

Another successful textbook dealing with the influence of natural selection on behavior.

JOHN R. KREBS AND NICHOLAS B. DAVIES (eds.). *Behavioural Ecology: An Evolutionary Approach.* Blackwell Science, Oxford, England. 1997.

A recent collection of chapters by specialists emphasizing not only function but also mechansims.

NIKO TINBERGEN. *The Animal in its World,* Volumes 1 and 2. Harvard University Press. 1972.

A collection of classic papers by one of the grandmasters of ecology.

GEORGE C. WILLIAMS. *Adaptation and Natural Selection.* Princeton University Press, Princeton, New Jersey. 1966.

Landmark book emphasizing the central role of selection on individuals.

JOURNALS:

Trends in Ecology and Evolution

Behavioral Ecology

Behavioral Ecology and Sociobiology

Three of many readable journals publishing original papers in the field.

Chapter 7

POPULATION ECOLOGY
The Waxing and Waning of Populations

Anthony R. Ives

> *In the case of every species, many different checks probably come into play, concurring in determining the average number or even the existence of the species.*
> —*Charles Darwin*

WHAT IS POPULATION ECOLOGY?

Have you ever wondered why starlings are so terribly common while other birds are so rare that they verge on extinction? Why some insect species appear and disappear from one year to the next, while other insect species are continuously abundant? Why dandelions occur throughout the world, while Venus's flytraps occur only in the marshy coastal areas of North and South Carolina? These are the questions of population ecology.

The goal of population ecology is to explain how and why populations change in abundance through time and through space. Population ecology's fundamental assumption is that changes in the abundance of populations can be explained in terms of the properties of individuals within populations and factors affecting those individuals. Individuals are affected by interactions with other individuals from the same species *(intraspecific interactions)* and by interactions with individuals from other species *(interspecific interactions)*. Also, individuals are affected by numerous environmental factors such as temperature, rainfall, and nutrient availability. In short, population ecology is the study of how intraspecific and interspecific interactions among individuals, and interactions between individuals and the environment, affect the abundance and distribution of species.

Definitions

What is a "population?" Simply, a **population** *is a collection of individuals from the same species that occupies some specified area.* A population of elephants might be all of the elephants in East Africa, and a population of grizzly bears might be all of the grizzly bears in Yellowstone National Park. There are no hard and fast rules for setting limits on what we call a particular population. Take, for example, the population of grizzly bears in Yellowstone National Park. The boundaries of the national park were set by politicians, and from the grizzly bear's point of view, defining its population in terms of human-drawn lines on a map is rather arbitrary. But in terms of managing Yellowstone National Park, defining the grizzly bears within the park as a population makes sense simply because this is the collection of grizzly bears that must be managed. This is not to say that grizzly bears do not move between the park and surrounding areas, and assuming that there is no such movement might be disastrous in terms of managing the grizzly bears within Yellowstone. Nonetheless, for the sake of management, it is reasonable to let the grizzly bears within the park boundaries be defined as a population.

The geographical extent of a population will depend on the species in question. All of the apple trees in the eastern part of North America could constitute a population, as could all the fruit flies in just one apple tree. One might be tempted to say that the geographical extent of a population of large organisms will always be greater than the geographical extent of a population of small organisms, but this is not always the case. For example, gorilla populations in Central Africa are restricted to quite small areas, since gorillas occupy small remnants of pristine tropical forests. In contrast, populations of desert locusts that swarm out of the Sahara to plague surrounding regions can cover millions of square kilometers.

The geographical extent of populations might depend not only on the species in question, but also on the question being asked about the species. When treating a person infected with *Plasmodium*, the protozoan that causes malaria, the population of interest is the *Plasmodium* in the person. However, if we are interested in controlling an outbreak of malaria in Central Africa, the population of interest is all the *Plasmodium* in all people in the region.

Because there are many ways to define a population, it is always important to first decide what the appropriate population is, since this decision affects how we ask and answer questions about it.

Population dynamics are the changes that occur in the number of individuals in a population through time. Questions about population dynamics include:

- What is the average abundance of a population over many years?
- What is the variability in the population abundance from one year to the next?
- How does the number of adults in a population change relative to the number of juveniles?
- Do changes in the abundance of a population occur simultaneously throughout the geographical range of the population, or does abundance increase in some areas while simultaneously decreasing in others?

This chapter is largely devoted to understanding the different processes that drive population dynamics.

QUESTIONS ASKED BY POPULATION ECOLOGISTS

Population ecologists ask more questions than can be discussed in a single chapter. Therefore, this chapter is organized around questions that can be linked in a reasonably logical fashion, with each question building on the previous ones (Box 7.1). The first set of questions involves just a single species. Even though no species lives in isolation from other species, population ecologists frequently focus their research around a single species, often either a species that, from a human point of view, is too

BOX 7.1. AUTHOR'S VIEWPOINT

I should confess several biases in my view of ecology. First, I am a theoretical ecologist. Mathematics has become increasingly important as a tool for understanding population ecology, and even the most experimentally oriented population ecologists need a working knowledge of mathematical theory. Therefore, I will not shy away from presenting ecological ideas in a mathematical way. However, to make things a little more accessible, I will confine most of the complicated mathematics to boxes.

Second, most of the experimental work I have done is on insects. I often think of populations as being very large and changing in abundance very quickly. Much of our basic understanding of population dynamics has been heavily influenced by researchers studying insects, because the life span of many insects is short enough that you can collect many generations worth of data in reasonably short time periods. This means that you can see the population dynamics. Although many of the examples I use involve insect populations, I will also discuss populations of plants, microbes, and other animals.

Third, ecologists differ in their view of what the science of ecology is. At one extreme, some ecologists think that the science of ecology is the search for fundamental laws of nature, like the laws of relativity in physics and the laws of thermodynamics in chemistry. At the opposite extreme, other ecologists think that the science of ecology is a collection of well-studied examples that illustrate what is possible in nature. They reason that because nature is so complex, the search for true laws is only going to lead to frustration; the only laws that are going to be universally true—with no exceptions—will have to be so broad that they are trivial, providing no real insights into how the world works. Faced with the problem that no laws are both nontrivial and universally correct, it makes most sense to describe patterns in nature through careful studies. Then, when a researcher embarks on a new study of an organism or a novel ecological problem, a library of examples will be available to show the researcher what to expect. Of course, there is a range of views between these extremes, but I personally lean towards the "ecology as a collection of well-studied examples" extreme. This means that throughout this chapter, I will often base arguments on examples rather than on supposed ecological laws. Don't think, though, that these examples are just rare oddities. I have picked examples that illustrate basic ideas that appear over and over in different ecological systems.

rare (i.e., an endangered species) or too common (i.e., a pest). After questions involving single species, the discussion will turn to questions involving interactions between two species: predation, competition, and mutualism. These interactions bind together the population dynamics of two species, so understanding the population dynamics of one requires understanding the population dynamics of the other. The next set of questions will introduce the complexities that space plays in interactions among species. **Spatial variability** refers to all of the variability that occurs from one geographical location to another within the range of a population. Spatial variability can include variability in the environment that affects species, or variability in the local abundance of species. The last set of questions in this section will ask what types of population dynamics can be produced when more than two species interact. Just as natural populations never really exist in isolation, neither do pairs of species coexist in isolation from other species. Trying to understand how the web of interactions among a group of species drives population dynamics is one of the major goals in ecology, and one of the major challenges.

What Limits Population Growth?

Many species of plants and animals are equipped with the reproductive potential for huge increases in population abundance. Tens of thousands of acorns can be produced by a single oak tree in one year; a spawning frog can produce thousands of eggs. A female rabbit may produce 30 offspring per year, and although this is less impressive than the reproductive potential of oaks and frogs, rabbit populations may still grow to huge sizes remarkably quickly. For example, a few European rabbits were introduced into Australia in 1859 in an ill-fated attempt to farm them for food and fur. Rabbits, in the absence of anything rabbit-like in Australia, did very well—too well. They soon escaped into the wild and did what rabbits do best. They multiplied, and by the early part of the twentieth century, the population of rabbits in Australia numbered hundreds of millions, despite every attempt by the Australian government to control them.

Given the huge reproductive potential of many species, what limits the increase in population abundance of different species? The specific answer to this question will be different for every species. Nevertheless, it is still possible to answer this question in a general way in terms of the reproduction and mortality of individuals within populations.

To discuss population growth, it is necessary to start with several definitions. The change in the number of individuals in a population over a specified time period is referred to as the **population growth rate**. Dividing the population growth rate by the initial number of individuals in the population gives the **per capita population growth rate**—the population growth rate per individual in the population. The time over which the population growth rate is measured will depend on the species. Human population growth rates are measured on a yearly basis; population growth rates for bacteria can be measured in hours. Population growth rates are often expressed in terms of changes in population density rather than population size. **Population density** is simply the number of individuals in the population divided by the area covered by the population. The reason for using density rather than size is that very often the frequency of interactions among individuals within a population depends on the density of the population rather than the number of individuals in the population. For example, the total human population in Canada is greater than that in Holland, but the population density of Holland is much greater than that of Canada. In terms of the frequency with which we would meet other people when walking across Canada vs. walking across Holland, population density plays a more important role than population size.

The per capita population growth rate depends on the **per capita reproduction rate** and the **per capita mortality rate**. These are the numbers of individuals that are born or die, respectively, over a specific time period divided by the number of individuals in the population at the start of the time period. The mathematical relationship between the population growth rate and the per capita reproduction and mortality rates can be expressed mathematically as follows. Let x_t denote the population density at time t. The time unit of t is the time unit used to measure the population growth rate—years, days, hours, etc. For example, x_{2000} could be the human population density in the year 2000. If the population density is known at time t, then the population density at time t+1 is

$$x_{t+1} = x_t + Bx_t - Dx_t = (1 + B - D) \bullet x_t = R \bullet x_t$$

where B and D are the per capita reproduction and mortality rates, respectively, and R is the per capita population growth rate. In words, this equation states that the population density at some time point (i.e., x_{t+1}) depends on the population density in the preceding time point (x_t) plus the density of newborn offspring (Bx_t) minus the density of dying individuals (Dx_t). Both B and D are multiplied by the population density x_t, because B and D are the per capita reproduction and mortality rates.

Exponential population growth

If the per capita reproduction and mortality rates of a population are unchanging, then it is easy to calculate the population density at any time in the future. This is done by iterating the equation $x_{t+1} = Rx_t$. Since this equation applies to any consecutive time points t+1 and t, it is true for t+2 and t+1, so $x_{t+2} = Rx_{t+1}$. Since you know that $x_{t+1} = Rx_t$, you can calculate the relationship between x_{t+2} and x_t as $x_{t+2} = Rx_{t+1} = RRx_t = R^2x_t$. Following this logic for T time units in the future,

$$x_{t+T} = (R^T) \bullet x_t$$

This equation produces **exponential population growth.**

The fact that the population growth is exponential is best seen by introducing yet another parameter, **r**, which is called the **intrinsic rate of increase** and is defined by the equation

$$e^r = R$$

Using the intrinsic rate of increase, the population density at time t+T can be computed from the population density at time t by the equation

$$x_{t+T} = (e^{rT}) \bullet x_t$$

Thus, the change in the population density over T time units depends exponentially on both r and T. When the per capita reproduction rate is greater than the per capita mortality rate (i.e., $B - D > 0$), then r is greater than zero, and population will increase exponentially. Conversely, when the per capita reproduction rate is less than the per capita mortality rate, then r is less than zero, and population will decline exponentially.

High intrinsic rates of increase can lead to huge population densities very rapidly. For example, pea aphids *(Acyrthosiphon pisum)* are a common pest of pea, bean, and alfalfa crops. The biology of pea aphids is geared for rapid reproduction. Pea aphids give live birth, and the young reach reproductive adulthood in only about eight days. Therefore, many generations occur over the course of a summer. Furthermore, during the summer months pea aphids reproduce asexually, meaning that adult females produce only female offspring and do so without being mated by males. Males only occur in pea aphid populations in the fall, when the sole sexual generation of the year occurs. Dispensing with the need for males during the summer doubles the pea aphid per capita reproduction rate—with asexual reproduction, all individuals in the population are capable of reproducing. Finally, asexual females produce offspring at a remarkable high rate of 4–6 per day.

These features of pea aphid biology lead to an intrinsic rate of increase of r = 0.52 per day, which translates into a daily increase in population density of about 170%. At this rate, starting with a single asexual female, in 10 days there will be about 180 aphids, in 15 days about 2,400 aphids, and in 20 days about 33,000 aphids. If this continues for 102 days, there will be about 10^{23} aphids, which is roughly enough aphids to cover the entire land surface of the Earth to a depth of two meters!

Species such as pea aphids with short generation times and high reproductive potentials have high intrinsic rates of increase, and are capable of reaching huge population densities rapidly. However, even species with relatively long generation times and low reproductive potentials can reach huge densities in a surprisingly short time. For example, the current human population is doubling about once every 40 years, an intrinsic rate of increase of 0.0175 per year. If the human population were to continue growing at this rate for 700 years, roughly 10^{15} people would inhabit the Earth. This number of people standing shoulder-to-shoulder, front-to-back would cover the entire land surface of the planet. Granted, the unabated pea aphid population could cover the Earth more quickly. Nonetheless, 700 years does not seem much time for a truly cataclysmic human population density to arise.

Density-dependent population growth

Of course, neither the pea aphid nor the human population could reach such densities. Figure 7.1A shows the actual growth of a pea aphid population on 16 alfalfa plants in a greenhouse. Exponential growth, represented by the dashed line, only occurs when the aphid density is low. Instead of growing exponentially, the population growth rate slows, with the population density actually declining after reaching a peak density of around 100 aphids/plant. Exponential growth is not maintained, because the per capita reproduction rate decreases and the per capita mortality rate increases as the aphid pop-

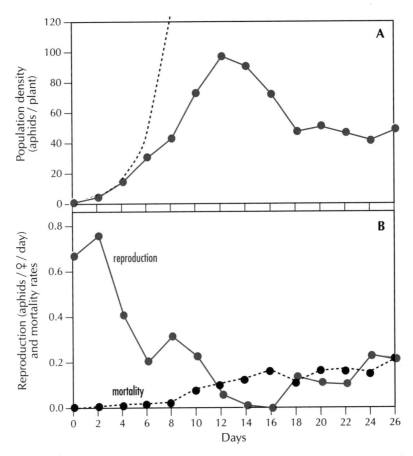

Figure 7.1. The population growth of pea aphids. To generate these data, two adult aphids were placed on each of 16 separated alfalfa plants. The numbers of living and dead aphids were counted every two days. **A.** The number of live aphids per plant for 26 days, with the dashed line giving the hypothetical case of exponential population growth. **B.** The per capita reproduction and mortality rates of aphids. In the experiment, all aphids were female and reproduced asexually. Note that when the per capita mortality rate exceeded the per capita reproduction rate (days 12–16), the population density decreased. These data were collected by Elizabeth L. Breuhl for an independent research project in introductory biology at the UW–Madison.

ulation density rises (Figure 7.1B). The decrease in the per capita reproduction rate and the increase in the per capita mortality rate occur because the pea aphids are sucking all of the nutrients out of the plants. As more nutrients are removed, less nutrients are available for each of the growing number of aphids. In addition, the plants themselves are weakened, and therefore plant production of nutrients is inhibited. On day 12 the per capita reproduction rate becomes less than the per capita mortality rate, thereby resulting in a decrease in the population density. On day 18, however, the per capita reproduction rate rebounds to the level of the per capita mortality rate. When the per capita reproduction rate equals the per capita mortality rate, reproduction balances mortality. Therefore, the population density remains approximately constant.

The pea aphid population gives an example of **density-dependent population growth** in which the per capita reproduction and mortality rates depend on population density. When the per capita population growth rate is density dependent, the intrinsic rate of increase r can no longer simply be defined in terms of the per capita reproduction and mortality rates, B and D, since these rates change with density. Therefore, the intrinsic rate of increase is defined as $e^r = 1 + B - D$ for the case when the difference between B and D is at its maximum. The intrinsic rate of increase gives the *greatest possible* per capita growth rate of a population.

In general, the per capita growth rate of any population must depend on population density; if it did not, the population would either increase exponentially (if r > 0) or decline exponentially (if r < 0). You could argue that it is possible for r to equal zero, thereby leading to no change in the population density. However, for r to equal *exactly* zero is infinitely unlikely, and when r is even slightly different from zero, the population density will eventually either increase to infinity or decrease to zero.

In natural populations, the effect of density on population growth is most easily observed when populations are growing from very low initial densities. As an example, Maria Alliende and John Harper studied the willow population of Newborough Warren in

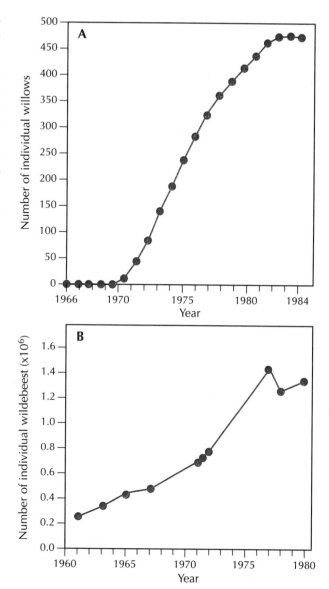

Figure 7.2. Population dynamics showing limits to population growth. **A.** The number of willow plants at Newborough Warren, Anglesey, Wales, following a myxomatosis outbreak that caused a crash in the rabbit population (Alliende and Harper 1989). **B.** The population growth of wildebeest in the Serengeti following an outbreak of rinderpest (Sinclair and Norton-Griffiths 1982).

Wales, United Kingdom (Alliende and Harper 1989). In the 1950s the rabbit population in the area crashed due to an outbreak of the disease myxomatosis. The willow population, freed from the nibbling of rabbits, increased rapidly (Figure 7.2A). However, as the population density increased, the per capita population growth rate decreased, and the population eventually plateaued at a size of about 500. At this population size, the entire area of suitable ground was covered with willows, and there was not enough space for new seedlings to establish.

Another example of rapid population growth followed by a plateau is the increase in the wildebeest population on the Serengeti plains of Africa (Figure 7.2 B). In the early 1960s there was an outbreak of rinderpest, a disease that killed large numbers of wildebeest and other ungulates (Sinclair and Norton-Griffiths 1982). Once the epidemic was over, the wildebeest population rebounded rapidly, initially showing nearly exponential population growth. However, eventually the population was limited by the abundance of grass on the Serengeti. With too little grass per wildebeest during the dry season, wildebeests suffered from reduced body condition and increased mortality.

Perhaps the most remarkable population growth curve is that of the human population (Figure 7.3A). The human population has been growing at a nearly exponential rate for the last 200 years. A particularly striking feature of human population growth is that for the last 50 years or more, per capita reproduction rates have been dropping (Figure 7.3B). This drop has been largely driven by social and economic forces that increase the benefits of family planning. During this time, the per capita mortality rate has also dropped due to the combined effects of improved nutrition, sanitation, and medical care. Thus, despite the drop in the per capita reproduction rate, the human per capita population growth rate has decreased only slightly. The drop in the per capita reproduction rate has been offset by the drop in the per capita mortality rate.

At some point in the future, however, the per capita human population growth rate must drop to zero, since the Earth could not conceivably support 10^{15} humans in 700 years. The human population will plateau when the per capita reproduction and mortality rates equal each other. The critical question is whether this will occur by the per capita reproduction rate dropping to match the per capita mortality rate, or by the per capita morality rate increasing to match the per capita reproduction rate. The answer to this question will determine whether the human population of the future will suffer lower or higher per capita mortality rates than it does today. Having lower per capita mortality rates is certainly better. But for this to happen, the per capita reproduction rate must drop.

Limits to population growth

Although the per capita population growth rates of all species must be density dependent, it is often very difficult to determine how and why per capita reproduction and mortality rates depend on density. For animals, density dependence may involve com-

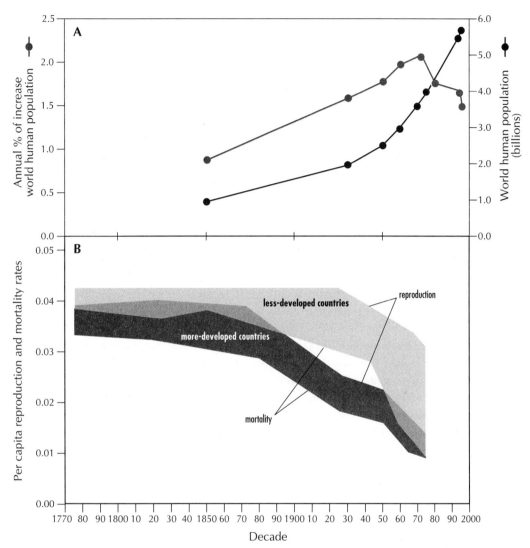

Figure 7.3. Human population growth. **A.** The size of the human population from 1770 to the present. **B.** The per capita reproduction and mortality rates for more-developed countries (Western Europe, United States, Canada, and Japan) and less-developed countries. The difference between per capita reproduction and mortality rates, shown by the shaded areas, gives the per capita population growth rate. Data for **A** are given in Ehrlich et al. 1973, and *World Population Data Sheets* from the Population Reference Bureau. **B** redrawn from Keyfitz 1989.

petition for food, territories, and space. For plants and microbes, density dependence may involve competition for sunlight, nutrients, water, and space. Density dependence may also involve predators and diseases, if predation rates and the spread of disease depend on population density. Also, density dependence is rarely the result of just a single factor. Instead, different factors may limit population growth at different times of year, or from one year to the next. Specific factors limiting population growth will be different for different species, and studying how density affects per capita population growth rates for different species is a key research area in population ecology.

What Regulates Populations?

What factors determine the patterns of population dynamics for a given species? *Regulate,* as defined in the dictionary, means to control or direct according to a rule or principle. This implies that something is doing the control, and that the actions of the controller have some purpose. In fact, **population regulation** is used by ecologists in a much more neutral way simply to describe the processes that affect the patterns of population dynamics. Although the language of ecology and popular opinion often implies purpose to the patterns of nature, there is no reason to suppose that a purpose exists. For example, a popular concept of "the balance of nature" implies that natural processes act in a purposeful way to create greater stability and harmony in the natural world. In fact, no scientific evidence supports this. The study of population regulation is just the study of patterns of population dynamics and the factors that create these patterns.

The range of population dynamical patterns shown by different species is large. Figure 7.4 gives examples for four species. The first species, *Dendrolimus* sp., is a common beetle that feeds on pine trees in Germany (Figure 7.4A). Over 60 generations, the population ranged in density from less than 10 to over 100,000 per square meter, showing at times huge changes from one year to the next. In contrast to the huge fluctuations in density of *Dendrolimus* sp., the population density of great tits *(Parus major)*, a common bird in the deciduous forests of Europe, varied much less (Figure 7.4B). The contrast between the highly variable population fluctuations of *Dendrolimus* sp. and the less dramatic population changes in great tits might seem at first to be a consequence solely of the fact that one is an insect and the other a bird. However, even similar species may show quite different population dynamics. Figures 7.4C and D show the population densities of red deer and Soay sheep on two islands in the North Atlantic. The yearly changes in Soay sheep densities are clearly more severe than those in the red deer population. The question of population regulation is why do these species differ.

Extrinsic versus intrinsic factors

Population regulation is at the center of one of the oldest and most contentious debates in ecology. This debate has taken a number of different forms over the last 60 years, but the basic argument rests on whether intrinsic or extrinsic factors are the most important in population regulation. **Intrinsic factors** are those factors that directly involve the individuals within a population. They include such things as competition among individuals for food or other resources, cannibalism of juveniles by adults (a common occurrence in many groups of animals such as insects and fish), and the shading out of seedlings by

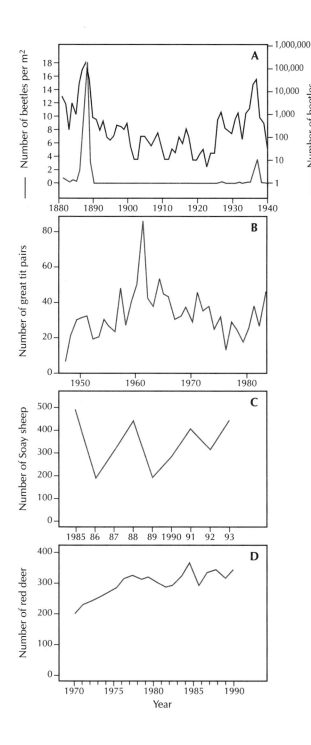

Figure 7.4. Examples of population dynamics from several different types of species. **A.** The population dynamics for the pine-feeding beetle *Dendrolimus* sp. (colored line), with the same data graphed on a $\log_{10}$ scale (black line). Redrawn from Varley 1949. **B.** The population dynamics of great tits in Marley Wood, Oxford, U.K. Redrawn from McCleery and Perrins 1985. **C** and **D.** Respectively, the numbers of Soay sheep on the island of Hirta, St. Kilda, and the number of red deer on the island of Rum, Inner Hebrides. Redrawn from Clutton-Brock et al. 1997.

large, established plants. The critical thing about intrinsic factors is that they may depend on the density of the population. Thus, intrinsic factors are ultimately responsible for density-dependent per capita reproduction and mortality rates.

Extrinsic factors are those factors that are external to a population yet nonetheless influence changes in the population's density through time. Extrinsic factors typically include environmental factors such as temperature, rainfall, and soil type. Other species may also be considered extrinsic factors. To classify another species as extrinsic, however, the effect of that species must not depend on the density of the species in question. The other species must act in a **density-independent** fashion; otherwise it would be considered an intrinsic factor. This is best explained with an example. Suppose a bird species depends on the seeds from a particular tree, so that when seed production is high, bird mortality is low, but when seed production is low, bird mortality is high. If the number of seeds consumed by the birds is never enough to affect the regeneration of the tree population, then seed production can be considered an extrinsic factor affecting the bird population. Just like temperature, seed production affects the bird population, but the bird population doesn't affect seed production.

Why have there been heated scientific debates about the relative importance of extrinsic and intrinsic factors? We would suspect that some populations are largely influenced by extrinsic factors, others largely by intrinsic factors, and the vast majority lie somewhere in between. This is in fact the case. Even though the debate over population regulation has been argued in terms of the relative importance of extrinsic vs. intrinsic factors, the real argument has been about what population regulation is. What particular properties observed in population dynamics are the most important? The debate is argued in terms of extrinsic vs. intrinsic factors because the relative importance of these factors depends upon the particular questions of interest about population dynamics. A real example will help clarify this debate.

The population dynamics of great tits

From 1956 to 1978, several researchers studied the population of great tits on the island of Vlieland off the coast of Holland (Tinbergen et al. 1985). They estimated per capita reproduction and mortality rates by banding birds and censusing the population each year. During ten of the years they experimentally reduced the population size by removing up to 60% of the breeding birds or their fledglings. Removing birds made it possible to measure reproduction and mortality at lowered population densities, thereby making it easier to study the causes of density dependence.

Data from this study can be analyzed to address three distinctly different questions about the population dynamics of great tits.

1. What factors are most important in limiting the per capita population growth rate?

The potential reproduction rate of great tits is quite high. In several years during the study, breeding pairs of great tits produced on average more than 10 fledglings, thereby leading to a per capita reproduction rate of more than 5 offspring per year. However, mortality was also quite high. Even in the best years, mortality of fledglings over the winter was 70%, and mortality of adults was 40%. Therefore, the single greatest factor limiting the per capita population growth rate was the mortality of fledglings. The possible causes of fledgling mortality are numerous: freezing, starvation, predation, accident. Identifying actual causes of fledgling mortality is difficult. Nonetheless, even without knowing them, it is still possible to pinpoint fledgling mortality rather than adult mortality as the most important factor limiting the per capita population growth rate.

> 2. What factors explain most of the year-to-year fluctuations in the per capita population growth rate?

Mortality of fledglings varied from year to year between 70% and 95%, and mortality of adults varied between 40% and 80%. Much of this variability can be explained by the year-to-year variability in seed production by tree species on the island. During the winter, great tits rely on the seeds produced by black alders, birch, and other species, and during the winters following summers with low seed production, mortality of both fledglings and adults was much higher. Therefore, the best predictor of whether the per capita population growth rate for a given year is going to be relatively high or low would be the abundance of seeds produced in the summer.

> 3. What factors explain how the per capita population growth rate depends on density?

Although the greatest amount of year-to-year variability in fledgling mortality was explained by the abundance of seeds produced in the summer, fledgling density had an additional effect on fledgling mortality. Breeding pairs of great tits produce two broods of offspring per summer. When fledgling densities from the first brood were high, the subsequent mortality of fledglings from the second brood was also high. Fledgling mortality is thus density dependent. The reason fledgling mortality depended on fledgling density is probably competition among fledglings for food. This could explain why the younger and smaller fledglings from the second broods had a particularly hard time when densities of the older and larger fledglings from the first broods were high. The importance of identifying density-dependent factors is that they will determine at what density the per capita reproduction and mortality rates balance each other out. Therefore, density-dependent factors determine the average population density expected to occur over the course of many years.

The population dynamics of great tits on Vlieland Island can be summarized as follows. The great tits have quite high per capita reproduction rates, but the per capita population growth rate is limited by mortality that is particularly high among fledglings. The year-to-year fluctuations in the per capita population growth rate are explained largely by year-to-year fluctuations in the seed production by several dominant tree species, which affects fledgling and adult mortality during the winter. Finally, fledgling mortality is density dependent, with high fledgling densities leading to increased mortality probably due to competition.

In answering the three questions above, note that the most important factors need not be the same for each question. For example, the density-dependent mortality of fledglings did not explain very much of the year-to-year variation in mortality. Even when the density of fledglings was low, if the seed production was very low during the summer, then fledgling mortality was very high. In fact, the analysis of the data only showed the effect of fledgling density on fledgling mortality *after* the overwhelmingly strong effect of seed production on fledgling mortality was statistically factored out.

Resolving the debate over extrinsic vs. intrinsic factors and population regulation

To return to the debate over the relative importance of extrinsic vs. intrinsic factors in population dynamics, the crux of the debate is the question being asked. Population ecologists most often pose the following three questions about population dynamics.

1. What factors are most important in limiting the per capita population growth rate?
2. What factors explain most of the year-to-year fluctuations in the per capita population growth rate?
3. What factors explain how the per capita population growth rate depends on density?

Which question you focus on depends very much on the goal of your study. For example, to predict what would happen if a bird species not found in Hawai`i were accidentally introduced, you might want to know the greatest source of mortality in its native country (question 1). If this source of mortality were not found in Hawai`i, then you might expect the population to increase in size rapidly. However, to predict the density of a bird population next year, you might want to know how much seed production occurred this year and therefore how much food would be available to the birds (question 2). Finally, to understand what sets the average density of a bird population over the course of many years, you would look for density-dependent factors (question 3).

The relative importance of extrinsic vs. intrinsic factors depends on which of these three questions you wish to explore. Extrinsic factors may be more important in answering the first two questions, but this is not necessarily the case. Intrinsic factors will generally be more important in answering the third question. Rather than debate the general importance of extrinsic and intrinsic factors in population dynamics, most ecologists now more carefully state the objectives of their studies, because the importance of extrinsic vs. intrinsic factors depends on these objectives.

Different population ecologists use "population regulation" differently. Some use the concept to refer only to density-dependent factors that affect population dynamics, and in this usage only question 3 is applicable to population regulation. Other population ecologists use population regulation to describe generally any factors that affect year-to-year fluctuations in population densities. In this usage, both questions 2 and 3 involve population regulation. The argument for confining the term population regulation only to density-dependent factors is that only these factors can stop population densities from growing or declining exponentially. The density-dependent factors set the density at which the per capita reproduction and mortality rates equal each other, and thereby ultimately set the long-term average population density. On the other hand, the argument for allowing population regulation to encompass density-independent factors is that density-independent factors very often have more impact on year-to-year fluctuations in population densities than do density-dependent factors. Although not universally accepted, today's preferred usage is for population regulation to refer only to density-dependent factors that influence population dynamics.

What Types Of Population Dynamics Are Possible In Simple, One-Species Systems?

No species in nature exists in isolation from any other species, but population ecologists often focus on the population dynamics of single-species systems for two reasons. First, many ecological studies are conducted most effectively by focusing on a single species, as in the study on great tits. Researchers don't ignore other species entirely; however, for the pragmatic reason that they can only study so much at the same time, ecologists limit investigation of other species and often lump other species with temperature, rainfall, and the like as extrinsic factors affecting the population dynamics of the primary study species. Second, the population dynamics of even single-species systems can show a wealth of complex patterns. Population dynamics of two- and many-species systems become increasingly complex. For the sake of investigating population dynamics, it makes sense to start with the simplest case of a single species.

A simple model of population dynamics

The best way to explore the types of population dynamics that can occur in single-species systems is to employ a simple mathematical model. The simple model used here might describe an insect or an annual plant, in which the adult population density in one year is a good predictor of the adult population in the next year. (Complexities associated with species whose populations are divided into non-reproducing juveniles and reproducing adults are discussed later in the chapter.)

Using a mathematical model to explore different types of population dynamics may seem unsatisfyingly abstract. However, the basic ideas illustrated by the model are essential for understanding population dynamics. Although abstract, the simple model more easily explains population dynamics than would a real biological example. Real examples contain unavoidable intricacies that, while fascinating in their own right, only compound the inherent difficulties of explaining population dynamics.

To construct the model, let x_t be the density of a population in year t, and let x_{t+1} be the population density in the following year. These can be related mathematically by

$$x_{t+1} = f(x_t)$$

where f() denotes a function that can be used to calculate x_{t+1} if x_t is known. When discussing exponential population growth, the model of the population dynamics was $x_{t+1} = (1 + B - D)x_t$. This is just a particular form of f(), and many other forms are possible. Since $f(x_t)$ gives the population density in year t+1 for any value of t, f() contains the information that determines population dynamics.

It might seem strange to talk about some function f() without saying what the function is exactly. Unlike physics and chemistry, which may use explicit equations like $E = mc^2$ and $PV = nRT$, population ecology often uses general functions to describe rough shapes without specifying mathematical formulas. This is done because every species is likely to have a different function f() that describes its population dynamics. Rather than specify f() in any detail, it is often more useful to ask what general characteristics of f() are important for different types of population dynamics.

Many things affecting population dynamics are variable and unpredictable from year to year. For example, the great tit population was affected by seed production, which in turn was affected by weather and other factors. Therefore, the function f() encapsulates both predictable and unpredictable factors that affect population dynamics. Many models in ecology ignore the unpredictable factors. These models are called **deterministic models**, because the population density in year t+1 can be determined using only the density in year t. Components of population dynamics that are unpredictable are called stochastic, and models including these factors are called **stochastic models**. Because they are simpler, the following discussion will first address deterministic models, and then turn to stochastic models.

Deterministic population dynamics

Although it is possible to use the general characteristics of an unspecified function f() to investigate population dynamics, specifying a particular function f() makes it easier to explain population dynamics in a textbook. This makes it possible to give concrete examples of different types of population dynamics. As a particular function, consider

$$f(x_t) = e^{r(1 - x_t/K)} \bullet x_t$$

This function is used frequently to model insect populations. The parameter r is the intrinsic rate of increase, which gives the maximum per capita population growth rate when the hypothetical population is at very low density. This can be seen by letting x_t be very small, in which case $(1-x_t/K)$ equals approximately one, and $f(x_t)$ equals approximately $e^r x_t$. Therefore, when the density is very low, population growth is approximately exponential with intrinsic rate of increase r. The parameter K gives the **equilibrium population density**, which is also sometimes called the equilibrium point. To see the meaning of K, suppose the population density x_t equals K. In this case, x_{t+1} = $f(x_t)$ = $f(K)$ = $e^{r(1-K/K)}K = e^0 K = K$. In other words, if the population density in year t is K, then the population density in the following year is also K. K is called the equilibrium population density, because if the population density were started at K, it would remain at K indefinitely.

In this model, the per capita population growth rate is density dependent. When the density is close to zero, the per capita population growth rate is e^r, which is greater than one. As the density increases to K, the per capita population growth rate decreases to one provided $r > 0$. Furthermore, the greater the intrinsic rate of increase r, the stronger the density dependence of the per capita population growth rate. This is because, for populations with the same equilibrium K, large values of r must be coupled with strong density dependence so that the per capita population growth rate drops to one by the time the density equals K.

Population dynamics with a stable equilibrium point

Figure 7.5 depicts population dynamics produced by the simple model for six different values of the intrinsic rate of increase, r. For all examples, the value of K is one. The left-hand panel of each figure shows the relationship between $f(x_t)$ and x_t. In each, the solid line gives $f(x_t)$, and the dashed line gives the 1:1 line along which the values of x_{t+1} (vertical axis) and x_t (horizontal axis) are the same. The equilibrium K is located where $f(x_t)$ crosses the 1:1 line, since at this point, $x_{t+1} = f(x_t) = x_t$. The right-hand panel of each figure shows how the population density changes through time, starting with a population density of $x_0 = 0.2$ at time $t = 0$.

First consider the example in Figure 7.5A in which $r = 0.2$. For values of x_t below K, $f(x_t)$ is greater than x_t; this is seen by the fact that $f(x_t)$ lies above the 1:1 line. Therefore, when x_t is less than K, the population density increases. This is seen directly in the right-hand panel of Figure 7.5A. When x_t is greater than K, $f(x_t)$ is less than x_t, since it lies below the 1:1 line. Therefore, $x_{t+1} = f(x_t)$ is less than x_t and the population density decreases. Because the population density increases when it is below K and decreases when it is above K, the population density will always move towards the equilibrium K through time regardless of its initial value.

When $r = 1.0$ (Figure 7.5B), the function $f(x_t)$ rises more quickly from low values of x_t and is nearly horizontal near the value of K. Compared to the case when $r = 0.2$,

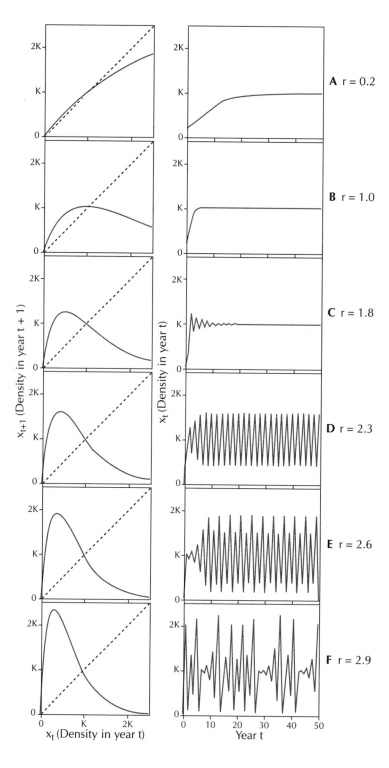

Figure 7.5. The population dynamics given by a simple model: $x_{t+1} = x_t\, e^{r(1-x_t/K)}$. The left-hand panel graphs x_{t+1} vs. x_t, and the right-hand panel gives the population density x_t vs. time. **A** through **F** correspond to increasing values of r. The dashed lines in the left-hand panels are the 1:1 lines along which $x_{t+1} = x_t$. In all graphs the value of K is 1.0, so the function $x_{t+1} = x_t\, e^{r(1-x_t/K)}$ crosses the dashed line at $x_t = 1.0$.

x_{t+1} (Density in year t + 1)

x_t (Density in year t)

x_t (Density in year t)

Year t

A r = 0.2

B r = 1.0

C r = 1.8

D r = 2.3

E r = 2.6

F r = 2.9

when r = 1.0 the population density moves towards K much more rapidly, as seen directly in the right-hand panel. To give a numerical example, consider the second points (x_1) in the right-hand panels of Figures 7.5 A and B. When r = 0.2, x_1 = 0.23, while when r = 1.0, x_1 = 0.45.

When r = 1.8, the population growth rate is high when the population density is low and low when the population density is high, and this produces a strong hump shape in $f(x_t)$ (Figure 7.5C). As a consequence, the population density alternates through time above and below the equilibrium K. This is seen in the right-hand panel of Figure 7.5C. Even though there are alternating peaks and troughs in the population density through time, the magnitudes of the peaks and troughs diminish, and ultimately the population density approaches K.

These three examples (Figure 7.5 A–C) illustrate two important features of population dynamics. First, the equilibrium population density K is **stable**, meaning that regardless of the initial population density, eventually the population density will approach K. Second, the population density approaches the stable equilibrium point more quickly when r = 1.0 (Figure 7.5B) than when either r = 0.2 or r = 1.8 (Figures 7.5A and C). For example, when r = 1.0, starting from an initial population density of x_0 = 0.2, it takes 4 years for the population density to get to K ± 1%. However, when r = 0.2 and r = 1.8, it takes 28 and 13 years, respectively. Thus, the population dynamics when r = 1.0 are said to be more **resilient** than when r = 0.2 or r = 1.8. For deterministic models in ecology, resilience is defined in terms of how rapidly a stable equilibrium point is approached through time.

Population dynamics with an unstable equilibrium point

More complex population dynamics occur as the intrinsic rate of increase r gets larger and density dependence gets correspondingly stronger. When r = 2.3, the population dynamics show perpetual peaks and troughs (Figure 7.5D). In this case, K is still the equilibrium population density, because if the population density were to start exactly at K, it would remain there forever. However, the equilibrium point K is unstable. If the population density were to start very close to K but not exactly at K, then population densities in subsequent years would alternate between increasingly high peaks and low troughs. To give a numerical example, suppose that the initial population density were x_0 = 0.99. Then the population densities in the following five years would be 1.01, 0.98, 1.02, 0.97, and 1.04. The magnitudes of the peaks and troughs in this alternating cycle will continue to increase until the pattern shown in the right-hand panel of Figure 7.5D is reached. Although the equilibrium point K is unstable, the alternating cycle shown in the right-hand panel of Figure 7.5D is stable, because regardless of the initial population density, an alternating cycle between the same two values will develop. This is called a **stable two-point cycle**.

Greater values of r lead to stable four-point cycles in which, regardless of the initial population density, the population eventually will cycle among four densities that are repeated perpetually (Figure 7.5E). Even greater values of r lead to complex patterns that show no apparent cycles (Figure 7.5F). In this last case, the population dynamics are chaotic. **Chaos** is a mathematical term referring to population dynamics in which patterns never repeat themselves because there are no stable equilibrium points and no stable cycles. The mathematical properties of chaos were discovered in the 1970s, and since then have fascinated ecologists and mathematicians alike (Appendix 7.1).

The lesson from this model is that even a very simple description of the relationship between population density in one year and the next can produce complex patterns such as stable cycles and chaos. Since the factors governing the population dynamics of real populations are likely to be far more complicated than this simple model, real populations clearly have the potential to show complex population dynamics.

In the absence of density dependence, population growth is exponential, leading to either exponential increases or exponential decreases in population densities. Density dependence is essential for keeping population densities bounded above zero and below infinity. However, the examples in Figure 7.5 show that patterns of density dependence do more than just bound population densities; they determine the general characteristics of population dynamics. For functions like that in Figure 7.5A, density dependence is weak, in that both the increase in population density from low initial values and the decrease in population density from high initial values are slow. The case in Figure 7.5F is the opposite extreme. At low population density, reproduction greatly exceeds mortality to create an explosive increase in density, and at high density, mortality greatly exceeds reproduction to create the drastic collapse in density. Thus, Figure 7.5 depicts very different patterns of density dependence that drive population dynamics.

From an ecological perspective, the important message is that strong density dependence may produce perpetually fluctuating population densities even in the absence of extrinsic factors. The study on great tits discussed earlier found that most of the year-to-year fluctuations in population density were driven by fluctuations in seed production, an extrinsic factor affecting the great tit population. The simple mathematical model shows that fluctuations in population densities may also be driven exclusively by intrinsic, density-dependent factors. Even when no extrinsic factors affect populations, as in the deterministic model, populations may nonetheless exhibit perpetual year-to-year fluctuations.

The main use of simple population models like the one analyzed above is to illustrate the possible types of population dynamics. Simple models make useful teaching tools. But are the types of population dynamics exhibited by the simple model realistic for natural populations? In particular, is density dependence in real populations strong enough to create hump-shaped functions of $f(x_t)$ that produce stable cycles and chaos? The answer, in short, is that hugely hump-shaped functions are potentially realistic, although they are rarely observed in nature. Many insects have the very high per capita reproduction rates necessary to produce huge humps; nonetheless, these high per capita reproduction rates are countered by constantly high per capita mortality rates. Since the mortality rates are high even at low population density, they produce low intrinsic rates of increase. Therefore, despite high per capita reproduction rates, the intrinsic rate of increase may be too low, and the strength of density dependence too weak, to create hump-shaped functions of $f(x_t)$. General surveys of data from single species of insects show only a few potential examples of strongly hump-shaped $f()$ functions, with none strong enough to create chaotic population dynamics (Hassell et al. 1976).

Given that density dependence sufficiently strong to drive population cycles and chaos is only rarely observed in nature, why spend so much time discussing them with the simple model? While very strong density dependence is uncommon for single species considered by themselves, interactions among two or more species can produce stronger density dependence, because the reproduction and mortality rates of species depend not only on their own densities, but also on the densities of other species. Multispecies density dependence may lead to more complex types of cycles, which will be discussed later in the chapter.

Stability and resilience in deterministic systems

Two ideas illustrated by the single-species model, stability and resilience, are of general importance in the study of population dynamics. **Stability** describes how population dynamics evolve through time. An equilibrium point is stable if population densities move towards that point through time. A two-point cycle is stable if population densities approach an alternating pattern between two values. In general, any cycle in population dynamics is stable if that cycle is eventually followed regardless of the initial population density.

Resilience describes how quickly a stable equilibrium point or stable cycle is approached through time. Therefore, resilience has meaning only in terms of something that is stable. The ideas of stability and resilience are relevant not only for single-species systems, but also for systems of interacting species. Since many of the questions in population ecology revolve around these ideas in one way or another, they will reappear throughout this chapter.

Stochastic population dynamics

Environmental variability affects all species on Earth, and for many species it is impossible to predict how population densities will change from one year to the next without knowing weather conditions, food abundance, and a myriad of other extrinsic factors. Even when precise predictions of population densities from one year to the next are impossible, it may still be possible to set a range of values over which the population density is likely to vary and to predict the chances that the population density will lie within some specified range. The range of densities traversed by a population in a naturally variable environment depends on the interplay between extrinsic and intrinsic factors affecting population dynamics. Understanding this interplay is central to understanding real population dynamics.

To illustrate different types of stochastic population dynamics, a stochastic model can be created by adding extrinsic variability, ε_t, to the deterministic model:

$$f(x_t) = e^{\varepsilon_t + r(1 - x_t/K)} \cdot x_t$$

In this function, the intrinsic rate of increase is given by the sum $r + \varepsilon_t$, since the per capita population growth rate is $e^{r + \varepsilon_t}$ when the population density is low. This breaks the intrinsic rate of increase into two components. The average intrinsic rate of increase over many years is given by r, while variability in the intrinsic rate of increase from year to year is given by ε_t. For example, ε_t for great tits would depend on variability in seed production and variability in any other extrinsic factors that affect reproduction and mortality.

In the stochastic model, ε_t is what is known as a random variable. A **random variable** is a variable that is not described by one particular value, but instead is described by a probability distribution of values. To explain this, suppose you could measure the intrinsic rate of increase of great tits for 1,000 consecutive years. After 1,000 years you calculate the average intrinsic rate of increase, r. Subtracting r from the actual yearly values of the intrinsic rate of increase gives ε_t. To characterize the distribution of ε_t, you could count up the number of times that ε_t fell between −1.0 and −0.9, the number of times ε_t fell between −0.9 and −0.8, etc. If you did this, you might get a histogram like the one shown in Figure 7.6. You could (hypothetically) do this again for one billion years' worth of data. If you were to do this for an infinite number of years, you would get what is called a **probability density function**, which is shown by the smooth curve in Figure 7.6. The

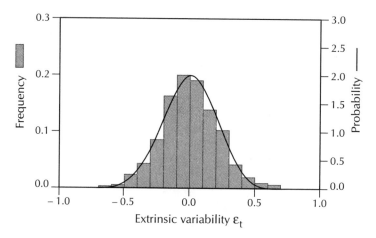

Figure 7.6. The distribution of the variability in extrinsic factors that affect population growth rates. ε_t is the deviation in the intrinsic rate of increase from its average value of r. The line gives the theoretical probability density function for the random variable ε_t. The bars give the results of selecting 1,000 values of ε_t using a computer random number generator that selects values according to a Normal probability density function. Note that because the bars give the realization of the random variable, they do not perfectly match the probability density function even after 1,000 values are selected.

probability density function is the theoretical distribution of probabilities that underlie a random variable. Thus, the random variable ε_t is defined by its probability density function. In this particular case, the probability density function of ε_t follows a Normal probability distribution.

What population dynamics are produced by the stochastic model? Figure 7.7 gives examples of stochastic population dynamics, changing the values of r as in Figure 7.5 while keeping the distribution of ε_t and the value of K the same. In the first of each set of panels, the solid line shows the function $f(x_t)$ excluding stochastic variability; these lines are the same as those in Figure 7.5 for the corresponding values of r. The points in these panels are 50 consecutive population densities obtained from the full stochastic population model starting at $x_0 = 0.2$. The variability of these points from the solid line is caused by the stochasticity produced by ε_t. The second panel shows the same 50 population densities graphed through time. The third panel shows both a histogram of population densities constructed from one thousand years of data and the probability density function for the population density. The probability density function for the population density is often called the **stationary population distribution**, because the probability density function does not change through time.

Figure 7.7 shows the effects of increasing the value of r on the population dynamics. Since the probability density function of ε_t in all examples is the same, there is the same variability caused by extrinsic factors. Therefore, differences among the figures are due solely to the intrinsic factors—specifically, differences in density dependence at different values of r.

When r is low (Figure 7.7A), the population density in one year tends to be similar to the population density in the previous year. This is because the function $f(x_t)$ lies close to the 1:1 line. The weak density dependence when $r = 0.2$ leads to moderately high variance in the stationary population distribution, for the following reason. If ε_t takes a relatively high value and pushes the population density above K, density dependence is too weak to bring the population density back down to K in the following year. Conversely, when ε_t takes a relatively low value and depresses the population density below K, density dependence is too weak to bring the population density back up to K in the following year. In the absence of a strong tendency to move towards K, the population density is free to wander among high and low values. This pattern is related to the population dynamics produced by the deterministic model with $r = 0.2$ (Figure 7.5A). In the deterministic model the equilibrium point K is not very resilient, so population densities away from K approach K slowly. The addition of extrinsic variability in the stochastic model keeps pushing the population density away from K, and the weak density dependence does not bring it back towards K very quickly.

In contrast, when $r = 1.0$ (Figure 7.7B), populations that are pushed away from K by extrinsic variability are more quickly returned by density dependence. Thus, compared to the case with $r = 0.2$, the stationary population distribution when $r = 1.0$ has lower variance. The same pattern of density dependence that produces low variance of the sta-

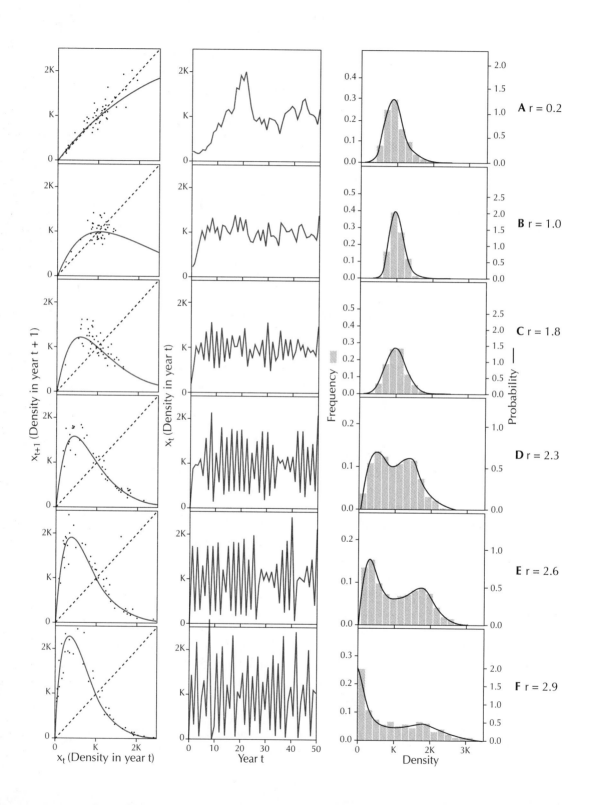

tionary population distribution in the stochastic model also produces high resilience of the stable equilibrium point in the deterministic model (Figure 7.5B).

The larger value of r = 1.8 in Figure 7.7C makes the population density tend to alternate between high and low values, although this pattern is not regular owing to the extrinsic variability. This creates higher variance in the stationary population distribution than when r = 1.0. Again, the higher variance of the stationary population distribution is caused by the same density-dependent pattern that makes the population dynamics in the deterministic model less resilient when r = 1.8 than when r = 1.0.

Finally, in the sequence of Figures 7.7C –F, higher values of r and stronger density dependence lead to higher variances in the stationary population distribution. In the corresponding sequence in the deterministic model (Figures 7.5C –F), starting from a stable equilibrium point (Figure 7.5C) the increase in density dependence produced a stable two-point cycle (Figure 7.5D), then a stable four-point cycle (Figure 7.5E), and finally chaos (Figure 7.5F). In the deterministic model, then, strong density dependence created intrinsic variability in the population dynamics. In a similar way, the strong density dependence in the stochastic model creates intrinsic variability in the population dynamics. This intrinsic variability is combined with the extrinsic variability caused by stochastic variation in ε_t. Thus, the variance of the stationary population distribution reflects the joint effects of extrinsic and intrinsic variability.

Stability, resilience, and resistance in stochastic systems

In stochastic population dynamics, stability is defined similarly to the way it is defined in deterministic population dynamics. The probability distributions for the population densities given in the third panels in Figure 7.7 are **stable stationary population distributions**. The stationary population distribution is akin to the equilibrium point in a deterministic model, but because the population dynamics are stochastic, the population density must be described using a probability density function, rather than as a single point. A stationary population distribution will be stable if the population dynamics eventually describe this distribution regardless of where the population density is started.

In practice, most stationary population distributions encountered for real organisms are stable (Dennis and Constantino 1988). However, if a population is on its way to extinction, then its population distribution is neither stable nor stationary, because the distribution of probable population densities will change as the population gets closer

Figure 7.7. (Facing page.) The population dynamics produced by a simple stochastic model: $x_{t+1} = x_t e^{\varepsilon_t + r(1-x_t/K)}$. The left-hand panels graph x_{t+1} vs. x_t, giving both the relationship in the absence of stochastic variability (solid line) and 50 points from a simulated data set. The dashed lines give the 1:1 lines along which $x_{t+1} = x_t$. The central panels give the 50 simulated data points x_t graphed vs. year t. The right-hand panels give the probability density function for the population dynamics and the frequency distribution of population densities in a simulation of 1,000 years of data. **A** through **F** correspond to increasing values of r, and the value of K is 1.0 throughout.

and closer to extinction. The exact time at which the population goes extinct can never be predicted precisely, but the probability that the population will go extinct can be predicted as the population density continues to drop.

Resilience in stochastic systems can be thought of as the variability in the population density for a given level of extrinsic variability (Ives 1995). For the values of r in Figure 7.7, r = 1.0 (Figure 7.7B) gives the most resilient population dynamics, because the variance of the stationary population distribution is lower than for any other value of r. Note that resilience is governed by intrinsic factors, since it is dictated by density dependence.

Finally, **resistance** is defined as the sensitivity of the per capita population growth rate to extrinsic variability. In all of the cases in Figure 7.7, the distribution of extrinsic variability ε_t is the same; each experiences the same variability in the intrinsic rate of increase from one year to the next; i.e., each has the same resistance. Although resistance is the same for all cases in the model, natural populations show very different levels of resistance. For example, insect populations often have low resistance because many insects are very sensitive to sudden freezes that occur in early spring or late fall. Low levels of resistance lead to high variances in population densities simply because per capita population growth rates are highly variable. In contrast, populations of large, long-lived species often have high resistance. The per capita population growth rate of red deer shows less year-to-year variability than the growth rate of most insects, because red deer are less sensitive to natural environmental fluctuations than are most insects.

The dynamics of real populations

Although the analyses of these simple deterministic and stochastic models have been fairly abstract, the basic ideas illustrated by the models go a long way to understanding the dynamics of real populations. Returning to the examples of *Dendrolimus* sp., great tits, red deer, and Soay sheep (Figure 7.4), why are the population dynamics of these four species so different? The population density of *Dendrolimus* sp. fluctuates wildly because the population dynamics are neither resistant nor resilient. Low resistance is seen in the ten-fold increases and decreases in population density that can occur between successive years. Low resilience is seen in the fact that these ten-fold increases and decreases both occur at low and high population densities; in other words, there is neither a regular increase in densities from low values nor a regular decrease in densities from high values. The actual intrinsic and extrinsic factors that are responsible for the highly erratic population dynamics are unknown, although they probably include predation, sensitivity to climatic variation, competition with other insects that also feed on pine needles, and the ability of large populations to strip trees of needles over wide stretches of forest (Varley 1949). In contrast, resistance in the great tit population is greater, with at most a two-fold increase or decrease in population density between successive years. Resilience is also greater, with increases in population density likely when the density is low, and decreases in population density likely when the density is high. These combine to produce relatively low variances in the great tit population density (McCleery and Perrins 1985).

Since both red deer and Soay sheep are large mammals, we would expect their population dynamics to show the same resistance (Clutton-Brock et al. 1997). The difference between their population dynamics turns out to be mainly differences in resilience due to differing patterns of density dependence. Soay sheep differ from red deer in two important ways. First, Soay sheep have a higher intrinsic rate of increase than red deer, because sheep reach reproductive maturity earlier than red deer (1 year vs. 3–4 years); they produce twins much more often; and they always have a higher proportion of females breeding each year. Second, the per capita reproduction rate of Soay sheep is density independent, while red deer per capita reproduction drops with increasing density. With this high intrinsic rate of increase and without density-dependent reproduction, Soay sheep populations build rapidly to a density that exceeds their winter food supply. With the winter food supply eaten up, winter mortality is high. In contrast, the red deer population with a lower intrinsic rate of increase grows more slowly, and as the density increases, the per capita reproduction rate decreases. Therefore, densities never get high enough for mass die-offs during the winter.

Why are per capita reproduction rates density dependent in red deer but not in Soay sheep? This is largely due to the timing of birthing. Soay lambs are born in April and finish suckling in June, leaving time during the summer for mothers to feed and recover their fat reserves. Entering the fall in good body condition, mothers conceive and produce lambs in the following year. In other words, because plentiful food in the summer allows ewes to recover from lactating, population density has little effect on the proportion of females who conceive. Food shortages produced by high population densities in the winter cause starvation of yearlings and adults alike. In contrast, red deer give birth in June, and owing to their slow maturation, calves suckle for most of the summer. If the population density is high and the food abundance low during the late summer and early fall, mothers don't have a chance to recover from birthing and lactating, and fail to conceive in the fall. This produces density dependence in the per capita reproduction rate. Red deer are not smarter than Soay sheep, nor do they intentionally reduce their reproduction rates in anticipation of food shortages. Instead, the reduction in reproduction rates is simply the result of red deer having less time to recover from birthing, and therefore experiencing food shortages before they conceive, rather than afterwards.

Summary

Variation in population densities through time depends on both extrinsic and intrinsic factors affecting the year-to-year changes in reproduction and mortality. If extrinsic factors such as temperature and rainfall cause large fluctuations in reproduction or mortality from one year to the next, the population density will naturally vary considerably.

However, the magnitude of variability in the population density will also depend on how reproduction and mortality vary with population density. If reproduction and mortality are only weakly influenced by population density (Figure 7.7A), then population densities tend to wander between high and low values. If reproduction and mortality are very strongly influenced by population densities (Figures 7.7D–F), large amounts of population variability can be driven by intrinsic factors. Density dependence of intermediate strength produces more resilient population dynamics (Figure 7.7B).

The analyses of simple models illustrated in a general way what types of population dynamics are possible for single species. However, when studying populations of real organisms, you need to know the specific extrinsic and intrinsic factors acting on the population dynamics. The variety of species on Earth makes any comprehensive list of factors affecting population dynamics impossible. For example, the factors affecting willows in Wales (rainfall, soil nutrients, shading from other trees, etc.) will be different from those affecting wildebeests in East Africa (abundance of grass, predation by lions, diseases, etc.). Much of the work of population ecologists involves identifying the important factors in the population dynamics of different species, and determining which of these factors could aid in saving an endangered species, suppressing an agricultural pest, or simply understanding why some species are rare or common from one year to the next.

Estimating Population Growth Rates for Species with Age Structure

So far, the basics of population dynamics have been explained using a model for species that have non-overlapping generations. A critical assumption of the model is that the population density in one year depends on the population density in the preceding year. For many species, however, changes in population densities depend not just on the total density of individuals in the population, but also on the *ages* of individuals in the population. For example, fledgling and adult great tits have different per capita mortality rates, and obviously adults can reproduce while fledglings cannot. Therefore, changes in great tit population densities will depend on the relative numbers of fledglings and adults in the population.

To study species in which per capita reproduction and mortality rates depend on age, ecologists often divide a population into different **age classes**. How the population is divided will vary from one species to another. The simplest division is between reproductive adults and non-reproductive juveniles. This is a reasonable approach to take with great tits, because once birds reach adulthood, the number of offspring they produce per year does not depend on age. Also, the chance that an adult dies over the course of the next year is independent of its age. Since neither per capita reproduction nor adult mortality rates depend on age, the main distinction within a great tit population is between fledglings and adults. In contrast, per capita reproduction and mortality rates in human populations depend strongly on age, with per capita reproduction rates highest among young to middle-aged adults, and per capita mortality rates highest among the elderly. It is therefore reasonable to divide a human population into yearly age groups, particularly since huge amounts of detailed data exist on age-specific human reproduction and mortality rates. Models that divide populations into different age classes are called **age-structured models**.

To understand age-structured models, consider the northern spotted owl, *Strix occidentalis caurina,* (Noon and McKelvey 1996). The northern spotted owl is a medium-sized owl that feeds primarily on rodents. It inhabits the coniferous forests of the Pacific Northwest of the United States, where it depends on old-growth forest. This requirement has led to one of the most contentious environmental debates in the United States. Old-growth forest is particularly profitable to logging companies, and many rural communities in the Pacific Northwest depend on logging. In 1990, after an intense legal battle, the United States Fish and Wildlife Service listed the northern spotted owl as a threatened subspecies, and it is therefore protected under the Endangered Species Act. Although we know that northern spotted owls need old-growth forest, the scientific problem is determining how much of that habitat is necessary for the protection of the northern spotted owl population. The first step is assessing the state of the northern spotted owl population. Extensive surveys in the 1980s found about 2,200 breeding pairs, and the total number of breeding pairs was estimated to be about double that. But is the population size increasing, decreasing, or staying roughly constant?

To implement conservation efforts, an immediate assessment of the state of the northern spotted owl was needed. Northern spotted owls can live for many years, so it was therefore not practical to determine whether owl populations were decreasing simply by censusing them through time. Instead, researchers set about measuring reproduction and mortality directly, beginning by dividing the population into three age classes: juveniles (one year old), subadults (two years old), and adults (three years old and older). These age classes were selected because northern spotted owls do not breed until their third year, and the per capita mortality rates of juveniles and subadults differ. Researchers collected data on the reproduction and mortality of the different age classes by observing banded birds over several years in several populations. As frequently occurs in ecological studies, different populations produced different estimates of the per capita reproduction and mortality rates. Some of the variation among estimates from different populations undoubtedly reflects real differences among populations; however, additional variation is likely the result of the difficulties of obtaining accurate estimates of per capita reproduction and mortality rates. In order to give an overall picture of the northern spotted owl populations throughout the Pacific Northwest, researchers combined the data from several studies to give composite estimates of per capita reproduction and mortality rates.

Box 7.2. Estimating the Population of the Northern Spotted Owl

To estimate the per capita population growth rate of the northern spotted owl, the population can be divided into juveniles, subadults, and reproductive adults. A model of the population dynamics can be built by letting J_t, S_t, and A_t denote the numbers of juveniles, subadults, and adults in the population in year t. Furthermore, let b be the number of female owls produced per adult female each year, and let s_0, s_1, and s_2 be fractions of juveniles, subadults, and adults surviving each year, respectively. The numbers of adults, subadults, and juveniles in year t+1 can be calculated as:

$$A_{t+1} = s_1 S_t + s_2 A_t$$
$$S_{t+1} = s_0 J_t$$
$$J_{t+1} = b s_1 S_t + b s_2 A_t$$

In words, this set of equations says that the number of adults in year t+1 is the number of subadults and adults that survive from year t. The number of subadults in year t+1 is just the number of juveniles from year t that survive. Finally, the number of juveniles in year t+1 is the birth rate multiplied by the number of adults in year t+1, which from the first equation is $s_1 S_t + s_2 A_t$.

The per capita population growth rate of the owls depends on the relative numbers of juveniles, subadults, and adults in the population. For example, for a total population size of 1000, if 980 are adults, 10 subadults, and 10 juveniles, the per capita population growth rate will be greater than if there are 980 juveniles, 10 subadults, and 10 adults, simply because of the difference in the number of reproducing birds. For calculating the per capita population growth rate, however, this is not much of a problem. Regardless of the initial proportions of juveniles, subadults and adults, after a few years the proportions will stabilize at a distribution determined only by the parameters b, s_0, s_1, and s_2. Thus, given that you know the values of b, s_0, s_1, and s_2, you can calculate this stable age distribution and then calculate the population growth

rate. Assuming that the population is at its **stable age distribution** is reasonable for predicting the change in population density for many years into the future, because regardless of the initial age distribution, in several years the proportions of juveniles, subadults, and adults will stabilize at the stable age distribution.

To make this more concrete, consider the values of b, s_0, s_1, and s_2 for northern spotted owls: b = 0.24, s_0 = 0.11, s_1 = 0.71, and s_2 = 0.94. At the stable age distribution, the proportions of juveniles : subadults : adults are 0.189 : 0.022 : 0.789. Thus, suppose in a population of 1000 owls, 189 are juveniles, 22 are subadults, and 789 are adults. In the next year the number of juveniles will be $J_{t+1} = b s_1 S_t + b s_2 A_t = 0.24 \times 0.71 \times 22 + 0.24 \times 0.94 \times 789 = 182$. Similarly, the numbers of subadults and adults will be $S_{t+1} = s_0 J_t = 0.1 \times 189 = 21$ and $A_{t+1} = s_1 S_t + s_2 A_t = 0.71 \times 22 + 0.94 \times 789 = 757$. Thus, the total population size in year t+1 will be 182 + 21 + 757 = 960, as predicted from the per capita population growth rate of 0.96. Furthermore, the proportions of juveniles : subadults : adults are 182/960 : 21/960 : 757/960, which equal 0.189 : 0.022 : 0.789, the stable age distribution.

Of course, these calculations were made when the stable age distribution was already known. Calculating the stable age distribution in the first place can be done mathematically in a straightforward way using linear algebra. A good reference giving all the details is Caswell (1989).

The same general approach can be used for organisms with more complicated types of age structure—for example, humans. In this case, there must be equations describing reproduction and mortality for each age class . Regardless of the number of age classes, though, there will always be a stable age distribution, and calculating the population growth rate at the stable age distribution is relatively easy (using computers, of course).

Reproduction was estimated as 0.24 female offspring produced per breeding female adult per year. Since breeding pairs of northern spotted owls are monogamous, the number of breeding males is necessarily the same as breeding females, and therefore the accounting can be done on just the females. The annual survivals of juvenile, subadult, and adult owls were 0.11, 0.71, and 0.94, respectively, representing 89%, 29%, and 6% annual mortality. The estimate of juvenile mortality is extremely high. Much of the problem faced by juvenile spotted owls is that in order to breed, they need to find a mate and establish a large territory. Territories of northern spotted owl pairs can be larger than 5,000 hectares, and they generally must include old-growth forest. High juvenile mortality occurs while the young birds search for potential territories and are killed by predators.

These estimates of reproduction and mortality can be combined to estimate the per capita population growth rate of the northern spotted owl; the analysis is described in Box 7.2. The estimated per capita population growth rate is 0.96. In other words, each year the population is predicted to decrease by 4%. At this rate, it would take only 17 years for the population size of northern spotted owls to decrease by 50%! Of course, there is always uncertainty when estimating per capita population growth rates for populations in the wild. Nonetheless, several researchers using different ways of collecting and analyzing the data all came to similarly bleak predictions. Clearly, the northern spotted owl population is in trouble.

The predictions for population declines of the northern spotted owl are not just of academic interest. They are central to a legal and political battle involving a multimillion-dollar-a-year logging industry, the economic livelihood of small logging communities, the most powerful conservation organizations in the United States, and about 5,000 northern spotted owls. Much of the battle hinges on two questions: what, exactly, is the current per capita population growth rate of northern spotted owls, and how will logging practices affect the per capita population growth rate in the future? In the late 1980s, the United States National Forest Service was sued because it failed to produce scientifically valid answers to these questions and to design a plan for protecting the northern spotted owl under the Endangered Species Act. With a recent compromise imposed by the United States government, the legal battle over the northern spotted owl has been reduced to skirmishes. Nonetheless, the future of the northern spotted owl is still uncertain.

One of the greatest sources of uncertainty about the future of the northern spotted owl is how the per capita population growth rate depends on density. The analyses described above ignore potential density dependence in reproduction and mortality. Since density dependence is ignored, the estimates of changes in the population density of northern spotted owls are based on the assumption of exponential decline. Unfortunately, it is very difficult to get direct estimates of density dependence for these birds. Without knowing patterns of density dependence, it is difficult to predict what will happen as the population density of owls continues to decrease. If per capita reproduction rates increase and per capita mortality rates decrease quickly as the population density decreased, and if old-growth forest were no longer logged, then the population could stabilize at a density not too much below its current level. However, if reproduction and mortality do not show strong responses to lowered population densities, then the northern spotted owl could drop to much lower densities, or even to extinction, despite the protection of old-growth forests.

Without direct estimates of density dependence, researchers have used models to predict how owl populations will change in the future. These models analyze the mortality of juvenile owls by asking how the spatial arrangement of old-growth and new-growth forest affects their chances of finding vacant breeding territories. Thus, the models explicitly link density-dependent juvenile mortality with old-growth forest to predict how logging affects the persistence of the northern spotted owl (Lamberson et al. 1994).

What Population Dynamics Arise from Interactions between Predators and Prey?

In broad ecological terms, **predation** is used to refer to any interaction between two species in which one is benefited and the other is harmed. Thus, predation includes not only classic predator-prey interactions such as those between lions and wildebeests, but also parasite-host interactions such as those between malaria and humans, and herbivore-plant interactions such as those between rabbits and dandelions. From the perspective of population dynamics, *benefit* and *harm* in predator-prey interactions have precise meanings that depend on how the population density of one species influences the population growth rate of another. If the population density of the predator is suddenly increased, then the per capita population growth rate of the prey will decrease. Conversely, if the population density of the prey is suddenly increased, then the per capita population growth rate of the predator will increase.

It is important to recognize that the response to changes in predator or prey density is measured by changes in the per capita population growth rate, *not* necessarily the population size. This distinction is necessary, because things other than predation can affect the population size of the predator and prey. For example, consider predation of rabbits by foxes. If there is a very mild winter, rabbit survival will be high, leading to a high population growth rate and an abundance of rabbits by the end of the summer. This occurs despite the fact that foxes are taking advantage of the abundance of rabbits, killing greater numbers than normal. The impact of the foxes on the rabbit population must be measured by the decrease in the rabbit population growth rate caused by the foxes. Even though the rabbit population is increasing, it is still possible to measure the effect of fox predation by comparing the actual rabbit per capita population growth rate to the per capita population growth rate that would have occurred had there been no foxes.

Why don't predators kill all their prey?

The world is full of predators, and no species is immune to attack from predators or parasites. Given the abundance of predators, why don't they kill all their prey? This question is sometimes answered as follows. If predators were to drive their prey extinct, then predator extinction would follow. Therefore, predators do not drive their prey extinct because this would cause their own extinction. This answer is sometimes extended to say that the predators actually act in the best interest of the prey by removing the "old and sick." However, this answer is wrong, because predators do not act this way, nor should they. Take the example of moose and wolves on Isle Royale in Lake Superior (Mech 1966). Moose arrived on the island probably around 1905, and in the abundance of an untouched food supply, the population exploded to more than 1,000 by the 1930s. The moose apparently stripped the island of their preferred food plants, and in the 1940s the population dropped to several hundred. Around 1948 timber wolves became established on the island, probably traveling over the ice from the Canadian mainland during the winter. The wolf population rose from a handful in the 1950s to more than 50, with wolves killing roughly 25% of the moose population each year. With this predation pressure, the moose population remained well under 1,000 despite the recovery of their preferred food plants. You might say that the moose population was healthier because the wolves kept the moose population below the level at which they over-exploited their food supply. Moreover, wolves killed primarily the old, the sick, and young calves. Wolf packs regularly "tested" moose, killing only 1 in 13 of the individuals they confronted. Doesn't this mean that the wolves both improved the conditions for the moose population and, by only killing the "sick, old and weak," improved the quality of the moose population on which they rely? Aren't the wolves acting in the best interest of both wolves and moose?

No. The problem with this argument is that it ascribes a purpose to the wolves' actions, a purpose that has no scientific support. Wolves kill only the old, young, and weak because these are the only moose they can kill without undue risk to themselves. A strong moose can maim and even kill a wolf. When hunting less-formidable prey like white-tailed deer, wolves kill even the healthy and strong. The reason wolves do not drive the moose extinct on Isle Royale is because they cannot. Furthermore, by considering this problem more closely, it is clear that if wolves could drive moose extinct, they probably would. Predators faced with the choice between killing prey and starving, or having their offspring starve, should always kill the prey. Of course, there are always some exceptions, but by and large the reason that predators do not kill all of their prey is because they cannot, not because they choose not to.

The question why a given predator cannot drive its prey extinct is surprisingly difficult to answer. One might immediately look for explanations involving the defenses that prey mount against predators. For example, almost all plant species produce chemical compounds that deter at least some herbivores, and many mammals like moose defend themselves, or like gazelles are swift of foot, or like ground squirrels are never far from a protective hole. Although these explanations help explain why predators don't kill all their prey, they cannot provide the whole story. The whole story must include information on the population dynamics of predators and prey.

A model of predator-prey population dynamics

To demonstrate the importance of predator-prey population dynamics in explaining why a predator cannot kill all of its prey, consider the simplest possible model of predator-prey interactions, the Nicholson-Bailey model (Nicholson 1933). The model is given in mathematical terms in Box 7.3, but can be described in words as follows. To present the model in concrete terms, suppose the prey in the model is a species of whitefly *(Trialeurodes vaporariorum)*. Whiteflies are common insects that feed on a variety of plants and are often pests of cotton crops and of vegetables grown in greenhouses. The predator is *Encarsia formosa*, a tiny wasp that attacks whiteflies by injecting an egg into them. The egg then hatches into a grub-like juvenile, which feeds on and grows within the still-living whitefly. When the juvenile gets large enough, it kills the whitefly and then metamorphoses into an adult. Insect predators such as *Encarsia formosa* are called parasitoids, because they share characteristics of both parasites and predators. Like parasites, they spend part of their lives within living prey (hosts), but like predators they invariably kill their prey, rather than leaving them alive as do many parasites.

The model assumes that whiteflies, in the absence of parasitoids, have exponential population growth. In other words, whitefly reproduction and mortality (due to factors other than predation) do not depend on density. How the model incorporates predation is best described from the viewpoint of an individual whitefly. The chance that an individual whitefly is attacked during a short period of time (say an hour) by a *particular* parasitoid in the parasitoid population is called the **instantaneous attack rate**. The model assumes that the instantaneous attack rate does not depend on the densities of either whiteflies or parasitoids. Therefore, the chance that an individual whitefly is attacked by any parasitoid in the population increases proportionally with the size of the parasitoid population. Furthermore, the total number of whiteflies killed by parasitoids each generation depends on the chance that each individual whitefly is attacked (by any of the parasitoids), and the total number of whiteflies in the population. Therefore, although the instantaneous attack rate is density independent, the total number of whiteflies killed by parasitoids increases with both the number of parasitoids and the number of whiteflies.

BOX 7.3. THE NICHOLSON-BAILEY MODEL

The simplest possible interactions between parasitoid and prey are described by the Nicholson-Bailey model (Nicholson and Bailey 1933), one of the first mathematical models used in ecology. The objective of the model was to determine the basic types of population dynamics that can be produced by parasitoid-prey interactions. In the model, the prey density is given by x_t, and the parasitoid density is y_t. Suppose that in the absence of the parasitoid, the population dynamics of the prey are given by $x_{t+1} = x_t e^r$. This produces exponential growth. Although exponential growth is unrealistic, since the objective of the model is to understand parasitoid-prey dynamics, it is reasonable to assume that the prey is not influenced by any other factors than predation. Suppose the parasitoid searches randomly for prey; it wanders throughout space and only finds prey by stumbling into them. If the instantaneous attack rate is a, and if predators spend a total of h time units searching for prey during their lifetimes, then the fraction of prey that escape the parasitoids is e^{-ahy_t}. The mathematical explanation for the occurrence of $-ahy_t$ in an exponent has to do with the process of random search. Even when the density of parasitoids is high and they each search a large area, there is still a chance that some prey escape; e^{-ahy_t} decreases with increases in the product ahy_t, but never goes to zero. Since a fraction e^{-ahy_t} of the prey escape parasitism, a fraction $(1-e^{-ahy_t})$ are killed. Finally, of the prey that are killed, suppose a fraction c produce parasitoids. Because parasitized prey may die before the parasitoid emerges, c may be less than 1.

This description of parasitoid-prey interactions leads to the set of equations

$$x_{t+1} = x_t e^{r-ahy_t}$$
$$y_{t+1} = cx_t(1 - e^{-ahy_t})$$

A mathematical analysis of these equations shows that there is never a stable point or stable cycles. Instead, they produce cycles of ever-increasing amplitude, and eventually either the parasitoid goes extinct or the prey goes extinct, with the parasitoid obviously going extinct in the following generation.

The population dynamics produced by this model are illustrated in Figure 7.8. The striking feature of the population dynamics is the cycle of parasitoid and whitefly densities. This is an example of a **predator-prey cycle**. This predator-prey cycle is produced by the time lag between changes in whitefly density and changes in parasitoid density. When whitefly densities are high, the parasitoid population growth rate increases in response to the abundance of prey. Eventually, when parasitoid densities are high enough, the whitefly population growth rate becomes negative. As whitefly densities drop, the parasitoid population growth rate becomes negative, causing decreases in the parasitoid density. When the parasitoid population density drops sufficiently low, the population growth rate of the whitefly becomes positive, and whitefly densities start to rise. Eventually parasitoid densities begin to rise, and the cycle repeats itself. As time progresses, the amplitude of the cycle increases, and the parasitoid goes extinct after 20 generations.

The importance of the time lag in creating the predator-prey cycles is seen in Figure 7.8—when whitefly densities increase, parasitoid densities also increase, but only after two generations. The reason for the time lag is that the parasitoid population can only begin to increase after the whitefly population has reached moderate population densities. Similarly, the crash in the parasitoid population occurs after the crash in the whitefly population, because only when the whitefly population density is low does the per capita population growth rate of the parasitoid drop low enough to create a crash in the parasitoid density. In other more complex models of predator-prey interactions, predator-prey cycles do not increase in amplitude; in these models, prey and predator densities can cycle indefinitely. Nonetheless, the cause of the cycles—the time lag between changes in prey and predator densities—is the same.

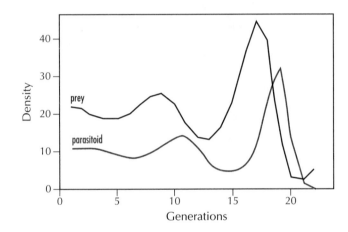

Figure 7.8. The population dynamics produced by the Nicholson-Bailey parasitoid-prey model.

Examples of real predator-prey population dynamics

Are the population dynamics shown by the simple model realistic? That depends on what is meant by realistic. Figure 7.9A shows the actual population dynamics of the whitefly *Trialeurodes vaporariorum* and the parasitoid *Encarsia formosa* in a study in a greenhouse (Burnett 1958). The match between these data and the model output in Figure 7.8 is remarkably close. However, the study was very artificial. Parasitoids were only allowed to attack prey for a 48-hour period. Parasitoids were then removed, and the next "generation" of whiteflies and parasitoids was produced by adding twice the number of whiteflies that were not parasitized, and by replacing parasitized whiteflies with new parasitoids. Thus, the reproduction of unparasitized whiteflies was experimentally set to 2, and the generations of whiteflies and parasitoids were carefully regimented. The only thing not experimentally controlled was the instantaneous attack rate of parasitoids, although the environment in which parasitoids searched for whiteflies was artificially simple. That the model produces very similar population dynamics to the experiment is reassuring—but not surprising, given the control that the experimenter exerted on the population dynamics.

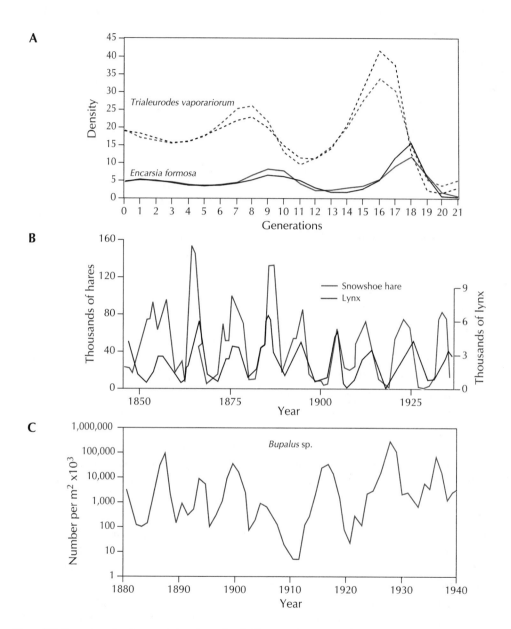

Figure 7.9. Examples of cyclic population dynamics. **A.** The greenhouse whitefly *Trialeurodes vaporariorum,* and its parasitoid *Encarsia formosa* (Burnett 1958). Data are plotted with black lines, while the output from the Nicholson-Bailey model (Figure 7.8) is graphed with colored lines. **B.** Snowshoe hares (colored line) and lynx (black line) populations inferred from the numbers of pelts sold to the Hudson's Bay Company (MacLulich 1937). **C.** The pine looper *Bupalus* sp. in Germany (Varley 1949). These data were collected at the same time as the data from *Dendrolimus* sp. in Figure 7.4A. Note they are graphed on a $\log_{10}$ scale. All graphs redrawn from original sources.

Figure 7.9B shows a natural example of a predator-prey cycle, that of Canada lynx and snowshoe hare inferred from the numbers of pelts of both species sold to the Hudson's Bay Company by trappers. The hare and lynx populations show distinct cycles that, unlike the whitefly-parasitoid example, do not produce extinction. Over much of the geographical range of snowshoe hares, lynx are the dominant predator and are responsible for most mortality in the snowshoe hare population. However, the smoking gun implicating lynx or other predators as the driving force behind the snowshoe hare cycles has yet to be convincingly found (Krebs et al. 1995). An alternative hypothesis is that the snowshoe hare cycles are driven by herbivore-plant interactions. Snowshoe hares feed on several species of plant that produce toxic compounds in their leaves if they suffer defoliation for several consecutive years. High hare densities promote the production of these toxic compounds, which then cause negative hare population growth rates. In addition to promoting toxic compounds, hares can reduce the density of their food plants. The effects of hares on plants can potentially lead to snowshoe hare-plant cycles in which the "predators" are snowshoe hares and the prey are plants. According to this hypothesis, the lynx cycles (Figure 7.9B) are passive, with changes in lynx densities being driven by the cycles in hare densities, but not participating in driving the cycles themselves.

Of course, these two hypotheses are not mutually exclusive, and the observed cycles could be produced by the three-way interaction among lynx, snowshoe hares, and plants. Evidence to support the combined effects of all three is that predation rates on hares are highest when plant resources are of poor quality, because this forces hares to spend more time finding food and puts them at greater risk of being caught and eaten. Additional factors could complicate matters further, such as parasites that build up in the hare population and make them more susceptible to predation. Regardless of the factors actually driving the cycles in snowshoe hare densities, the important thing is that they almost certainly involve some form of predator-prey interaction (including herbivore-plant and parasite-host interactions).

As a final example of predator-prey cycles, Figure 7.9C shows data from 60 generations of *Bupalus* sp., a pine looper in Germany. The cycles shown in these data are somewhat irregular, although distinct peaks occur roughly every 11 years. No data on any predators or parasitoids of *Bupalus* sp. were collected during the 60 years, so it is impossible to know the exact causes of these cycles. Nonetheless, the 11 generation cycles are too long to be produced solely through single-species density dependence; the cycles produced in the single-species models all show strong alternating peaks and troughs from one year to the next (Figures 7.5 and 7.7). Although predation on *Bupalus* sp. was not studied directly, the suggestion of predator-prey interactions given by cyclic population dynamics is strong enough to warrant more research to determine whether predators are in fact driving the cycles.

Do predator-prey cycles occur frequently in nature?

The implication of the simple model in Figure 7.8 and the examples in Figure 7.9 is that predator-prey interactions produce cycles, and if the cycles are severe enough, they can lead to the extinction of prey and/or predator. However, clear examples of predator-prey cycles are not common in nature, and many species exist in the world despite suffering chronically severe predation. Why cycles are not shown by more predators and their prey in nature is one of the central questions in population ecology, and it is still largely unanswered.

One explanation for the limited number of examples of predator-prey cycles in nature is simply that detecting these cycles requires many years of carefully collected data, and this type of data set is rare. In support of this, recent analyses of long-term data sets from forest insects have shown more examples of cyclic dynamics than people previously suspected (Turchin 1990). Unfortunately, such examples as the *Bupalus* study (Figure 7.9C) only involve observations on prey species, rather than observations on both prey and predators, so ascribing the cycles definitively to predation is impossible. Despite the growing evidence of cycles in long-term data sets, if predator-prey cycles were common in nature, examples would probably be far more numerous.

Another explanation is that almost every species is attacked by several species of predator, and most predator species attack more than one species of prey. This diffusion of interactions may stop the cycling of any predator-prey pair. In support of this argument is the fact that many of the clear predator-prey cycles in nature, such as that observed for snowshoe hare and lynx (Figure 7.9B), involve strong interactions between relatively small numbers of species. In addition, several rodent species show population cycles in northern parts of their ranges but not in southern parts, and the numbers of predator species are higher in the southern parts of their ranges. However, this explanation for lack of cycles is unlikely to hold in all situations, because examples exist in which even strongly interacting pairs of predators and prey do not show cycles.

Another explanation is that some features of predator-prey interactions dampen out any tendency for the population dynamics to cycle. In the very simple model of predator-prey interactions, the population dynamics are cyclic (Figure 7.8). Are these cyclic population dynamics the product of the very simple assumptions of the model? What realistic complexities could be added to the model that would stop cycles from occurring? The real utility of the simple model is that it shows, under the simplest assumptions, the types of population dynamics that can occur in predator-prey interactions. The model is useful not because it mimics the dynamics of natural populations, but because it fails, and exploring why it fails may reveal what is going on in nature.

Features of predator-prey interactions that may act to dampen potential cycles can be divided into three general categories.

- Density-dependence in the prey population growth rate may dampen predator-prey cycles in much the same way that density-dependence leads to resilient stable equilibria in systems with only a single species (Figure 7.5B). If prey reproduction and/or survival decrease with increasing prey density due to factors not involving the predator, this would act against the formation of peaks and troughs in the prey density during a predator-prey cycle.

- The efficiency of the predator in finding and attacking prey could decrease at low prey density. In the simple model, the instantaneous attack rate of predators on prey is assumed to be density independent, so a constant fraction of prey are attacked regardless of their density. If, however, prey became relatively more difficult to find and attack at low density, this would act to counteract deep troughs developing in the prey density during predator-prey cycles.

- Predator efficiency may decrease at high predator density, leading to a decrease in the instantaneous attack rate. In this case, the rapid increase in predator population density in response to high prey densities would be impossible. Therefore, the predator population could not reach high enough densities to drive prey densities to low values. This effect of predator density on predator efficiency could occur if, for example, all predator individuals tended to search for prey in the same areas. The resulting buildup in the predator density would concentrate more predators into the same areas, thereby depressing the local density of prey in those areas and decreasing predator efficiency.

Although predator-prey interactions are clearly more complex than single-species systems, the basic objective in understanding why cycles do or do not occur is similar to the problem of understanding resilience in single-species systems. Predator-prey cycles are created by strong interactions between predators and prey. Even though cycles can occur when the instantaneous attack rate and the prey per capita population growth rate are density independent, the per capita mortality rate of the prey depends on predator density, and the per capita reproduction rate of the predator depends on prey density. Therefore, predator-prey cycles represent an intrinsic source variability in population densities. Features which diminish cycles decrease the intrinsic variability in population dynamics and increase the resilience of the system. Understanding predator-prey cycles and understanding the resilience of predator-prey systems go hand in hand.

Summary

Predator-prey interactions have the potential to drive cycles in population dynamics, possibly cycles so extreme that they lead to the extinction of predators, or prey and predators. However, despite several clear examples in nature, cyclic population dynamics seem to be uncommon in predator-prey systems. Several possible explanations for this include the following: long-term cycles really are common in nature, but we don't have enough data sets sufficiently long to show them; cycles in nature are blurred by the combined interactions among many prey and predator species; and simple models showing cyclic population dynamics ignore important features of prey and predator population dynamics acting to dampen possible cycles. Disentangling the possible explanations for why cycles do or do not occur in natural predator-prey systems is a major challenge for population ecologists. Even when such cycles clearly exist, it is difficult to understand their cause—the cycles in lynx and snowshoe hares densities (Figure 7.9B) have been studied for decades by many different researchers, yet the causes of these cycles are not unambiguously resolved.

Why Don't Some Species Outcompete Other Species and Drive Them to Extinction?

As is the case for predation, interspecific competition is defined in terms of how population densities of one species affect the population growth rate of another. **Interspecific competition** occurs between two species when an increase in the population density of either species causes a decrease in the per capita population growth rate of the other. (Because it is cumbersome to refer to "interspecific competition" when two or more species are clearly involved, for shorthand ecologists typically use "competition" without the "interspecific" modifier.) The actual mode of competition between two species can be varied and diverse. For example, competition between two bird species can occur when individuals from different species contest the same nest site. In this case, competition takes the form of direct confrontations between individuals; this is often called **interference** or **contest competition**. Competition between two species of mice could occur when they feed on the same types of seeds. Here, competition occurs not through direct confrontation, but instead through the depletion of a food resource shared in common; this is often referred to as **exploitation** or **scramble competition**. Plants can compete for water and nutrients by removing them from the ground so they are unavailable to neighboring plants. Above ground, plants may also shade out individuals from other species. Some plants even engage in the more aggressive practice of producing compounds toxic to other species; these compounds are leached into the soil, killing neighboring plants and seedlings. Space itself can be the object of competition. On rocky ocean shores, mussels, barnacles, and other species attach to rock surfaces and filter food that is brought in by ocean currents. Here, food is not limited, but attachment sites are. Since many species share the same resources—food, nutrients, space, nesting sites, etc.—the potential exists for rampant competition in nature.

Experimental studies of interspecific competition

Numerous experimental studies have been performed to investigate competition among species. Typical experiments involve varying the densities of two potentially competing species, and then measuring the population growth rate, or some surrogate, of both species. Surrogates for measuring population growth rates include separate measurements of survival, fecundity, and growth rates of individuals in the populations. Ecologists rely on these surrogates when it is difficult to measure all of the components that together make up the per capita population growth rate. Thus, in investigating competition between two plant species, the measure of competition might be a combination of seedling survival to flowering and the number of seeds produced. This excludes the success of the seeds in finding good germination sites, and so misses part of the life cycle of the plant. Nonetheless, the effects of competition on adult survival and seed production probably incorporate most of the possible effects of competition on population growth rates.

Experiments are frequently conducted at small spatial scales. For example, many studies on competition among plants are conducted in plant pots or small field plots measuring a few tens of centimeters in diameter. These small scales are necessitated by the large number of different densities over which the species must be varied in order to reveal the effect of density on per capita population growth rates (or surrogate measures). Designing competition studies inevitably involves a trade-off between the degree of realism and the detail with which the effects of competition can be measured. Different researchers make different choices in this trade-off, and very often organisms are chosen for studies based on how severe the trade-off is likely to be. For example, weedy annual plants are much easier for studies on competition than redwood trees, because the small size and rapid growth of annual weeds allows reasonably natural studies to be conducted over a short period of time at a small spatial scale—say, over the course of a summer in one-meter-square field plots. In contrast, a redwood tree can be hundreds of years old and have a trunk well over one meter in diameter.

Most detailed experimental studies reveal competition occurring at least sometimes among potentially competing species. It is difficult to generalize from this observation, however, to the frequency with which populations experience competition in nature. For obvious reasons, ecologists typically don't set off to study competition between a pair of species unless they suspect that competition occurs. Nonetheless, these studies do show a pattern that probably is generalizable: in most studies, competition is asymmetrical, with one species being less affected by competition than the other (Schoener 1983). This leads to another of the central questions in population ecology. Given that all species are at least potentially affected by competition, and given that competition is generally asymmetrical, what stops some species from reaching high enough densities to drive other species to extinction? This leads to the related question of whether there is a limit on the number of species that can coexist in the same habitat. Given the potential for some species to be excluded through competition with other species that are superior competitors, what explains the large number of species that coexist in many ecological communities?

An example of the competitive exclusion principle

The possibility that some species are able to exclude other species from a habitat through competition is illustrated by the following example (Bess et al. 1961). When commercial fruit production was starting in Hawai`i, the Oriental fruit fly was accidentally introduced, and it caused considerable damage to the developing fruit industry. In response, between 1947 and 1952 the Hawai`i Agriculture Department imported 32 parasitoid species that attack the Oriental fruit fly in its native range in Asia. Of these, three species in the genus *Opius* became established in sequence (Figure 7.10). All three of these species are similar and have generation times of roughly three weeks, comparable to that of the fruit fly. *Opius longicaudatus* was the first of the three to become established, and it increased to the point of parasitizing roughly 20% of the fruit fly population. *O. vandenboschi* followed *O. longicaudatus*, appearing to replace it slowly during late 1949 and early 1950, while increasing the parasitism rate on the fruit fly to 40%. Finally, *O. oophilus* replaced *O. vandenboschi* as the main parasitoid, eventually reaching 80% parasitism while the other two species declined to non-existence. Thus, *O. oophilus* competitively excluded two previously established parasitoids.

Two features of this example are noteworthy. First, competitive replacement is shown clearly because the parasitoids were monitored as they were introduced into Hawai`i. If we were to go to Hawai`i now and found only *O. oophilus,* we would be completely unaware that *O. oophilus* had previously competitively replaced two other species. The point is that the potential for competitive exclusion could be common even though actual competitive exclusion is rarely seen, simply because it has already occurred. This

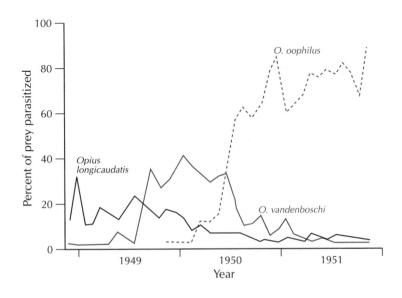

Figure 7.10. Replacement of two parasitoids of the Oriental fruit fly, *Opius longicaudatus* and *O. vandenboschi,* by a third species, *O. oophilus,* in Hawai`i. Population densities of the parasitoids are given in terms of the percent of prey parasitized by each species. Redrawn from Bess et al. 1961.

phenomenon is sometimes called "the ghost of competition past," since the effects of competition on a community (specifically, what species actually occur) depend on past competitive events that are now unknown.

Second, the only prey attacked by the three parasitoids on Hawai'i is the Oriental fruit fly. Therefore, the parasitoids compete for a single resource. Each successive *Opius* species achieved a greater level of parasitism on the Oriental fruit fly, from 20% for *O. longicaudatus* to 80% for *O. oophilus*. Thus, *O. oophilus* was the most efficient predator of the three species, with the ability to suppress the fruit fly population to a level below which the two other parasitoids could not maintain a positive population growth rate. This is an example of the **competitive exclusion principle**, which states that two or more species cannot coexist if they use the same limiting resources in the same way. A **limiting resource** is one that, if increased, will increase the population growth rate of a species. The Oriental fruit fly is a limiting resource of the *Opius* parasitoids, because the population sizes of the parasitoids are tied closely to the availability of their prey. Stated another way, the competitive exclusion principle says that if species depend on the same limiting resources, whichever species can use the resource more efficiently will drive the others to extinction. Stated yet another way, the competitive exclusion principle says that two species cannot share exactly the same **fundamental niche**. The fundamental niche is a way of referring to all of the requirements of a species; it is defined by the upper and lower bounds of all factors—for example, temperature and food abundance—that the species requires to persist (see Chapter Five).

Resource partitioning

The value of the competitive exclusion principle is not that it says when one species will drive others extinct, but instead that it shows how similar species can coexist: some of the resources they use can be different, or they can use the same resources in a somewhat different way. The general term for species using different resources is **resource partitioning**. An example illustrating resource partitioning involves two bird species, the medium ground finch (*Geospiza fortis*) and the small ground finch (*Geospiza fuliginosa*), two evolutionarily closely related species of finch that occur in the Galapagos archipelago (Grant 1986). As their common names suggest, the medium ground finch is larger than the small ground finch, and has a larger beak. The larger beak makes the medium ground finch better able to crack open large seeds than the small ground finch. Not surprisingly, larger, more energy-rich seeds make up a larger portion of the medium ground finch's diet than of the small ground finch's diet. Although the small ground finch would seem to be at a disadvantage, its smaller size means that it can do well on a diet of smaller seeds. Although the medium ground finch can eat smaller seeds, its larger body size means that it needs the greater energy source provided by larger seeds. These two species still compete, because they both eat medium-sized seeds. Thus, large populations of the medium ground finch consume all the medium seeds, leading to less food and a decreased population growth rate of the small ground finch. Similarly, large populations of the small ground finch decrease the population growth rate of the medium ground finch. However, regardless of how high the density of either species rises, seeds will always be available to the other species—either large seeds for the medium ground finch or small seeds for the small ground finch.

The popular literature often claims that resource partitioning is an example of how species learn to "live in harmony" with each other. Like the idea that predators are good for populations of their prey, this is another concept that, though warm and comforting, is scientifically unfounded. Species do not partition resources through mutual agreement. Instead, species are constrained by the inability to be masters-of-all-trades. In the example of the medium and small ground finches, the small ground finch cannot crack large seeds, and the large ground finch cannot survive on a diet of small seeds. Therefore, neither species can be the superior competitor for both large and small seeds. Resource partitioning arises because there is a trade-off, either physiological, morphological, or behavioral, between being efficient at acquiring one type of resource and being able to acquire many different types of resources. Although "living in harmony" is not the reason why species partition resources, coexistence is the outcome of resource partitioning nonetheless.

Resource partitioning between diatoms

David Tilman discovered an example of resource partitioning in a detailed study of two diatoms, *Asterionella formosa* and *Cyclotella meneghiniana* (Tilman 1977). Diatoms are single-celled organisms that make up one of the major photosynthetic groups in lakes and oceans. Diatoms need both phosphate to produce RNA and other compounds, and silicate to produce shells. The relative required amounts of these nutrients, however, varies from species to species. In particular, *A. formosa* requires relatively higher silicate concentrations, while *C. meneghiniana* requires relatively higher phosphate concentrations. Tilman conducted 76 separate experiments pitting *A. formosa* against *C. meneghiniana* in small laboratory containers where he could vary the relative concentrations of phosphate and silicate in the water (Figure 7.11A). As expected, when phosphate concentrations were low relative to silicate, *C. meneghiniana*, with its requirement for higher phosphate concentrations, was outcompeted by *A. formosa*. However, the reverse occurred when silicate concentrations were low. At intermediate ratios of silicate and phosphate, however, both species coexisted—*A. formosa* is limited by silicate while *C. meneghiniana* is limited by phosphate. Because they are limited by different nutrients, they cannot drive each other extinct. This doesn't occur just in the lab; in Lake Michigan, areas in which the water had relatively higher concentrations of phosphate were dominated by *A. formosa*, areas with relatively higher concentrations of silicate were dominated by *C. meneghiniana*, and areas with intermediate ratios of silicate-to-phosphate concentrations had both species (Figure 7.11B). Finding the same patterns in the field and in the lab strongly suggests that both patterns have the same cause, and that resource partitioning makes coexistence of diatoms possible in nature.

This diatom example highlights a particularly useful way of looking at interspecific competition and resource partitioning. In explaining how competing species coexist, we need to consider not only competition between individuals of different species, but also competition between individuals from the same species. In the range of concentrations of phosphate and silicate allowing coexistence of both diatoms, the maximum density of *A. formosa* is set when they draw down silicate concentrations to a level that

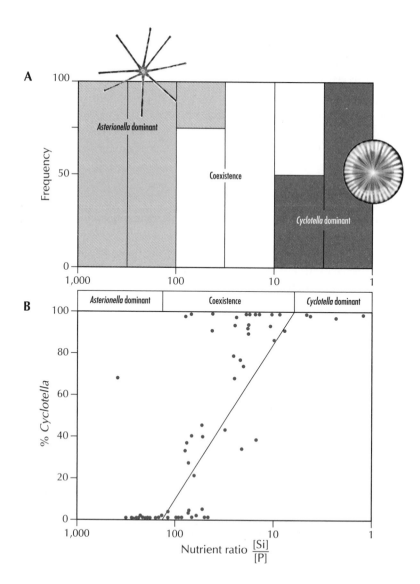

Figure 7.11. Competition and coexistence between two diatoms, *Asterionella* and *Cyclotella*. **A.** The frequency with which laboratory experiments were dominated by *Asterionella*, dominated by *Cyclotella*, or produced coexistence of both species at different relative concentrations of silicate and phosphate. **B.** The percent *Cyclotella* in samples of *Asterionella* and *Cyclotella* taken from Lake Michigan waters that differed in the relative concentration of silicate and phosphate. Redrawn from Tilman 1977.

their per capita population growth rate equals one. Individuals of *A. formosa* are still using phosphate, but because their population density is limited by silicate, they cannot reach high enough densities to reduce phosphate to sufficiently low concentrations to exclude *C. meneghiniana*. Stated another way, intraspecific competition among *A. formosa* individuals for silicate makes it impossible for *A. formosa* to reach high enough densities to exclude *C. meneghiniana*. Similarly, intraspecific competition among *C. meneghiniana* individuals for phosphate makes it impossible for their population density to become high enough to consume enough phosphate to exclude *A. formosa*. Stated even more simply, the two diatoms coexist because intraspecific competition is stronger than interspecific competition. This is the mechanism by which resource partitioning allows species to coexist. When the similarity in resource use by individuals from the same species is greater than the similarity of resource use by individuals from different species, intraspecific competition is greater than interspecific competition, and coexistence is possible.

Coexistence of fugitive species

In addition to partitioning different types of resources, species may also coexist on the same limiting resources if they use the resources in a different way. Good examples are the species of grassland plants that are found in small patches of disturbed ground, e.g., areas created by ground squirrels digging holes for protection or badgers digging holes to capture ground squirrels (Platt and Weis 1985). These plants have very light, windblown seeds that are produced in copious quantities. Almost as soon as disturbances occur, seeds from these plants reach the sites and germinate. The fast-growing seedlings reach adulthood and produce seeds in their first year. Slowly, plants from the surrounding prairie spread over the disturbed ground. These late-comers outcompete those plants that were first to arrive, but not before the early plants have produced great quantities of seeds that can spread to other, newly disturbed sites. The plants that make a living through dispersing into disturbed areas are examples of **fugitive species**. Fugitive species coexist with species that can beat them in head-to-head competition by always staying one step ahead of the competitive dominants. These fugitive species use the same resources as the plants that eventually outcompete them, but they use them differently, by dispersing rapidly into newly-disturbed sites.

As in the case of resource partitioning among small and medium ground finches, fugitive species illustrate how trade-offs lead to the coexistence of species. Fugitive plant species must have small, light seeds that can be spread easily in the wind. This means that the seeds cannot carry much in the way of provisions with them. In contrast, non-fugitive species can have large seeds containing a nutrient store that increases the hardiness and competitive ability of the seedling. The necessary trade-off between small, highly dispersive seeds and large, heavily provisioned seeds gives contrasting advantages to fugitive and non-fugitive species, thereby making it possible for both to coexist.

The population dynamics of competing species

Like predation, competition is an intrinsic factor that affects the population dynamics of coexisting species. In general, interspecific competition decreases the resilience of population dynamics for both competing species. If one species becomes relatively rare, the other species will be freed from interspecific competition and therefore become relatively common. In turn, competition with the now-common species will slow the rate at which the rare species can once again attain high densities. Thus, interspecific competition makes population dynamics of individual species less resilient by perpetuating the rarity of a species when it is rare, and perpetuating the commonness of a species when it is common.

On the other hand, rather than looking at each species separately, if we look instead at the sum of the densities of two species, the effects of interspecific competition on both species combined can be quite different. When one species is rare, the other is likely to be common. Therefore, the sum of their densities is likely to vary less than either of their densities taken separately. In terms of individual species, interspecific competition decreases resilience, while in terms of the species taken together, interspecific competition may have no effect on resilience if changes in the density of one species are balanced by opposite changes in the density of the other. This problem brings up the issue of how one views populations of interacting species. In this example, a population ecologist might view the population densities of each species separately, while a community ecologist might lump the two species together (see Chapter Eight). Thus, the population ecologist would conclude that interspecific competition decreases resilience, while the community ecologist would conclude that interspecific competition has no effect on resilience. As is the case in many situations, arguments can be avoided by making clear exactly what we are talking about—in this case, whether we are talking about the resilience of individual species or the resilience of two species taken together.

Summary

Interspecific competition poses the problem stated by the competitive exclusion principle: two or more species cannot coexist if they use the same limiting resources in the same way. Since many species coexist in nature, many studies have been directed towards finding out how they do so. Resource partitioning is probably the most widespread factor explaining coexistence of competitors. Species may also exist as fugitives, surviving despite being poor competitors by being able to disperse rapidly and take advantage of areas before better competitors arrive. Other scenarios for coexistence of competing species are possible. For example, predators that attack two competing species could play a role in allowing coexistence. Regardless of the particular mechanism, the essential requirement for coexistence is that species be sufficiently different in some way to prevent either species from reaching a high enough density to drive the other extinct.

How Important Are Mutualistic Interactions in Nature?

In the same manner as predation and competition, mutualism may be defined in terms of population growth rates. **Mutualism** occurs when an increase in the density of either member of a species pair increases the per capita population growth rate of the other species. A survey of the scientific literature shows that the population dynamics of mutualism are far less studied than the population dynamics of predation and competition. Most research on mutualism focuses on determining how the species benefit each other, and how the mutualism could have evolved. The relatively small amount of research on the population dynamics of mutualism might indicate that the study area is young and still going through initial phases of describing the natural history of mutualistic interactions. Alternatively, predation and competition could figure more importantly in the population dynamics of organisms that researchers have chosen to study. Regardless, numerous mutualistic interactions are very important in ecological communities and ecosystems, and below are a few examples.

Pollination and seed dispersal

The most familiar and possibly the most important mutualism in nature is between plants and the animals that disperse the plants' pollen and seeds. Flowering plants constitute the most common type of plant on Earth. By co-opting the aid of insects and other animals, flowering plants have found an admirable solution to the problems of being sedentary. Animals carry pollen between flowers and disperse seeds to distant sites. The benefits to the animals that aid plants are numerous. Plants provide both nectar to attract pollinators and fleshy fruits to attract seed dispersers. Many plants are wholly dependent on animals as pollinators and seed dispersers, and many animals are wholly dependent on the nectar or fruits that plants produce.

Despite this seemingly rosy state of affairs between plant and animal mutualists, evolution has lead to numerous ways in which plants and animals take advantage of their mutualistic partners. For example, plant nectar provides only a dilute concentration of sugar, forcing pollinators to visit many plants and thereby disperse pollen to many recipients. Some plants even take advantage of pollinators without providing any benefits, for example by withholding nectar, or by attracting unsuspecting insects not with a reward, but instead with a chemical compound mimicking the insects' sex pheromones. Animals also take advantage of mutualistic plants. Some bees, rather than climb down the flower to collect nectar and thereby get covered in pollen, instead chew through the sides of flowers to steal the nectar. Even though mutualism is predicated on both species benefiting from their interactions, mutualism should not be confused with altruism. Mutualism must be viewed from the perspective of natural selection, which favors those characteristics in both plants and animals leading to increased survival and fecundity. Mutualisms only arise from the mutual self-interest of plants and animals acting through natural selection.

Digestion of cellulose

Many mutualistic interactions occur between microbes and animals that eat relatively indigestible food. Cellulose, the compound that makes up the bulk of structural material in plants, is largely indigestible to almost all organisms except certain bacteria and protists. When a cow chews its cud, it is grinding up plant material so that the cellulose-digesting microbes in its rumen have easier access to the cellulose. Similarly, termites have an array of protists that join the termites' own cellulose-digesting enzymes to decompose the wood that termites feed on. In these mutualisms, the animals take advantage of the diverse chemical reactions unique to their microbes, while the microbes take advantage of the favorable environment of the animals gut and the regular servings of food they receive.

Nitrogen fixation

Another mutualism occurs between bacteria and plants and involves nitrogen fixation. Although nitrogen makes up 78% of the atmosphere, plants cannot use it in its gaseous form. In fact, many plants live with constant nitrogen shortages—nitrogen is the prime ingredient of most fertilizers. Several groups of prokaryotes (a few bacteria and cyanobacteria) have the requisite biochemistry to convert atmospheric nitrogen into nitrate or ammonium, forms of nitrogen that plants can use readily. With the obvious benefits that nitrogen fixing has for plants, several groups of plants have developed specialized structures for culturing nitrogen fixers. Legumes, including peas, clover and beans, develop nodules in their roots that provide the perfect environment for nitrogen-fixing bacteria. The frequency of plant species having specialized structures to culture nitrogen-fixers increases in habitats with soils that have low concentrations of nitrogen. This is particularly true in tropical areas with heavily leached soils, where co-opting the help of nitrogen-fixers is critical to many tropical trees.

Mycorrhizal fungi

A final example of mutualism involves plants and fungi. Most species of higher plants have their root systems surrounded by the hyphae of specialized fungi, called mycorrhizal fungi. The fungi are particularly efficient in absorbing water and nutrients (nitrogen, phosphorous, and calcium) from soils. These nutrients are passed to the roots of the plants, and in some species the hyphae of the fungi actually penetrate the root cells. In return for water and nutrients, the plants provide the fungi with at least some of the carbohydrates they require. These mycorrhizal associations are extremely important for trees and grasses. When non-native pines were introduced into both Puerto Rico and Australia in attempts to establish commercial plantations, the trees grew very slowly until receiving applications of soil containing the appropriate mycorrhizal fungal species to help the pines.

The population dynamics of mutualism

From a population-dynamics point of view, mutualistic interactions increase population growth rates of both species, thereby leading to higher densities of both. In fact, very simple models of mutualism, analogous to the simple model of predation presented previously, "lead to silly solutions in which both populations undergo unbounded exponential growth, in an orgy of mutual benefaction" (May 1981). Of course, this doesn't occur in nature. Despite mutual benefaction, other factors such as intraspecific competition, interspecific competition, or predation step in to stop run-away population growth. Although Bob May pleaded in 1981 for more work on the theory of mutualistic population dynamics, little has been done, and despite the clear ecological importance of mutualisms, we have only a rudimentary understanding of how mutualisms can influence population dynamics.

How Does Space Influence Population Dynamics?

A central tenet of population ecology is that population dynamics can be understood in terms of the interactions among individuals from the same and from different species. Competition involves individuals using resources, and predation involves individual predators attacking individual prey. Because population dynamics are the product of individual actions, population dynamics depend on how individuals experience the world around them. Thus, if the environment varies spatially, then individuals in different places will experience the world differently. These differences, averaged over the entire population, can have profound effects on population dynamics. A comprehensive summary of the diverse effects of spatial variation would take several books. Only two types of spatial processes will be presented here: *aggregation* and *metapopulation dynamics*.

Aggregation and competition among carrion flies

Aggregation refers to the clustering of individuals in particular locations in space. How aggregation affects competition can be seen in the example of carrion flies. The community of carrion flies that breed from small mammal carcasses in New Jersey is dominated two main groups: *Phaenicia coeruleiviridis*, a green-bottle fly, and a group of three species of flesh flies in the genus *Sarcophaga* (Ives 1991). The *Sarcophaga* species are ecologically similar. The females of two of the species are indistinguishable, and all of the taxonomic differences between these two species lie in the male penises. Because of the ecological similarities within the *Sarcophaga* group, we can lump them together in order to ask how they coexist with *P. coeruleiviridis* (Plate 7.1).

Competition among carrion-fly maggots for the limiting resource of a carcass can be intense, and carcasses are often stripped to skin and bone in just a few days. When initial maggot densities are high, the carcass is consumed before all maggots have had their fill, and the maggots either die, or, if they are large enough, go through metamorphosis successfully and become small adults. Adult size is important for population dynamics, because the smallest females produce only 10% as many eggs as the largest females. Since maggots from all carrion fly species eat the same fleshy tissue, there is little opportunity for resource partitioning within carcasses. Nonetheless, *P. coeruleiviridis* and the *Sarcophagas* coexist.

The explanation for coexistence may be aggregation. Aggregation of carrion flies among carcasses is easy to see. Suppose you spent a Sunday afternoon spreading mouse carcasses in a field, and then sat and watched female carrion flies come and lay eggs on them. You would find that one carcass might be visited by a large number of *P. coeruleiviridis* females, while another carcass might attract very few. This pattern of variability in the number of *P. coeruleiviridis* females among carcasses represents **intraspecific aggregation.** If you watched *Sarcophaga* females, you would find that they too are intraspecifically aggregated. However, those carcasses that attract large numbers of *P. coeruleiviridis* females are not necessarily those that attract large numbers of *Sarcophaga* females. In other words, *P. coeruleiviridis* and *Sarcophaga* females are intraspecifically aggregating on different subsets of carcasses, so there is no **interspecific aggregation**. An experiment in which female flies were trapped at thirty mouse carcasses showed these patterns quantitatively (Figure 7.12).

Aggregation affects competition by changing the average densities of competitors experienced by individuals from both species. In those carcasses which attract large numbers of *P. coeruleiviridis* females, maggots experience intense intraspecific competition. Conversely, in those carcasses which attract few *P. coeruleiviridis* females, maggots experience little intraspecific competition. From the perspective of the maggots as a group, however, on average the increase in intraspecific competition in carcasses attracting many females outweighs the decrease in intraspecific competition in carcasses attracting few females. This is because more maggots occur on those carcasses with greater intraspecific competition, and therefore more maggots experience increased intraspecific competition than decreased intraspecific competition. In other words, from the perspective of *P. coeruleiviridis* maggots, intraspecific aggregation increases the average *experienced* density of their fellow maggots.

For the data in Figure 7.12, intraspecific aggregation of *P. coeruleiviridis* increases their experienced density of conspecific maggots by 80% above the actual density. Intraspecific aggregation of *Sarcophaga* increases their experienced density of conspecific maggots by 240%. However, because there is no interspecific aggregation, the density of *Sarcophaga* maggots experienced by *P. coeruleiviridis* maggots is the same as the actual density, and the converse is true from the perspective of *Sarcophaga* maggots. Thus, intraspecific aggregation of *P. coeruleiviridis* and *Sarcophaga* females increases intraspecific competition experienced by both species. Although interspecific competition could be affected by interspecific aggregation, interspecific aggregation did not occur. The net result is that by accentuating intraspecific competition relative to interspecific competition, the pattern of carrion fly aggregation makes coexistence more likely.

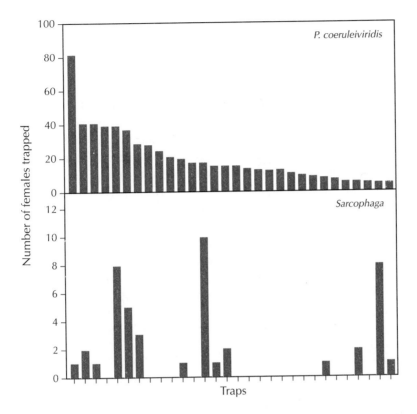

Figure 7.12. The number of *Phaenicia coeruleiviridis* and *Sarcophaga* spp. females captured at 30 traps baited with fresh mouse carcasses. In the top panel traps are ordered according to the number of *P. coeruleiviridis* females captured (not according to their actual location). The order of traps in the *Sarcophaga* panel is the same as in the *P. coeruleiviridis* panel. Redrawn from Ives (1988).

The aggregation shown by carrion flies is a type of resource partitioning, in that different species are using different carcasses. However, unlike many cases of resource partitioning in which the resources differ in some obvious way, the carcasses seem to be all equally good for maggots. Experiments comparing the survival and size of maggots reared from attractive vs. non-attractive carcasses showed no differences in the quality of carcasses for maggot growth. The reason some carcasses attract one species while other carcasses attract other species seems to be differences in the microbial communities that grow on carcasses. Carrion flies are faced with a challenge in finding carcasses, because carcasses are extremely rare in nature. To find carcasses, carrion flies detect chemicals produced during the initial phases of carcass decay, chemicals that are very specific to the decay process and therefore distinguishable from the huge array of volatile chemicals in the air. This specificity has made carrion flies very efficient at finding carcasses, but also susceptible to small differences in the microbial communities among carcasses. Different species of carrion fly perceive the environment and its resources differently, and this leads to aggregative patterns in their distributions among carcasses that facilitate coexistence of species. The pattern of different species aggregating on different carcasses is not the result of some mutual agreement among carrion flies to partition up their resources. Instead, it is the unavoidable consequence of differences in how they find carcasses.

Aggregation among plants

Aggregation figures prominently in competition among many kinds of species. Joy Bergelson found this to be the case for two plant species, common groundsel (*Senecio vulgaris*) and bluegrass (*Poa annua*) (Bergelson 1996). Both species go through two generations per year, with seeds produced by the spring generation germinating in the summer, and seeds from the summer generation overwintering until they germinate in the spring. Although they have similar generation schedules, the two plants are very different in structure. Groundsel is tall and thin, while bluegrass grows as a bushy mass (Plate 7.2). This difference in structure gives bluegrass a competitive edge that occurs *among* generations, rather than within each generation. After producing seeds during the spring generation, bluegrass plants die, leaving a mass of dry grass litter where the plant stood. If groundsel seeds germinate under this mass, the seedlings cannot break through the litter and grow. Thus, the mass of litter produced by the spring generation of bluegrass causes a strong competitive effect on the summer generation of groundsel. The summer generation of bluegrass suffers less than the groundsel, since its seedlings are better able to break through the mass of litter.

Aggregation comes into play to help the summer generation of groundsel. If the spring generation of bluegrass is aggregated in space with clusters of plants close together, then the gaps between the bluegrass clusters provide openings for the summer generation of groundsel. To demonstrate how important these gaps are, Bergelson planted plots of both bluegrass and groundsel in the spring. In half of the plots she spread the bluegrass seeds evenly, while in the other half she spread the bluegrass seeds in an aggregated fashion. She planted groundsel evenly in all plots. As expected, groundsel in the summer generation did much better in the plots with aggregated bluegrass distributions, because there were plenty of gaps in which the groundsel seeds could germinate and grow.

So what could cause bluegrass plants to be aggregated in space? Grazing by mammal herbivores often kills groups of bluegrass plants that are close to each other. In addition, although bluegrass is wind-dispersed, it is a relatively short plant with rather large seeds, so most seeds are dropped around the base of the adult plant. Thus, the gaps in bluegrass cover created by herbivores are not rapidly filled by dispersing bluegrass seeds. This type of aggregative pattern is frequently found among plants and may play a central role in plant competition.

Metapopulation dynamics of mite predators and prey

Metapopulation dynamics refer to the situation in which populations are subdivided into smaller groups (subpopulations). Being partially isolated from each other, the population dynamics within subpopulations are somewhat independent of the dynamics in other subpopulations, but not completely so. The dynamics of subpopulations are linked together by the dispersal of individuals among subpopulations. Understanding the dynamics of all subpopulations together—the metapopulation—requires understanding how dispersal among subpopulations influences the overall population dynamics.

Metapopulation dynamics are most easily explained using an example like the classic experiment on predator-prey interactions conducted by Charles Huffaker (Huffaker 1958). Huffaker studied the predator-prey interactions between a predatory mite and its prey, a herbivorous mite that feeds on citrus fruit. Huffaker set up arrays of oranges in a laboratory, introduced prey and predator mites, and observed the population dynamics. Initially, he started with a small array of six oranges. In this simple array, the predatory mite population reached high densities, eventually leading to the extinction of its prey (Figure 7.13A). Thus, as in the simple model of predator-prey interactions (Figure 7.8), high-amplitude cycles in predator-prey dynamics led to rapid extinction.

Huffaker went on to study population dynamics in a more complicated array of 120 oranges. He used Vaseline to make a network of barriers on the board, separating different groups of oranges. He also set up vertical sticks as "launch pads" for the prey mite. These mites "balloon" by setting out a thin line of silk (similar to the silk of spiders) that can be caught by air currents. Thus, the prey mites with launch pads could move among oranges more easily than the non-ballooning predatory mites.

The population dynamics in this more complicated array of oranges are quite different from those found in the simple array (Figure 7.13B). Over the course of eight months (roughly 12 generations of prey mites), the populations showed three cycles. Although the mites eventually went extinct, they persisted for much longer than in the simple array. The explanation for the increased persistence is seen in the spatial pattern of prey and predators. The distribution of prey and predators among the oranges shows a mosaic pattern, with some oranges containing both prey and predators, while other oranges contain only prey or predators. Even though predators eventually find and exterminate prey populations from a given orange, the prey are continuously dispersing to other oranges that are free of predators. Thus, this is a large population-dynamics cat-and-mouse game. Because the prey can move more readily among oranges, their populations can establish and grow on oranges in the absence of predators, at least for a while.

The greater persistence of mites is the result of metapopulation dynamics. The subpopulations are the prey and predators on individual oranges, while the metapopulation is the populations of prey and predators in the entire array of oranges. The subpopulations themselves have unstable dynamics. If both prey and predator were put together on an orange, the predator would kill all the prey. However, the persistence of the metapopulation occurs because the dynamics of the subpopulations are not all synchronized. While the subpopulations of prey on some oranges are being exterminated by predators, subpopulations of prey on other oranges are growing in the absence of predators. As long as enough asynchrony is maintained among subpopulations, the metapopulations will

Figure 7.13. The population dynamics of a prey mite (black lines) and a predatory mite (colored lines) in arrays of oranges. **A.** The population dynamics on six oranges. **B.** The population dynamics on 120 oranges with Vaseline barriers separating oranges, and stick launch pads. In both cases, the amount of orange exposed to the orange-feeding prey mites was the same. In **A** the entire orange was exposed to feeding, while in **B** only $1/20$ of each orange was exposed. The set of panels in **B** gives the spatial patterns of prey and predator densities. Shading represents the density of prey mites on each orange: white, 0–5 mites; pale color, 6–25 mites; gray, 26–75 mites; black, greater than 75 mites. Presence of the predatory mite (1–8 individuals) is indicated by colored dots. Redrawn from Huffaker (1958).

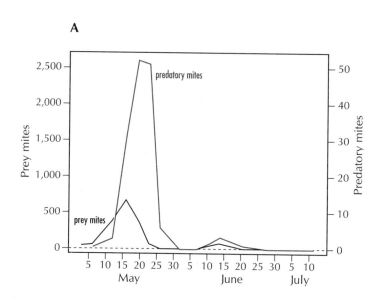

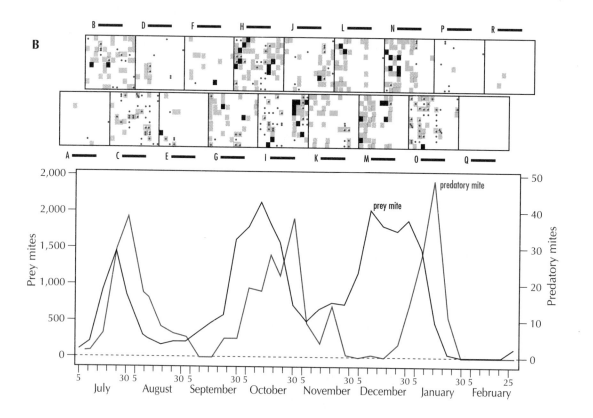

persist. The collapse of Huffaker's metapopulations after the third population cycle occurred because predators synchronously exterminated prey from all oranges.

Metapopulation dynamics of rare species

Metapopulation dynamics don't just apply to predators and prey, and the importance of metapopulation dynamics in nature is probably huge. For example, metapopulation dynamics are important for fugitive species such as the Karner blue butterfly *(Lycaeides melissa samuelis)*. Historically, this butterfly occurred as a metapopulation spread over wide areas of sandy habitats in the north-central United States (Andow, Baker and Lane 1994). The caterpillars feed exclusively on lupines, and lupines only occur in areas that have recently been disturbed, either by fire or by human activity. Although a Karner blue subpopulation may flourish in a large patch of lupines, the subpopulation will inevitably go extinct, because after several years lupines will be replaced naturally by other plant species. However, even though any one subpopulation of butterfly inevitably goes extinct, the metapopulation persists because all subpopulations do not go extinct at the same time; there are always subpopulations that provide colonists to establish new sub-populations when new patches of lupines arise.

As is the case of the Karner blue butterfly, metapopulation dynamics are often important for species that are rare and depend on specialized habitats. It is not surprising then, that many threatened and endangered species exhibit metapopulation dynamics. The Karner blue butterfly itself is threatened, and it illustrates the difficulty of protecting species that have metapopulation dynamics. Because any one subpopulation of Karner blue butterfly is destined to extinction, simply protecting the habitat in which the butterfly is found is not enough. Instead, it is necessary to protect habitat where the butterfly is not found, or even create new habitat through managed burning of prairies. Furthermore, protecting only one subpopulation is not enough, because the persistence of the metapopulation requires sufficiently many subpopulations located sufficiently close together to allow the establishment of new subpopulations through dispersal. Thus, protecting many rare and threatened species requires a thorough scientific understanding the spatial structure of their population dynamics and how dispersal affects the overall persistence of the species.

Summary

Aggregation and metapopulation dynamics depend on the spatial and temporal scale at which ecological interactions occur. For species in which aggregation is important, population dynamics depend on how individuals are distributed at small spatial scales— maggots among carcasses or plants within small field plots. For species in which metapopulation dynamics are important, understanding population dynamics will be impossible if one focuses only on small areas, because the overall metapopulation dynamics are governed in part by dispersal among small areas. This presents a considerable challenge to ecologists. Experimental studies are generally limited in spatial extent. In fact, in a survey of experimental studies on interactions among species, 50% were conducted in areas of less the one square meter (Kareiva and Andersen 1988). Understanding broad-scale patterns in the population dynamics of interacting species is one of the great challenges in population ecology.

Can the Population Dynamics of Communities of Species Be Understood from Pairwise Interactions between the Species?

Population ecologists generally think of population dynamics of interacting species in pairwise terms. Predation can lead to population cycles of one prey and one predator, competition can lead to the exclusion of one species by another, mutualism can lead to higher densities of mutualistic pairs of species. When confronted with a group of interacting species, the population ecology mindset is to divide the community up into pairs of interactions: Who eats whom? Who competes with whom? Who benefits whom? But any community may have hundreds or thousands of species, and the total number of possible interactions greatly exceeds the number of species. Is it possible to understand the dynamics of species in a community in terms of pairwise interactions between species? Although this is a question about communities, a population ecologist might approach it very differently than a community ecologist (Chapter Eight). The focus of a population ecologist is on the individual species.

Whether the population dynamics of communities can be understood in terms of pairwise interactions between species depends on the number of strong interactions relative to the number of weak interactions that can be safely ignored. If relatively few strong interactions dominate the population dynamics, then studying pairwise interactions may not be fruitless. Since the relative frequency of strong vs. weak pairwise interactions between species may vary from one community to another, a first step toward understanding population dynamics of communities of organisms is learning how many pairwise interactions are strong enough to affect the population dynamics of the species involved.

Interactions among species in the rocky intertidal

The strengths of interactions between species are most clearly revealed when a community is subjected to some form of perturbation, either natural or experimental. Tim Wootton used this approach to study the interactions between species occurring in the rocky intertidal area of Tatoosh Island on the Pacific coast of Washington (Wootton 1994). The community consisted of goose barnacles, acorn barnacles, and mussels, all of which are filter-feeders that compete for attachment sites on rocky surfaces; *Nucella* snails that are predators of barnacles; the starfish *Leptasterias hexactis* that eats *Nucella* snails and acorn barnacles; and three bird species (glaucous-winged gulls, black oyster-catchers, and northwestern crows) that eat everything else except acorn barnacles (Plate 7.3). Figure 7.14A gives a web of possible interactions among species.

In an initial set of experiments, Wootton perturbed the community by eliminating bird predation. Using ten study plots, he covered half with mesh cages that he attached to the rock. After one and two years, he censused both covered and exposed plots to determine the relative densities of goose barnacles, acorn barnacles, mussels, and *Nucella* snails. In covered plots without bird predation, the density of goose barnacles increased relative to the exposed plots, while the densities of the other three groups of species decreased. Why did this occur? Three possible explanations are consistent with the observed changes in the densities of species:

- *Hypothesis One* (H1). The main effect of birds could be on goose barnacles (Figure 7.14B). According to this hypothesis, elimination of bird predation allowed the density of goose barnacles to increase, and this in turn reduced the densities of acorn barnacles and mussels through competition for attachment sites on the rock surface. The decrease in *Nucella* snails then occurred because of the decrease in acorn barnacles which are their prey. This hypothesis supposes that acorn barnacles are more important for *Nucella* snails than goose barnacles. Finally, the increase in starfish density with the elimination of bird predation had no effect on the other species.
- *Hypothesis Two* (H2). The main effect of birds could be through predation on starfish, which in turn have their main effect by feeding on *Nucella* snails (Figure 7.14C). According to this hypothesis, the removal of bird predation allows the population of starfish to increase, causing a decrease in *Nucella* snails. This in turn reduces Nucella predation on goose barnacles, allowing the goose barnacle population to increase. Finally, the increase in goose barnacles causes decreases in acorn barnacles and mussels through competition for attachment sites.
- *Hypothesis Three* (H3). As in the second hypothesis, the main effect of birds could be through predation on starfish, but the main effect of starfish could be on acorn barnacles (Figure 7.14D). In this case, the increase in starfish density decreases the density of acorn barnacles. With fewer prey, this decreases the density of *Nucella* snails, and with the reduction in *Nucella* predation, the population density of goose barnacles increases. Finally, the increase in goose barnacles decreases the density of mussels through competition.

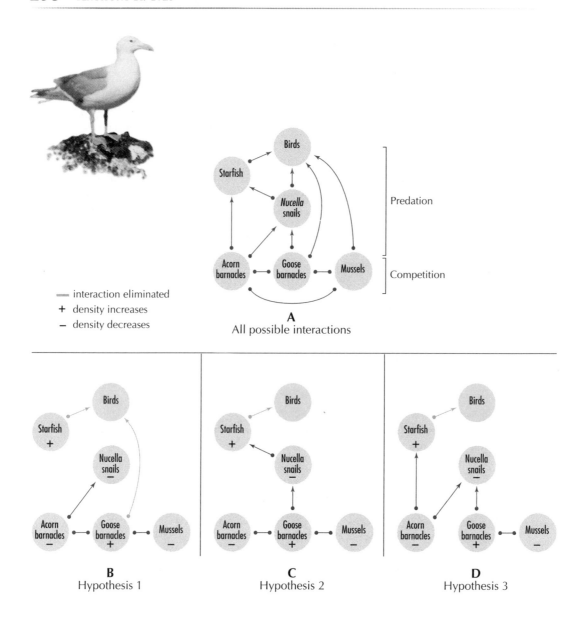

— interaction eliminated
+ density increases
— density decreases

A
All possible interactions

B
Hypothesis 1

C
Hypothesis 2

D
Hypothesis 3

Figure 7.14. Interaction webs for species in the rocky intertidal of Tatoosh Island (Wootton 1994). **A.** Total possible interactions among species. The lines connecting the circles depict the type on interaction. A colored line with two circles on each end depicts competition, in which both species have a negative effect on each other. Predation is shown by a line with a circle on the prey end (since the predator has a negative effect on the prey) and an arrow on the predator end. **B–D.** Different hypotheses about the strengths of interactions among species that explain the experimental results when bird predation was eliminated. In the experiment, predation by birds was removed, indicated by the gray lines. The resulting increases and decreases in population densities of the other species are shown by pluses and minuses, respectively. Each hypothetical pattern of interactions shown in **B–D** is consistent with the observed changes in population densities following elimination of bird predation.

To determine which of these hypotheses is more likely to be correct, Wootton conducted a set of additional experiments to investigate which interactions between species were the strongest. He conducted three separate sets of experiments, manipulating the densities of *Nucella* snails, acorn barnacles, and goose barnacles, respectively. Manipulating *Nucella* snails made it possible to determine whether they have a greater effect on goose barnacles than on acorn barnacles (H1 vs. H2). Manipulating acorn barnacles showed whether they have a negative effect on goose barnacles (H1 and H2 vs. H3) and a positive effect on *Nucella* snails (H1 and H3 vs. H2). Manipulating goose barnacles revealed whether they have a negative effect on acorn barnacles (H1 and H2 vs. H3) and a positive effect on *Nucella* snails (H1 vs. H2 and H3). By analyzing the strengths of interactions revealed by these experiments, Wootton found that Hypothesis 1 (Figure 7.14B) was best supported by the data.

This study illustrates the complexities involved in understanding how entire communities respond to perturbations. Even though the community was simple, consisting of only six groups of species, the strengths of many pairwise interactions between species needed to be known before the consequences of a perturbation, elimination of bird predation, could be understood. For larger, more complex groups of species, the challenges are daunting, but the rewards are appealing.

What Is the Relationship between Community Diversity and Stability?

The final question of this section has dogged ecology since its inception: Are more diverse communities more stable? This question has clear relevance to a plethora of environmental issues. If more diverse systems are more stable, this gives a compelling reason to preserve ecological diversity. With the loss of species one by one from a community, the community would become increasingly vulnerable to environmental perturbations. Borrowing Paul and Anne Ehrlich's analogy of rivets on an airplane (Ehrlich and Ehrlich 1981), with the removal of rivets one at a time, the plane will continue to fly, but it will become increasingly vulnerable to the stresses produced by turbulence. Eventually, and inevitably, enough rivets will be lost so that even a little turbulence will rip the plane apart. According to this view, the effects of the loss of species might not be seen immediately, but as more species are lost from a community, the community becomes ever more vulnerable to collapse.

Diversity and stability

Despite the general perception of the importance of diversity, scientists have not proven that more diverse communities are in fact always more stable. A major problem in addressing this issue is devising concepts of diversity and stability that are relevant for assessing the vulnerability of communities to environmental perturbations. The simplest attribute of **diversity** is the number of species in a community; this is known as

species richness. Species richness counts each species equally, regardless of how abundant each species is relative to the other species. Thus, for example, the species richness of a community containing ten species, with 1,000 individuals of each species, is the same as the species richness of a community with ten species in which one species is represented by 9,910 individuals while the other nine species have only ten individuals. These two communities are quite different, and several mathematical formulae for diversity account for the **evenness** of the number of individuals among different species. In these formulae, communities in which all species are equally abundant—communities with greater evenness—have greater diversity. Another factor that might be included in figuring the diversity of a community is the number of different types of organisms present. For example, a community containing grasses, annual and perennial herbs, shrubs, and trees might be identified as having greater diversity than a community containing only grasses, even though the number of species and the abundances of species are the same in the two communities. Although it seems natural to include the variety in the different types of organisms when measuring the diversity of communities, it is unclear exactly how to do this.

Devising a relevant concept for the stability of communities is equally difficult—if not more so. Throughout this chapter, stability has been used in the context of the population dynamics of individual species. For a deterministic single-species model, an equilibrium point is stable if, when population densities are moved away from the point, they then move back towards the equilibrium point through time. Similarly, a population cycle is stable if population densities are attracted to the cycle through time. The idea of stability can be applied to measurements other than the densities of individual species. For example, in the case of two similar competing species, one could ask whether the sum of their densities has a stable equilibrium point. Community composition might also be stable or unstable. If the community composition is stable, then the same species remain in the community through time, while if community composition is unstable, then different species enter or drop out of the community. Even when community composition is unstable, the number of species in the community (species richness) might be stable if the community contains the same number of species through time despite the species being different. Finally, even though the number of species in a community might change, measures of processes within the community, such as net primary productivity or the rate of nutrient cycling, may have stable equilibrium values that are maintained despite major changes in the number and kinds of species (see Chapter Four). The actual meaning of stability is the same in all of these cases, but stability is applied to different attributes of the system, from the densities of individual species to ecosystem-level processes of primary productivity and nutrient cycling.

An additional complication in addressing stability is that *stability* and *resilience* are used interchangeably by some ecologists, while they are given distinct definitions by others. (This chapter uses distinct definitions, defining resilience in deterministic systems as the rate at which population densities approach a stable equilibrium point or stable cycle through time.) Furthermore, many ecologists think of stability and resilience only in terms of deterministic population dynamics, rather than stochastic population dynamics. However, stability and resilience in stochastic systems may be defined in terms of probability distributions rather than equilibrium points and cycles.

The variety of ways in which diversity and stability can be measured make any discussion of the relationship between diversity and stability complicated, because this relationship will depend on the precise definitions of diversity and stability used. Rather than try to disentangle all of the evidence, a brief discussion of different views on the relationship between diversity and stability follows. This discussion enters the interfaces between population, community, and ecosystem ecology, because the concept of stability is often applied to properties other than population densities (see Chapters Four and Eight).

Stability and redundancy in the functions of different species

In 1955, Robert MacArthur (MacArthur 1955) articulated the prevailing view of the relationship between diversity and stability: diversity, taken as the number of species in a community, increases stability by providing parallel redundancy in the functions going on within communities. For example, suppose a community has many species of plants. If one species became extinct, other species would carry on photosynthesis. Or if one of many species of predacious insects went extinct, other species could step in to keep the prey in check. MacArthur emphasized the flow of energy through communities—plants capturing light energy through photosynthesis, and passing this energy to herbivores and then to carnivores that eat the herbivores or each other. The definition of stability applied here is based on the maintenance of the flow of energy through the community. The main conceptual idea is that if many species are doing essentially the same thing, then the loss of any particular species does not change how the community functions.

Diversity and the stability of point equilibria of individual species

The idea that more diverse systems are more stable was shaken by Bob May (1973). May used a different definition of stability, one that applies to population dynamics of individual species in the community. Specifically, May showed mathematically that the chances of having population dynamics with a stable equilibrium point become smaller and smaller as the number of interacting species in a community increases. This concept of stability focuses on the population dynamics of individual species, not on the maintenance of the functional properties of communities. The reason for May's result is that, as the number of species in a community increases, the sum of interaction strengths experienced by any one species increases. As demonstrated earlier in the chapter, strong interactions often (but not always) increase the intrinsic variability of population dynamics, thereby making equilibrium points for all species unstable.

Stability and the identity of particular species in a community

A third conceptual link between diversity and stability was made by Sam McNaughton (1977). McNaughton studied savanna grasses in Kenya that are subject to fire, drought, and grazing by numerous species of ungulates. He found that, following a drought, grass communities with more species returned more rapidly to their pre-drought biomass than communities with fewer species. Thus, the more diverse communities were more resilient. Here, resilience was measured in terms of the combined biomass of all plant species (i.e., the total dry weight of all plant material). He attributed the increased resilience of more diverse communities to the presence of several species of grass that were particularly resistant to the drought. The role of diversity in maintaining biomass resilience is simply to increase the chances that particular species which are resistant to perturbations are present in the community. The greater the number of species, the greater the chance that a species with some important, but possibly unanticipated, attribute will be present.

Compensation and the stability of combined species abundances

Finally, recent work by David Tilman (1996) emphasized the comparison between changes in the densities of individual species and changes in their combined densities. Tilman was conducting studies on prairie grasses. Fortuitously, in 1988 his field site experienced the worst drought in 100 years. Experimental plots that he had previously established differed widely in the number of species they contained. Tilman found that during and after the drought, those plots with the most species experienced smaller changes in total plant biomass. In addition, the changes in biomass of species taken separately was greater than the change in biomass of all species taken together. In other words, the combined biomass of plants in more diverse plots (measured by species richness) changed less, because changes that occurred in the biomasses of individual species tended to cancel each other out; the decrease in biomass of one species was at least partially compensated by increases in the biomass of other species. This concept of resilience of individual species vs. resilience of aggregates of species in part reconciles the contrast between the views of MacArthur and May. Even though greater diversity may decrease the resilience/stability of the population dynamics of individual species, it may increase the resilience/stability of aggregates of species, if changes in the densities of some species are compensated by changes in the densities of other species.

Summary

The issue of the relationship between diversity and stability is hardly resolved. The variety of different ways in which ecologists use the terms diversity and stability create not one relationship, but a whole suite of relationships between different pairs of definitions of diversity and stability. The appropriate pair of definitions will change from one problem to another. For example, when studying ways to control agricultural pests, one wants the pest population to be remain stable at a density below which crop damage occurs. Thus, the focus is on individual population dynamics. On the other hand, when studying algal blooms in lakes, what happens if there is a sudden influx of phosphorus? If herbivory on the algae by zooplankton is heavy, then an algal bloom will not occur despite the influx of nutrients. In this case, the stability and resilience of the system are gauged by nutrient fluxes and population sizes of entire groups of species, such as all algae or all zooplankton (see Chapter Four). As an added complication, stability in response to nutrient influxes does not guarantee stability in response to other perturbations, such as acidification through acid rain. In short, there is no satisfactory simple summary for the relationship between diversity and stability, because diversity and stability can be used to describe a variety of properties of real ecological systems.

THE ROLE OF EVOLUTION IN POPULATION ECOLOGY

The link between evolution and population ecology is two-directional. **Evolution** is responsible for characteristics of organisms that influence how they respond to environmental changes, how they use resources, and how they interact with other individuals of their own or another species. Evolution molds ecological interactions. Conversely, ecological interactions that drive population dynamics do so by changing reproduction and mortality rates. Reproduction and mortality of an individual organism determine its fitness under natural selection. Therefore, the same ecological interactions that drive population dynamics also drive evolution. Evolution in the past sets the stage for ecological interactions, and as ecological interactions play out, they determine the future direction of evolution.

Very often, ecological changes in population densities occur more rapidly than evolutionary changes in the genetic make-up of populations, and ecologists therefore often ignore evolutionary changes. However, evolutionary changes have occasionally occurred quickly enough to have noticeable effects on ecological interactions. Possibly the best example of this occurred during the release of the myxoma virus to control rabbit populations in Australia (Fenner and Myers 1978).

Following the introduction of European rabbits to Australia in 1859, and their subsequent population explosion, the Australian government tried various control tactics. The most successful was the introduction of rabbit myxoma virus from England in 1950. Myxoma virus is transmitted among rabbits by mosquitoes, and the particular strain of virus released into Australia was so lethal that it could kill more than 99.8% of infected rabbits. Following release, this strain of myxoma virus swept through much of Australia. After the epidemic ended, the rabbit populations started to rise once again, only to be swept by another myxomatosis epidemic. But this time only 90% of the rabbits were killed. During the third epidemic, the myxoma virus killed only 40–60% of rabbits in infected populations. This pattern continued to occur, with each successive epidemic killing a smaller proportion of rabbits. Thus, the effect of the myxoma epidemics on rabbit populations became weaker through time.

Two things caused the lessening of effect of myxoma epidemics on rabbit populations. First, the rabbits became more resistant to myxoma infection. A standard strain of myxoma virus was used to infect test rabbits from a population after each successive epidemic. As the number of epidemics experienced by the population increased, the rabbits become more resistant to the standard myxoma strain. Second, the virus itself became less virulent. Samples of the virus were analyzed in 1958 and 1963 by infecting standard laboratory rabbits. While the virus that was released killed more than 99.8% of test rabbits, strains of the virus present in 1958 killed roughly 90%, and strains present in 1963 killed roughly 80%. In addition, the time it took for the virus to kill the test rabbits changed. The strain first released killed test rabbits in less than 13 days on average, while in 1958 and 1963 the average times were 22 days and 28 days, respectively.

Evolution of resistance to the myxoma virus in rabbit populations is not surprising. Those rabbits that survived each successive epidemic were those with greater than average resistance. This greater resistance was inherited by their offspring, leading to a change in the genetic make-up of the rabbit population—i.e., evolution. Evolution of reduced virulence in the myxoma virus, however, is at first more puzzling. To understand this, remember that spreading from one rabbit to another is essential for the virus. The virus is vectored by mosquitoes. Therefore, to be spread from one rabbit to another, there must be time between the virus's build-up in the blood of the infected rabbit and the death of the rabbit, during which the virus can be picked up by a mosquito. If the virus is too virulent, the rabbit will die before it is picked up by a mosquito. Less virulent strains of the virus are more likely to spread, because they kill the rabbit more slowly. Since they are more likely to spread, they will increase through time and eventually become the most prevalent strains in the virus population. Thus, there is evolution for reduced virulence.

This striking example of evolution of parasite and host, and the impact of evolution on the ecological interaction between them, is largely due to two unusual circumstances. First, the myxoma virus was introduced into a rabbit population that had not experienced the virus for a century. Consequently, initial natural resistance in the rabbit population was very low. Second, the strain of virus introduced was chosen by scientists to be particularly virulent. These two circumstances meant that the parasite-host interaction was out of balance when the parasite was introduced. In general, however, evolutionary biologists assume that species are molded by the continuous, or near continuous, pressure of natural selection. Therefore, unless an anomalous event occurs, evolution holds the characteristics of individuals in a population static.

Of course, humans are particularly active in producing anomalous events. They introduce new species onto islands and continents, thereby subjecting those species to new selective forces and subjecting the native species to the selective forces produced by the introduced species. Humans cause all kinds of environmental changes by introducing toxins into the air and water, and by using pesticides in the explicit attempt to influence the population dynamics of pest species. Even the global climate is influenced by human activity. In the face of human activities that have such sudden and diverse effects on the environment, evolutionary changes may play an ever-increasing role in the population dynamics of species.

THE ROLE OF MODELS IN POPULATION ECOLOGY

In population ecology, as in other types of ecology such as ecosystem ecology (Chapter Four), mathematical models are used for three different purposes that have been illustrated by the different models presented throughout this chapter.

- *Qualitative understanding:* Models can be used to obtain a qualitative understanding of the types of population dynamics that can arise through interactions among species and the environment. Most of the models used in this chapter fall under this category. For example, the deterministic and stochastic models for a single species were used to illustrate how density-dependent population growth rates influence population dynamics. The simple model for a parasitoid and its prey is another example of a model used to understand how properties of population dynamics, in this case cycles, can be produced through interactions between species.
- *Quantitative predictions:* Models can be used to make quantitative predictions about changes in the abundance or distribution of species. An example is the model used to predict the population growth rate of northern spotted owls. Although this particular model did not include density dependence and therefore cannot be used to predict long-term changes in population density, it nonetheless gives a short-term prediction for the northern spotted owl population.
- *Analysis of data:* Models can be used to analyze observational and experimental data. For example, estimating the effect of aggregation on competition among carrion flies used a simple statistical model to calculate how aggregation changed the average density of competitors experienced by maggots within carcasses. This model can be extended to produce a mathematical formula that takes data directly from experiments, such as the data illustrated in Figure 7.12, and makes quantitative assessments of how much aggregation is needed for coexistence. Using models for data analysis is relatively rare in population ecology, although it holds huge potential. The goal is to have models that guide experimental research, and simultaneously have experiments that guide the development of models.

In ecology, mathematical models are not designed to be perfect representations of the real world, in the sense of incorporating all ecological processes that affect populations. Any such model would be too complex both to construct and to analyze. Mathematical models by definition are simplifications of reality, and the degree of simplification depends mainly on the model's purpose. The simple, single-species model used in this chapter is appropriate to give a qualitative understanding of population dynamics, but it is too simple to make quantitative predictions about the abundance of a species next year. To make quantitative predictions, more complexity is needed. The art of ecological modeling is not figuring out how to analyze models. Instead, the art is knowing what level of complexity in a model is appropriate—what needs to be included and what can really safely be left out. In general, the more simple the model for any given objective, the better, because a simple model more clearly reveals the reasons for its results.

Some ecologists complain that models in general are not useful because they are not realistic. However, models provide a wonderful tool to ask whether their degree of realism is appropriate. Building a series of models of increasing complexity often helps decide what processes must be included and what processes can be left out without influencing the conclusions. When applied in conjunction with experimental research, models can distinguish which processes are essential and which are inconsequential, and therefore can focus experimental research.

TECHNIQUES USED IN POPULATION ECOLOGY

Population ecology is ultimately based on counting. To characterize a population, we need to know its density, and how its density changes through time. The difficulty of counting organisms is probably the single greatest limitation of population ecology. Typically, no attempt is made to count all the individuals in a population. Instead, a subset of the population is sampled, and from this subsample the density of the entire population is estimated. Even sampling subsets of populations is generally arduous, particularly if you want to know how the density of the population varies from one location to another. Since population dynamics involve changes in the population density through time, sampling needs to be repeated many times, sometimes for years. Data sets exceeding more than a decade are extremely rare in population ecology. When one considers that many organisms have life spans exceeding a decade, the limited amount of data on population dynamics becomes obvious.

Rather than rely solely on accumulating data on population dynamics, experimentation provides a powerful tool with which to examine interactions within and among populations. Experiments can be used to measure the effects of intra- and interspecific competition, predation, mutualism, or anything else on population growth rates. By necessity, many experiments are short in duration and small in spatial extent. However, the ability to vary simultaneously population densities, resource abundances, and/or environmental factors can be used to investigate properties of ecological interactions that are unapproachable by non-experimental means. Because experiments can be designed to vary only one factor, for example the density of one species or the availability of one resource, they can be used to pin-point unambiguously the roles of different components of ecological systems. Since experiments are repeatable, it is possible to disentangle results due to the intentional experimental manipulations from those that arise by chance.

Some ecologists are very careful to make their experiments mimic reality as closely as possible. They argue that, unless the experiment is realistic, it tells nothing about ecology. The converse view is that experiments are by definition unrealistic, and that much more can be learned about ecology by intentionally making experiments unnatural. For example, what happens when the density of a population is increased to ten times its normal level? Interpreting any experiment done in this fashion must be done carefully. Nevertheless, "what if" scenarios can be very useful in understanding why the scenarios do not in fact occur in nature. It is probably best to acknowledge up front that any experiment is unrealistic, and then explicitly consider how the unrealism affects the results.

APPLICATIONS OF POPULATION ECOLOGY

Because population ecology asks why populations increase and decrease in size, and what factors can lead to their extinction, population ecology applies directly to the conservation of threatened species such as the northern spotted owl and the Karner blue butterfly. Another critical issue for conservation is the spread of non-native species among continents and onto islands that were formerly isolated from the biota of the rest of the world. Countless extinctions of species have occurred by the accidental or intentional introduction of species by humans. For example, of at least 84 species of birds native only to Hawai`i, today only 44 survive (Scott et al. 1988), and of the surviving species 18 are classified as endangered or threatened under the United States Endangered Species Act. Although habitat destruction has probably played some role in the loss of Hawaiian bird species, introductions of non-native species have been more devastating. Species introduced by Polynesians and, more recently, Europeans, include mongooses, rats, feral cats that prey on birds and bird eggs, and pigs and cattle that destroy bird habitat. Two introduced diseases, avian pox and avian malaria, have also taken a huge toll on the native species. Currently, the native birds are confined to elevations above 600 m, which is above the elevational range of the mosquito which vectors avian malaria. Understanding how non-native species are able to invade novel communities and proceed to drive native species to extinction through competition or predation is a problem of population ecology.

Population ecology also has many applications in agriculture and forestry, where the goal is to control pest species. Biological control, the use of natural enemies to control pests, is widespread. The first use of biological control was in the citrus industry of

California. In the 1880s, the cottony cushion scale, an insect that damages both fruit and trees, threatened to destroy the industry. In 1888, the vedalia, a predatory ladybird beetle, was introduced from Australia, and its population exploded in response to the abundance of scale insect prey. Sustenance of the citrus industry in California is largely credited to the vedalia's suppression of the cottony cushion scale. This was only the first example of successful biological control. As of 1988 (Waage and Greathead 1988), a total of 534 biological control programs had been successful in introducing insect predators and herbivores to control insect and plant pests. Very often these biological control programs are unsung successes, because the introduced species controls the pest without further input from scientists and farmers. In addition to these cases of biological control through introduced species, native natural enemies play critical roles in controlling pest species without any input from scientists and farmers. With growing concerns over the environmental and healthy risks of chemical pesticides, and with the growing incidence of pests evolving resistance to chemical pesticides, biological control is an extremely attractive alternative.

Population ecology has also been used to regulate commercial harvesting of natural populations. Some of the best examples involve fisheries management (Hilborn and Walters 1992). Understanding factors governing fish population growth rates allows estimation how many fish can be harvested each year while maintaining the fish stock for future harvests. To manage fisheries, it is necessary to know how fish population growth rates depend on density. If a moderate reduction in fish density through harvesting produces a large increase in the population growth rate, then fish can be harvested with the expectation that the population can be maintained. However, over-fishing can produce such reductions in fish density that the population cannot recover, or can only recover very slowly. This has happened in many fisheries throughout the world, such as the Peruvian anchovy fishery and the cod fishery on the Georges and Grand Banks off the Atlantic Coast of North America. The only way to prevent further examples of crashes in fisheries is to use good ecological science to develop strategies for managing fish harvesting, and then enforce these strategies with international fishing regulations.

LINKS TO OTHER KINDS OF ECOLOGY

Population ecology has strong links to the other kinds of ecology presented in this book. Because population ecology is based on the idea that population dynamics can be explained in terms of interactions among individuals and interactions between individuals and the environment, behavioral and physiological ecology play central roles. For example, understanding the efficiency of predators as they search for prey, and how their efficiency changes with prey density, overlaps the realm of behavioral ecology. However, a behavioral ecologist and a population ecologist might approach the study of a predator's foraging behavior in different ways. When trying to understand why predators behave the way they do from an evolutionary perspective, a behavioral ecologist might want to know the costs and benefits of different foraging behaviors. A population ecologist, on the other hand, might not be as interested in explaining why predators forage the way they do, but instead be interested in explaining the consequences of the foraging behavior on interactions between predator and prey populations. Given these different objectives, behavioral and population ecologists might design experiments on foraging behavior differently.

Physiological ecology is the study of interactions between individual organisms and the environment. Earlier in the chapter, the discussion of extrinsic factors that affect population dynamics largely glossed over what these extrinsic factors are and how they affect reproduction and mortality. Extrinsic factors are clearly important to understanding population ecology, and population ecologists typically spend a lot of time trying to figure out what the important extrinsic factors are. However, investigating the actual mechanism by which an extrinsic factor affects individual organisms falls to the physiological ecologist.

The link between population and community ecology is seen, for example, in the discussion of interactions among more than two species. Population ecologists often think of communities of interacting species in terms of predation, competition, and mutualism—interactions between pairs of species. Almost on faith, many population ecologists assume that understanding the dynamics of populations embedded in a community of interacting species is possible through studying pairwise interactions between species. This faith rests on the hope that interactions within communities are dominated by a limited number of strong interactions that are primarily responsible for the observed densities and dynamics of populations. Community ecologists, on the other hand, often focus on "emergent properties" of communities, properties that involve the aggregate of all species, rather than species taken separately. The tension between these two perspectives is currently generating a lot of interesting new research and has heightened interest in "community dynamics."

Landscape ecology and population ecology overlap extensively on issues involving the spatial distribution of organisms. Many of the same ideas, such as metapopulation dynamics, are important in both fields. But surprisingly, landscape and population ecologists don't often collaborate with each other, and there is as yet precious little cross-fertilization of ideas. Landscape and population ecology approaches to spatial problems have different traditions, with landscape ecology oriented more towards ecosystem management and population ecology more towards the management of individual species. Furthermore, landscape ecology relies more heavily on complex computer models than does population ecology, which is dominated by a pencil-and-paper style of modeling. The current schism between landscape and population ecology, however, is beginning to be bridged, and both fields are bound to benefit.

Ecosystem ecology might at first seem to be the most distant field from population ecology. Ecosystem ecology deals with nutrients and energy, while population ecology deals with densities of populations. Nevertheless, ties between ecosystem and population ecology are growing, fueled largely by overlap in objectives rather than overlap in perspectives. Many of the environmental problems today involve both ecosystem functioning and how the species within the ecosystems respond to environmental assaults. Ecosystem ecologists are learning that the identities of species do matter in understanding how ecosystem function changes in response to environmental perturbations. Conversely, population ecologists are learning that the way species respond to environmental perturbations can only be understood when nutrient and energy flows are taken into account.

Finally, understanding historical roots is important for any scientific field. Many of the debates in population ecology today began 50 years ago, as discussed in Chapter Two. Have we then learned nothing in 50 years? In fact, far more information and understanding is brought to bear in today's debates. That the same issues are being debated is a testament to how challenging ecological problems really are—many of which are highlighted by the numerous unanswered questions showcased throughout this book.

SUMMARY

Population ecology asks how interactions among individual organisms and between organisms and the environment affect the densities and dynamics of populations. Questions asked by population ecologists include:

- What regulates the population of a particular species?
- Can you predict changes in the density of a species, and whether the species is likely to go extinct?
- Why don't predators kill all their prey?
- Why don't superior competitors exclude inferior competitors from their domains?
- What would be the effect of removing one species from a community on the population densities of all the remaining species?

These questions are phrased abstractly in terms of such concepts as population regulation, predation, and competition. But population ecology is most compelling when these questions are applied to particular species. Is the northern spotted owl bound for extinction if logging practices are not changed? Can a predator be made more effective in controlling scale insects that attack orange trees in California? What would happen to the rocky intertidal community if starfish populations on the Pacific coast of the North America collapsed?

Population ecology has a strong tradition of application to important ecological and environmental problems, and applied research is becoming increasingly critical in today's world of endangered species and environmental uncertainties. Although this chapter has focused on questions that are set squarely within the field of population ecology, many ecological and environmental problems require the joint talents of researchers from all ecological fields, as well as scientists from other disciplines.

RECOMMENDED READING

MICHAEL BEGON, JOHN L. HARPER AND COLIN R. TOWNSEND. 1990. *Ecology: Individuals, Populations and Communities.* Blackwell Scientific Publications, Boston, Massachusetts.
>   *The author's favorite of the main comprehensive ecology texts available today.*

CHARLES J. KREBS. 1994. Ecology: *The Experimental Analysis of Distribution and Abundance.* Harper-Collins College Publishers, New York, New York.
>   *A good comprehensive ecology text.*

Robert E. Ricklefs. 1990. *Ecology.* W. H. Freeman and Company, New York, New York.
>   *Another good comprehensive ecology text.*

LEAH EDELSTEIN-KESHET. 1986. *Mathematical Models in Biology.* Random House, New York, New York.
>   *An excellent treatment of the more theoretical side of population ecology.*

NICK J. GOTELLI. 1995. *A Primer of Ecology.* Sinauer Associates, Sunderland, Massachusetts.
>   *A shorter and more elementary treatment of the more theoretical side of population ecology.*

NAOMI CAPPUCCINO, AND P. W. PRICE. 1995. Population Dynamics: *New Approaches and Synthesis.* Academic Press, San Diego.
>   *This is an excellent compilation of chapters by different authors on population dynamics, with most chapters emphasizing experimental approaches to population ecology.*

DAVID ATTENBOROUGH. 1995. *The Private Life of Plants: A Natural History of Plant Behaviour.* Princeton University Press, Princeton, New Jersey.
>   *An excellent and entertaining book on the natural history of plants.*

HOWARD E. EVAN. 1985. *The Pleasures of Entomology: Portraits of Insects and the People Who Study Them.* Smithsonian Institution Press, Washington, D.C.
>   *Of the many excellent and entertaining books on natural history, this is one of the author's favorites*

DAVID QUAMMEN. 1988. *The Flight of the Iguana: a Sidelong View of Science and Nature.* Doubleday, New York, New York.
>   *Good ecology must start with good natural history, and this is one of the author's favorite books on the subject.*

APPENDIX 7.1. CHAOS

In an ecological context, chaos was first discovered in models like the one used to generate Figure 7.5; models for organisms that have discrete, non-overlapping generations, as do insects with one generation per year. Although this is the simplest type of model in which chaos occurs, chaos actually arises more frequently in more complex models. For models with discrete, non-overlapping generations, chaos requires very strong density dependence, and is produced when potential exists for explosive population growth from low densities and for catastrophic population crashes from high densities. This type of population dynamic is ecologically extreme. For more complex models with many interacting species, however, chaos may arise under ecologically more realistic levels of density dependence. Since ecological systems hold potential for chaos, it is worth additional discussion.

In the simple model illustrated in Figure 7.5, the route by which the population dynamics become chaotic is through a set of cycles with ever-doubling periods (May and Oster 1976). Thus, when r = 1.8 (Figure 7.5C) the equilibrium point K is stable, but the population density approaches K alternating between values above and below K. When r = 2.3 (Figure 7.5D), the alternation between values above and below K becomes permanent, leading to a stable two-point cycle. This two-point cycle gives way to a four-point stable cycle when r = 2.6 (Figure 7.5E). Although not shown in the figures, when r equals approximately 2.7 the four-point cycle gives way to an eight-point stable cycle, and an additional increase in r leads to a 16-point cycle.

This pattern of period doubling, in which as r increases, stable cycles give way to stable cycles with double the period, can be illustrated by plotting the values of the stable cycles produced by the model against the value of r (see Figure A7.1). This figure was constructed by selecting a value of r for the model and then using an initial density of $x_0 = 0.9$ to calculate the population density in 500 years. Five hundred years is long enough so that if a stable equilibrium point or stable cycle exists, the population dynamics will be at this point or cycle. In the figure, the next 128 population densities (years 501 to 629) are plotted against the value of r. Therefore, if there is a stable equilibrium point, all 128 values will be the same (i.e., K). If there is a stable two-point cycle, then the population density will have two values. If there is a 32-point cycle, then the population density will have 32 points. The number of points in the figure for each value of r gives the period of the stable cycles.

The box figure shows a pattern of period doubling. As r increases above 2.6, the rate at which period doubling occurs continuously increases until the population densities are smeared across a range of values. What has happened is that period doubling—2, 4, 8, 16, 32, 64, 128, 256, 512, … —has eventually produced cycles with infinite period. In other words, the pattern of successive population densities through time never repeats itself. This is chaos. As r increases to about 3.1, chaotic dynamics give way to three-point cycles. These three-point cycles then go through a set of period doubling, with increas-

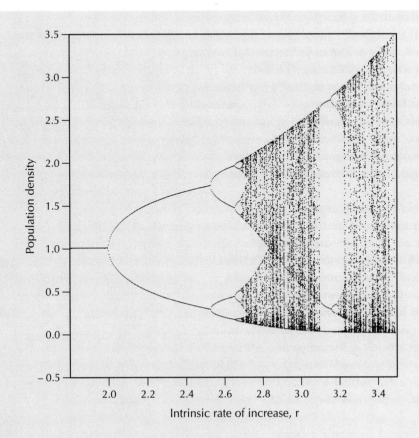

Figure A7.1. Period doubling for the model $x_{t+1} = x_t\, e^{r(1-x_t/K)}$. To construct this figure, the value of r was first selected, and the model was used to generate 500 years of data. Then, the population densities from year 501 to 629 were plotted vs. the value of r.

ing r leading to 6, 12, 24, 48, ... period cycles. Again chaotic dynamics are reached when period doubling of the 3-point cycles leads to cycles with infinite period. Although not shown in the figure, after this region of chaos, cycles of period 5 appear and go through a set of period doubling again leading to chaos. This pattern continues to repeat itself, going through all of the prime numbers, 7, 11, 13, etc.

A mathematically intriguing property of chaos is that this general pattern of period-doubling occurs for any model of single-species population dynamics, provided density dependence can be made strong enough. The main requirement for a model to produce chaos is that the peak in $f(x_t)$ can be made very high; i.e., the model has a large hump.

The most important mathematical attribute of chaos is the absence of any stable cycle in the population dynamics. This means that if you were to start two populations

Table A7.1. Dynamics of four-point cycles and chaos

Four-point cycle

time	density of population 1	density of population 2	difference between densities
0	0.4	0.45	−0.05
1	1.90	1.88	0.023
2	0.18	0.19	−0.0089
3	1.52	1.56	−0.038
4	0.39	0.36	0.028
5	1.90	1.90	0.0035
6	0.18	0.18	−0.0013

Chaos

time	density of population 1	density of population 2	difference between densities
0	0.4	0.41	−0.01
1	2.28	2.27	0.0098
2	0.056	0.057	−0.0014
3	0.86	0.88	−0.018
4	1.28	1.24	0.039
5	0.56	0.61	−0.048
6	2.00	1.89	0.11

governed by the same function $f(x_t)$ at very close but non-identical values of x_t, then through time the densities of the two populations would get farther and farther apart. This can be illustrated by comparing the dynamics of four-point cycles with chaos (Table A7.1). This example shows that the population dynamics in chaotic systems are sensitive to the initial population density. In other words, your ability to predict what the population density will be in 10 or 100 years depends on how accurately you know the population density today. Even if you know the population density to within 0.001% of its true value, you will still be unable to predict the population density in 100 years. This is true even though you know the exact form of the function $f(x_t)$, and even though there is no extrinsic variability.

But is this unpredictable nature of chaos important for real populations? No. Ecological systems have built-in unpredictability due to extrinsic factors that affect population dynamics. When taking account of stochastic extrinsic factors, chaotic population dynamics do not look much different from non-chaotic population dynamics. Chaos in stochastic systems can still occur; as in deterministic systems, it is defined in terms of the property that long-term population dynamics are sensitive to initial population densities, with the proviso that two populations with slightly different initial densities experience exactly the same pattern of stochastic extrinsic factors. As in deterministic models, chaos does increase the variability in population dynamics in stochastic models. However, creating unpredictability is not unique to chaos, since unpredictability is also created by extrinsic factors. Therefore, the fact that chaos produces cycles of infinite period is an interesting mathematical phenomenon, but this is not particularly interesting for ecologists. The important lesson for ecologists is that strong density dependence can lead to high levels of intrinsic population variability. This lesson applies for any increase in r above r = 1.0, not just for increases in r great enough to produce chaos.

Chapter 8

COMMUNITY ECOLOGY

The Issue At The Center

T. F. H. Allen

No two stands [of vegetation] can be identical, if sufficient detail is taken into account Moreover, they may differ in so many characters that is it impossible to take all of these into consideration. A subjective selection must, therefore, be made of the criteria to be used in characterization and comparison of stands.
—*P. Greig-Smith*

WHAT IS COMMUNITY ECOLOGY?

Look at a tract of vegetation or an assemblage of animals and we see patterns. The patterns are likely to involve a multiplicity of species, across a range of environmental conditions. Furthermore, there will be small-, medium- and large-sized organisms, and an array of species each with its own patterns in average size, mobility and longevity. Pattern abounds, but often we cannot quite put our finger on it. And on top of all this is an enormous number of reasonable explanations for what we see.

The challenge in community ecology is to avoid compromise on the richness of the system in convenient but sweeping simplifications. Often feeling that they must deal with too many things that interact in too many ways, community ecologists study nature as she presents her complexity on a country walk, or on an expedition to exotic places. Community ecology addresses the full mix of species and their interactions as we might find them in a woodlot or hedgerow. Communities invoke so many different types of relationships between their parts, and they occur across such ranges of environmental variation, that simple rules to capture community behavior do not work. Because of the heterogeneity inside communities, students of them must deal with fairly long chains of causality to account for what they see. In community ecology, it does not suffice to ignore all species except one or two, as in population biology. In community ecology, hidden physiological processes are secondary, as are unseen nutrient cycles. Because of its frontal assault on complexity, community ecology must use its own style of data collection, analysis and model building that stands in contrast to the methods employed over the rest of ecology.

In community ecology, one wishes to capture the various relationships between a large number of species and see the outcome in system behavior. Unfortunately, it is not possible to write helpful equations that link together more than a few species at a time. The limitation comes from the interaction of different reaction rates, one set for each species. For example, when a prey item is taken it dies immediately, whereas there is a time lag before the predator who eats the prey derives benefit; the prey population goes down immediately, whereas the predator population takes time to grow. A system of equations that links too many long lags will appear chaotic, in both the technical and commonplace meaning of the word **chaos** (see Chapter Seven). Community equations involving all the species in the community as separate entities will be so long and involved, that the reasons for the behavior the equations exhibit become as obscure as the system they represent.

Not all the difficulties with communities come from species interaction, i.e., from the biotic side of ecology. On the abiotic side there is environmental variation that changes the carrying capacity to which the species respond. To the extent populations occur in various habitats, those differences can be studied systematically by experimental design. The assumption is that environmental variation is orderly enough for the population to respond as a unit. All science is about making assumptions, so that need not be a problem, but this one is often hidden. In this case, assumptions which work so well for populations are of little use in accounting for communities. This is because community ecologists insist on giving account to more realistic complications of patchy environmental changes that may influence each species differently. In communities, we are interested precisely in how those assumptions of unified behaviors of populations fail.

Definitions

The definition of the term **community** is far from obvious. Several definitions have currency, some more deserving of attention than others. We present four ways to define communities, less as formal definitions per se, but more as classes of approach to communities:

- The community as a group of species.
- The concrete versus abstract community.
- The community as a collection of populations.
- The community as interaction through periodic occupancy of space.

The community as a local group of species

Here the focus is on collections of species in a place. The advantage of this definition is that it corresponds to the commonplace experience of sitting on a stump in the woods, or having a picnic in a prairie. However, this definition is not very useful for scientific purposes, because it treats the community as a mere collection of bits with little coherence or organization between them. The community as a thing in a place is a landscape-defined entity. This conception focuses on community through example, rather than the general class from which the example is taken.

The concrete versus abstract community

In his 1923 paper, G. E. Nichols refers to the above definition as he describes the use of the term association. "By some its use has been restricted to the concrete individual pieces of vegetation which we study in the field… By others these individual concrete communities have been viewed as merely 'examples' of an association." Wrestling with definitions of the plant "association" (read community), Nichols in October 1921 circulated a letter to some 85 members of the Ecological Society of America. In that letter, he recommended "That the term Plant Association be recognized as applicable both to the abstract vegetation concept and the concrete individual pieces of vegetation on which the concept is based." The idea of the abstract community occurred first in an unpublished response of Cooper to a questionnaire sent out a month before the circulated October 1921 letter (Nichols, 1923).

Cooper (1926) makes an analogy of communities as a "braided stream." In a braided stream the sediment drops out and causes the present stream channel to become filled. Eventually the stream diverts around the old channel, leaving a sand bar with the stream flowing around it, sometimes on both sides. The braiding comes from meanders going around and cutting through old meanders. The valley fills with sediment, so digging down into the sediment reveals the history of various channels active across the valley at any given time in the past. (Figure 8.1).

The analogy of the braided stream indicates that, much as water in a stream flows downhill, so the concrete examples of a community flow through time. The point of tension here is captured by the observation of Greek philosopher Heraclitus, "You cannot step into the same river twice." Heraclitus means that it is not the same material water when you step in the second time, even if there is water flowing in the same place. Vegetation in a place is ephemeral, in that organisms and species come and go as time passes. To borrow from Heraclitus, "You cannot go to see the same concrete community twice."

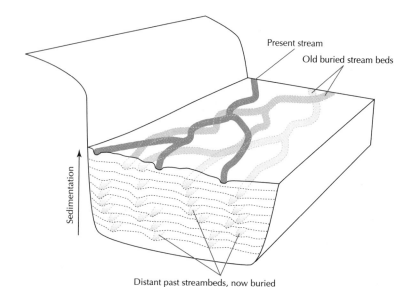

Figure 8.1. Diagram of a braided stream bed through time and space.

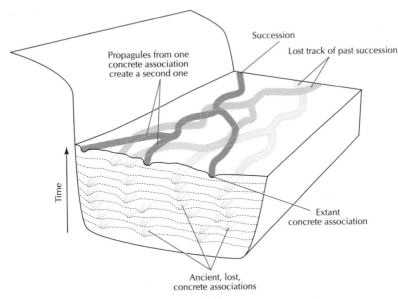

Figure 8.2. Diagram of a community expressed as a braided stream.

At any instant in time, each concrete example of a community corresponds to a single braid, a streamlet finger poking from the past to be caught in the cross section of the present (Figure 8.2). The braiding of the stream refers to the way that present examples of a community type come from prior examples that have split and conjoined in the past to give the present concrete example. Some of the separate braids of the present can be expected to coalesce in the future, while others will dribble to a halt. At any given instant in time, there exists the set of concrete communities of a type. The type of community, the abstract association, is the braided stream in its entirety. In the end, each changing exemplar loses its identity, often in an exchange of seeds between sites. The whole stream exists over millennia, in Cooper's assessment, so the identification of a general type is not tied to any place at any time. Cooper (1926) does not connect the phrases explicitly, but the whole braided stream is the "abstract association."

Nichols (1923) gives an example that appears to anticipate the braided stream analogy.

> We have in central Connecticut any number of pitch pine *(Pinus rigida)* communities, all of them essentially alike in their vegetation, but all (broadly speaking) topographically separated from one another and therefore organically disconnected. Now, to some ecologists each one of these pitch pine communities, by itself, is an association. To others the term "pitch pine association" implies a sort of pigeonhole to which all the individual pitch pine communities can be referred; the individual community is merely a concrete example of the abstract association. Still others would say that all the individual pitch pine communities in existence, considered as a single aggre-

gate and concrete whole, constitute the "pitch pine association," and that the individual pieces themselves are but parts or fragments of this association...

The abstract concept of pitch pine association includes by implication not only the pitch pine communities in existence, but all similar communities of the past and of the future as well [the entire braided stream]...

The concrete and the abstract communities are models that operate each at its own level of analysis. Both serve some purpose but both present difficulties. In the concrete examples, each one is unique and its particular configuration is a matter of happenstance. This is the position of Henry Gleason (1926) in his individualistic concept of the association. The explanation for the species composition of each woodlot is unique, and science only deals with patterns that repeat. The abstract association, then, is a valid object for scientific investigation, but the concrete association appears more problematic. One can point to a woodlot's uniqueness, but prediction of the composition of anything so particular as the concrete association is difficult.

On the other hand, one can go too far in pressing the reality of the abstract association, as did Frederic Clements in his assertions that the growing, changing vegetation was a superorganism (Clements 1905). One can fairly inquire as to where is its skin and what constitutes its genetic system. As the ultimate reference, Clements used the point where vegetation no longer changes, because it is at its most mature state. Fully mature vegetation is said to be *at climax*. At **climax**, the individuals of the species present replace themselves with their progeny in the same site, and the vegetation is more or less in compositional stasis. Unfortunately, the climax is a rare condition, not the point on which all vegetation in a region is converging. Climax is probably not a sufficiently general benchmark for the notion of the plant community.

Note how Cooper's braided stream analogy implies several levels operating and existing over widely different spans of time: the vegetation at a site; that same vegetation over successional stages; the existence of the whole type of association over extended time. (Figure 8.2). Faced with such richness of pattern and behavior, both Clements and Gleason sought some unequivocal reference around which to organize the structure and dynamics of communities. Clements picked the climax as an abstraction, removing successional stages from the discussion; Gleason picked the composition of the concrete association at an instant at a site, ignoring the climax as a unifier. Neither is wrong, but both are forced to give up something.

The community as a collection of populations

A common conception amongst animal ecologists is the notion of a community as a collection of interacting populations. Field zoologists often collect data by making measurements of populations as they steam past the sampling point. For example, ornithologists commonly collect bird data by standing at a series of points and listening. As the birds fly around, a given species scores a hit at a sampling point if the song of that species is heard in the time allotted for the sample. Animal data collection stands in contrast to plant community data collection, which often involves recording particular plants in a sample, not simply the presence or absence of a population. The identity of each individual is usually more important in plant data as opposed to animal data. For example, a botanist collecting herb data will often be on hands and knees, counting all the plants by species in a small sample area (Figure 8.3). *The community as a collection of populations works well for animal communities, but the concept is usually less satisfactory for plant communities.*

Animal ecologist data collection

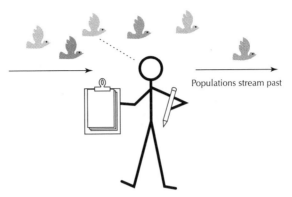

Populations stream past

Figure 8.3. Collecting data from communities.

Plant ecologist data collection

Individual plant organisms are counted

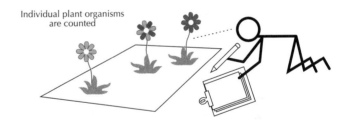

The central issue here is the cause of *community pattern*. If the physical environment is seen as the primary cause, then the model is said to employ *environmental determinism;* if the biota are seen as the primary cause, the model is one of *biotic determinism.* When, for whatever reason, biotic forces such as competition or predation are primarily responsible for the pattern, then the concept of the community as a collection of populations works well. Population processes explain much of what one sees. Since animals move, they are less determined by physical environment than are most plants—animal communities can often be cast successfully as collections of populations. The manner in which a plant is a member of its community is often not the same as the manner of its membership in its population. One can certainly refer to populations of trees, but the relationship of a tree to its fellow population members may be insignificant in its community interactions. The relationships between individual trees in forest communities are less with whom they breed, and more who is immediately adjacent—often a member of a different species. Usually little is to be gained by inserting the population as a coherent entity between the individual plant and its community. By contrast, the way that animal populations move across the landscape often

Animals in population

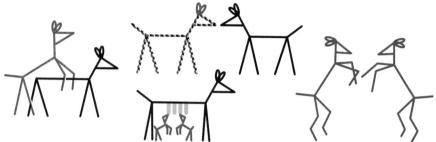

Figure 8.4. The concept of a community as a collection of populations works well for animal communities; however, the manner in which a plant is a member of its community is not often the same as the manner of its membership in its population.

Plants in community

makes the population a helpful conceptual device as a link between the individual animal and the community to which it belongs (Figure 8.4).

Interaction through periodic occupancy of space

Different species occupy the landscape at different scales. As a result, the very place that is occupied by members of different species is different, even if the two individuals are right next to each other. An ant and a wolf who happen on the same square meter occupy the surrounding space at very different scales, breaking down the notion of a multispecies community as a thing in a place. Ants may come and go from a square meter frequently, whereas a wolf may return only occasionally, if at all, to that exact spot (Figure 8.5). The way each species reads the habitat at different scales means that two species differently scaled cannot occupy the same place. A place big enough to service a far-ranging carnivore consists mostly of the land far beyond that required for a small rodent, and yet wolves prey on mice, inviting inclusion in the same community. The same dilemma also applies to plants in plant communities, as when a tree occupies a spot for centuries and saturates a large surrounding area with seeds, whereas a herb occupies its spot under the tree for as little as a growing season, and may distribute most of its seeds very locally. Heterogeneous spatial and temporal scales inside both plant and animal communities deny the community a particular place.

The way out of the dilemma is not to assign the community to any place in particular. And yet communities occupy space, so what to do? Allen and Hoekstra (1992) point to a solution recognizing that *the community is the interaction, not the physical presence, of organisms or populations in a place*. Think of each species not as tied to a place, but as occupying a particular spot only intermittently. Like the sea forming waves on a beach, each species occupies a site and then departs for a while, as does the sea between waves and between tides (Figure 8.6). The size and duration of the wave of occupancy of a site is different for each species' individual biology. If the community were a static thing, just a simple collection, then there would be no dilemma. But there is always some sort of interaction between community members, and it is that interaction that gives the community coherence. The community ecologist must at some point acknowledge a whole that is all the species' waves of landscape occupancy.

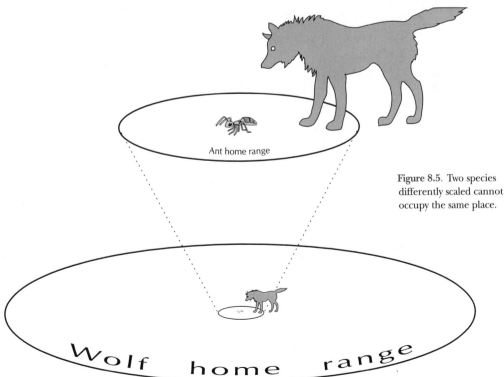

Figure 8.5. Two species differently scaled cannot occupy the same place.

The Edge of the Sea

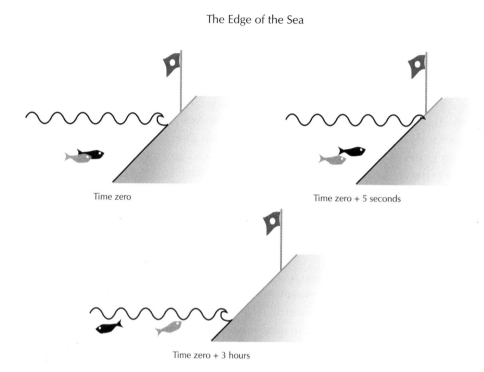

Time zero

Time zero + 5 seconds

Time zero + 3 hours

Figure 8.6. Like waves on a beach, each species occupies a site and then departs for a while, as does the sea between waves and between tides. The community is the interaction, not the physical presence, of organisms or populations in a place.

Note that in the intertidal zone by the sea each wave swells for only seconds, whereas the tide waves up and down the beach every six hours or so. One species might occupy a site for a short while but return quickly, like the individual waves, while another species may occupy the site for a long time and be absent for long periods, like the ebbing tide. It makes no more sense to talk of the exact position of the real edge of a community than it does to talk of the exact edge of the sea. Both are the edges of interacting periods of waves.

Periodic interaction in physics is called **wave interference**. *Wave interference patterns can appear static, but include dynamics inside them.* That fits community biology rather well. Notice also that very small changes in the period or direction of one of the waves in a physical system can produce a very different wave interference pattern (Figure 8.7). That too fits the community conception. If there is a change in amplitude (i.e., abundance of a critical species) or a shift in periodicity (duration of the temporary, local extinction of a competitor), then the biology of a community and its patterns in time are likely to change radically. For example, Robert Paine's work on the intertidal

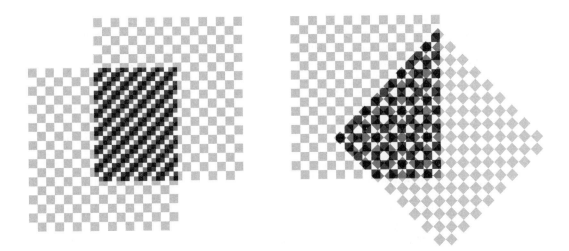

Figure 8.7. Wave interference model of the ecological community. The two checkerboard grids represent some aspect, such as life history or home range, of two species, distributed in time and/or space. The derived patterns (where the two species overlap) is the community—the relationship between the components. Note that changes in the period or direction of one of the waves (checkerboards) in a physical system can produce a very different wave interference pattern (community, represented by the patterns of black areas of overlap).

zone showed exactly that sort of radical change in community composition when he temporarily removed the predatory starfish (Plate 8.1), an example of a **keystone-species.** Without starfish intermittently opening space, the interaction between the other species changed completely (Figure 8.8). Also frequent stormy seas (waves in a literal sense) crashing on exposed parts of the coast caused the intertidal fauna to be different from the fauna in sheltered coves, which were disturbed infrequently. Storms and starfish both periodically open space for animals that may not be able to compete and hold onto a spot, but can move quickly in a wave of colonization onto recently denuded rocks (Box 8.1).

The community considered as a wave interference pattern need not be conceived as a thing in a place. This explains why the field observer can be so conscious of the presence of a community, but still have difficulty putting a boundary round it. The thing itself, *the community, is the dynamical wave interference between species.* Particularly in animal communities, the place is not occupied by the species all at the same time, as illustrated by the waves of ungulate herds that graze the Serengeti Plains (Box 8.2). The community as a wave interference pattern allows the occurrence of sharp community boundaries in vegetation in some places to be in stark contrast to the gradual changes from one community type to another that occur in other places. The edge of a community can be discrete and static or continuous and dynamic.

Box 8.1. Population conceptions and biotic determinism

The intertidal animal communities of Robert Paine (1966) are well conceived as collections of populations interacting and replacing one another in time and space. Paine's work represents a prime example of a biotically determined community. The lead players are a starfish *(Pisaster)* and a mussel *(Mytilus)* living on rocks in the intertidal zone. The walk-on parts are played by three species of acorn barnacles, a goose-necked barnacle, two species of limpet that graze algae, and a snail-like *Chiton* (Figure 8.8). The lead actors appear to influence the rest of the species, the mussel by crowding out all other species for space, and the starfish by opening up the space by grazing on the mussels and anything else in its way. The starfish cuts a swath through its prey populations, and wave action may also open up space. In stark terms, the system has one unchanging resource base—space—occupied by a superior competitor, which is mediated by a predator.

In the intertidal, one species *(Mytilus)* gains ascendancy over the others in the absence of a population of predators to keep them in check. Paine removed the starfish and saw the mussels take over the space of other sedentary animals and algae. The grazers moved away in the absence of their algal food source. The important interactions between animals in the intertidal zone are reproduction and occupancy of space on the one hand, and predation on the other. Both reproduction and predation are easily cast as population processes, so the population becomes the natural subunit inside Paine's animal communities. There is a component of physical disturbance, but that acts just like a starfish. There is a physical resource, space, but that is constant. The reasons for the success of Paine's conception of the animal intertidal community as a collection of populations is that *the system is unusually cleanly biotically determined in a remarkably homogeneous environment.* Yes, the environment varies strongly as the tide comes and goes, but that variation is regular, entirely predictable and resets the system by flushing it out. In the absence of unpre-

dictable environmental variation, biotic processes of populations (growth, competition and predation) come to explain the pattern almost entirely.

Paine's observation is not limited to animals. Robin Kimmerer showed the same overriding effect of biotic processes for mosses and liverworts on the walls of cliffs channeling the Kickapoo River in Wisconsin (Kimmerer & Allen 1982). The river floods in spring to various heights up the cliff, allowing floating logs to damage the vegetation to various heights with decreasing frequency higher up the cliff (Figure 8.9). The species most susceptible to damage is the best competitor for space, a liverwort *(Conocephalum)*. Indeed, susceptibility to floating log damage is related to superior competitive ability. The liverwort achieves overtopping of other species by growing as a crinkled sheet that is raised above the rock surface, attached by rhizoid threads. Floating logs rip the *Conocephalum* from the rock in great sheets. Above the normal high flood line we find uninterrupted sheets of liverwort, the result of uninterrupted competition. Immediately below the high flood line is a rich species diversity of mosses. In the lowest zone, which regularly gets flooded and often battered by floating logs, are pure stands of a moss *(Fissidens)*. It is the smallest moss and most closely pressed to the rock, essentially immune to log damage.

The point to note is that, like Paine's intertidal, Kimmerer's rock cliffs consist of only one overriding environmental resource—unchanging areas of rock surface. Space provides a constant **carrying capacity** as the context for the biotic processes of populations to create the simple banded pattern. By contrast, invoking populations of trees as an explanatory unit in forest communities gives little insight, and generally confuses the important issues in forest communities. Populations of trees is a weak concept because there is no single unchanging carrying capacity to bring biotic processes to the fore as unequivocal causal factors.

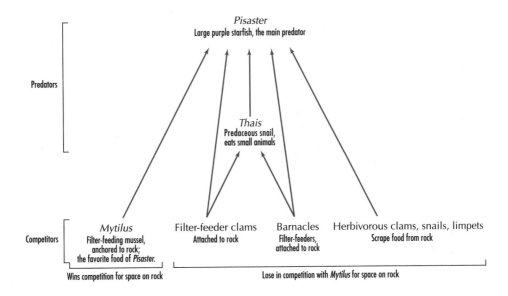

Figure 8.8. The interactions among invertebrate species living on rock faces along the Pacific coast of Washington and British Columbia.

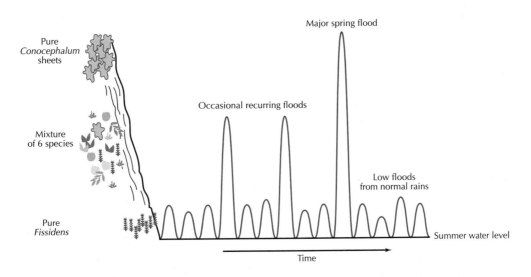

Figure 8.9. The effect of biotic processes for mosses and liverworts on the walls of cliffs channeling the Kickapoo River, Wisconsin.

Box 8.2. Animal communities in time over space

The Serengeti Plain of East Africa has impressive densities of grazing animals, mainly ungulates. Ungulates fall into two broad groups exemplified by horses and cows; these two groups are separated by a feature that reflects their evolutionary decent. Horses walk on one finger (on each leg), their hoof being the fingernail of the middle finger. Cows walk in two fingers; that is, each hoof embodies two "fingernails," and these animals are therefore referred to as "cloven hoofed."

Horses, to use a cliché, "eat like a horse," consuming large amounts of food, high quality if they can get it. Comparing horse to cow diets, horse feed has a higher crude fiber content. The passage of food through a horse is between 30 and 45 hours, whereas the passage through a cow takes 60 to 100 hours, so the horse makes up for shorter digestion time by eating larger quantities. Cloven-hoofed animals can benefit from low quality food by processing it until almost all the goodness is removed. Compare cow pies to horse hooey, and you can see the difference. Horses work like a combination of a mowing machine and a conveyer belt, passing enormous amounts of low grade food through quickly, making up for quality with quantity.

This book turns on the distinction between scale of an ecological entity and its type, and the way it is best to keep those two organizing principles separate. The above distinction within ungulates is one of type, but that leaves aside the separate issue of scale. For animals with comparable digestive systems, smaller animals require higher quality food with more protein. Beyond the fundamental difference in diet between horses and cloven hoofed grazers, there are the differences between the large and small cloven hoofed ungulates. On the East African plains, cloven-hoofed ungulates bracket the size of the zebra (a horse relative): buffalo are huge, bigger than a draft horse; wildebeest are three-quarters the size of zebra; topi, another cloven-hoofed animal, is half the size of a zebra, and the Thomson's gazelle is about the size of the ancestral horse (as large as a small greyhound). As animals increase in size, there are exponential changes in requirements for food and home range. Thus size and type of ungulates interact to produce an elaborate set of interrelationships with the environment. Put on top of that the dynamic interaction between the animals and their environment, each ungulate preferring a certain type of forage, and each having its distinctive effect on the plants. R. H. V. Bell summarizes the Serengeti grazers:

- The proportion of protein of the diet is lowest in zebra and highest in Thomson's gazelle.
- The food is toughest for zebra and most delicate in Thomson's gazelle.
- Zebras use the commonest and most accessible food, whereas Thomson's gazelle use the rarest and most inaccessible food.

The situation is further complicated by spatial and temporal pattern (Figure 8.10). The hilltops are dryest and have shorter vegetation. New veg-

Figure 8.10. Spatial and temporal pattern on the Serengeti.

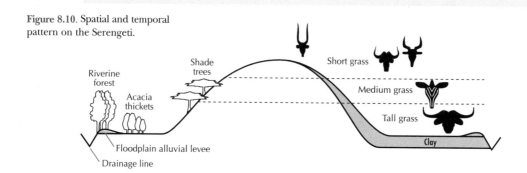

etation is plentiful all over the plain in the rainy season. At that point the grazers are independent of each others' effect on vegetation and grazers use the hilltops. As the dry season comes, there is a succession of animals driven downslope to the marshy areas where vegetation continues to grow longer after rain ceases. Movement upslope and downslope with the seasons is in order of size (with the exception that Wildebeest and Topi are reversed). The first to arrive downslope at the end of the rains are the biggest cloven-hoofed grazers, the buffalo. They graze the plants down to a size that the zebras can use, about waist-high. Then the zebras, with their voracious appetites and strong teeth, remove the bulk of the vegetation, leaving grass in the condition of a pasture. More slowly and thoroughly than the zebras, the topi and wildebeest work over the vegetation to produce a lawn. At that point, there is insufficient bulk food to support large animals, so along come the Thomson's gazelle. "Tommies" feed in relative isolation, working over the stubble, gleaning grains from the aftermath of the heavy grazers. Note that each species prepares the vegetation for the next; each comes in its place in the sequence. Clearly there is a grazing community here, but members in certain seasons may not overlap much in a given site with the other grazing species. For example, the Thomson's Gazelle cannot deal with tall, coarse vegetation to find seeds and high-protein shoots until the Zebras are gone, and probably the Wildebeest too.

So where is the boundary of the community? One cannot look at a given site at a given time, and say who is in the community. However, there is an animal community to be studied, with its distinctive phenomena and five major players, several of whom meet only rarely. Thus an animal community may well not yield to a characterization as a thing at an instant in a place. Even the edges of such concrete units as herds blur, as the density of animals decreases at the margin. Is a stray in or out of the closest herd? The observer must decide for the purposes of the investigation at hand. Making distinctions is scientist's responsibility—nature does not tell where distinctions lie.

QUESTIONS ASKED BY COMMUNITY ECOLOGISTS

Community ecologists ask their own style of question, and have their own methods for dealing with their systems. So what are the questions that might interest a community ecologist, and how might they vary across different community types?

Community Boundaries

Differences between plant and animal community conceptions are so great that often each will need its separate treatment, although there are, of course, some parallels. As to plant communities, D. A. Webb (1954) said, "The fact is that the pattern of variation shown in the distribution of species among **quadrats** (sampling units) on the Earth's surface chosen at random hovers in a tantalizing manner between the continuous and the discontinuous." Webb is saying that there is *regularity of association* between species, some of it reflected in sharp boundaries, but some of it occurring as gradual change of composition across the landscape. It is often very hard to put a boundary around the plant community. For animal communities, the notion of a boundary as a line across a space is clearly unworkable. In animal communities, we must ask about the very identity of the thing one is calling a community, when members come and go from the place in question.

Out of these concerns come questions such as "What is the meaning of sharp boundaries?" Are they historical, like lines where fires of the past stopped for whatever reason (see Chapter Two), or does the explanation lie in a discontinuity in some environmental factor, like soil type? And if there is no sharp delineation, what does it mean to recognize a thing which has no obvious boundary in space, like an assemblage of animal species that move across an area, as do the ungulates that crop a given part of the Serengeti Plain?

Dynamics Versus Structure

All of the above questions reflect the nature of the **pattern** directly. The pattern is not dynamic in itself, but rather it just *is*. Pattern depends on distinctions, and consists of ordered or repeated distinctions. It emphasizes quality rather than quantity, structure more than dynamics. Structure and dynamics are of equal importance in community ecology, so if there is a set of questions that probe system structure, there is another set that validly takes a more dynamical approach. The questions here probe at the level of processes that hold the whole community together. The processes generate the patterns seen in space and over time. Dynamical questions take the form of "what are the feedbacks that lead to mutual reinforcement of species presence?" It is unlikely that one species will determine the presence of another in a community context, but one can still ask several questions: How does the more abundant species become the setting for the lesser species? How regular is the change in species and their relative abundance over time? What are the causal agents for that change, and what explains the degree to which the sequences of species change are repeated in other patches of vegetation or in other animal communities? Conversely, one could also ask what causes some aspects of the patterns to be unique and fail to ever repeat. For example, we do see regular patterns of change in vegetation, and even give that process a name: **vegetational succession**. The successional sequence of species replacement is by no means always the same, however, even in different patches of vegetation in a local region. Indeed, the particular change in each patch is influenced by the constitution of the surrounding vegetation, a situation that is never going to be exactly the same for any two vegetational units (Figure 8.11). This is the point on which Gleason's individualistic concept turns in the Gleason/Clements debate. Thus succession is often repeated in general terms, but is never exactly the same. The questions vegetation scientists ask often probe the causes of what makes vegetation often similar but never the same.

How Do Ecologists Describe Communities?

The point of departure of community ecology is to deal with a blizzard of detailed differences in some coherent way, and so generate repeatable description upon which different observers can agree. Accordingly, a deal of effort is put into description of communities, usually in quantitative terms. The descriptive phase of ecology needs to be conducted properly, and usually elaborately in community ecology, so that the important questions that lead to explanations of the pattern can be posed and answered unequivocally. The question at the root of description is, "What is out there?" The question becomes more focused in particular settings as when John Curtis established the "Wisconsin School" of plant ecology. For the purposes of knowing what to restore and conserve, he asked the question, "What is the vegetation of Wisconsin?" And fifteen years and a dozen or so graduate students later his seminal book answered it (Curtis, 1959).

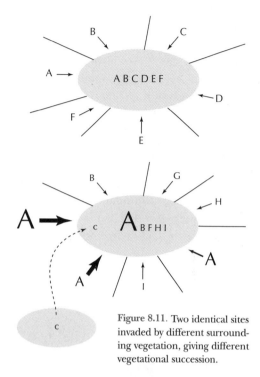

Figure 8.11. Two identical sites invaded by different surrounding vegetation, giving different vegetational succession.

What Causes Repeated Patterns in Vegetation?

Communities show pattern, albeit with variation and only degrees of similarity of pattern between sites. Community ecologists often ask questions about the explanations of those patterns and their repeated appearance at various sites. As we shall argue in the evolution section of this chapter, evolution appears to play a minor role. If communities are not importantly configured by evolution, what does give the patterns of association that correspond to communities? The answer is the existence of stable configurations of ready-made species. The history of global climate change and the geography of recovery from those events combine to produce a jumble of species in a given region at any particular instant in time. Given that species availability as potential players is largely a matter of biogeographic happenstance, many stable configurations of species association are possible. Furthermore, the possible stable configurations of species occur across a range of scales. Thus questions of the origin and mechanism behind pattern occur across significant changes of scale. Some examples will illustrate the point.

An example of a very small-scale stable configuration was described by A. S. Watt in 1947, when he explained various patterns of vegetation dynamics. Watt gives three examples of hummock building, each with its phases of growth, maturity, degeneration and hollow (often by erosion). A hummock is a slightly raised area; mole hills and ant nests are examples of small hummocks. The scale of these examples of hummocks is smaller than a meter. Watt also provides one large-scale example of a cycle in a forest, as do McCune and Whittaker as well. Each of these studies starts with a question, and the cycles and processes described below are the answers to those questions. McCune's paper

(McCune & Allen 1985) even has a question for a title: "Will similar forests occur on similar sites?" The answer was a surprising no! In the following examples, note the order of increasing size, nonetheless presenting a common class of original questions about the cause of the clear, repeated patterns of hummocks, waves of vegetation, or mosaic patches in forests.

Watt reports that the wind opens up relatively bare soil in a hollow, subsequently colonized by fescue grass. Particles of soil accumulate around the growing grass to fill the hollow and build a hummock. In the mature phase, when the hummock is about four centimeters high, various species of *Cladonia*, lichens with upright form, colonize the top of the hummock. The hummock gradually erodes, and the lichen species change to more crust-like flat forms. In the end, a hollow is eroded out, and the grass colonizes again. This cycle is repeated again and again (Figure 8.12).

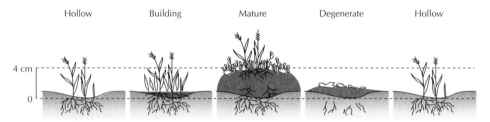

Figure 8.12. Cycle of small hummock formation.

A similar cyclical process of hummock building followed by wind erosion explains bands of heather *(Calluna)* about three feet wide that march across hillsides in upland Europe. Between the bands are the eroded phases where the heather has declined. Since the wind commonly comes from one direction, and the erosion occurs on the windward side of hummocks, the hummocks occur as bands that move across the hillside over decades (Figure 8.13). With just a little more protection, a species of lichen can join the hummock community, and another pattern occurs (Figure 8.1.).

Small hummock-building processes are also critical in the raising of the surface of bogs across Britain. *Sphagnum* moss builds the hummock, but the moss grows more slowly as it dries with increasing height above the bog surface. Various heath species inhabit the hummock, shade the moss, and dry the hummock. When high and dry, the heather succumbs and the hummock is vacant and not growing. However, as it grew it had left behind pools forming at its base. *Sphagnum* thrives in the pools and in turn forms new hummocks that overtop the old hummock, turning it into a pool, suitable once again for the growth of *Sphagnum*. Note that there is no reason to suggest that these species are evolved explicitly to deal with their role in these cyclical processes. It is much more likely that ready-made species happen to form a stable configuration of cyclical species replacement.

At larger scales, Watt points to the cycle called "gap phase" in forests. The death of large trees opens gaps that are filled by seedlings that become saplings, one or two of which become trees again (see Chapter One, Figure 1.1L). It is this process that Shugart's FORET model describes, a model described later in this chapter.

At a larger scale again, Bruce McCune identified various different, apparently stable configurations of forest trees occurring each in its own valley in the Bitterroot Mountains on the Idaho-Montana border (Figure 8.15). These distinctive forests appear to be insignificantly environmentally determined, and result mostly from historical accident, such as the patches of fire discussed in Chapter Two.

At a yet larger scale, Whittaker identified stable configurations of occupancy of the Great Smoky Mountains of Tennessee. The *cove forests* (forests in the valleys) are dominated by tulip poplars. These forests butt up against pine-dominated forest on low ridges,

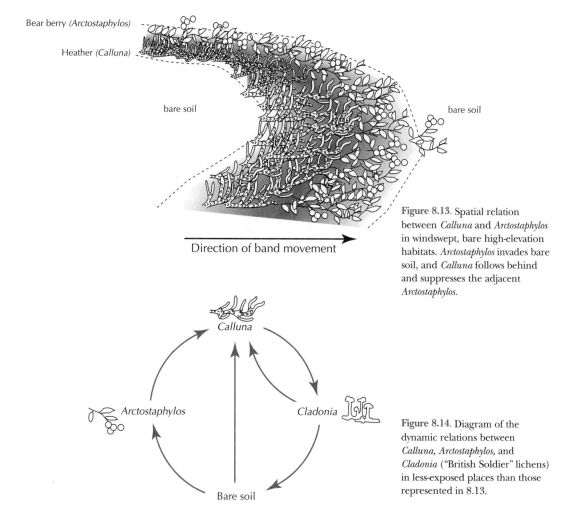

Figure 8.13. Spatial relation between *Calluna* and *Arctostaphylos* in windswept, bare high-elevation habitats. *Arctostaphylos* invades bare soil, and *Calluna* follows behind and suppresses the adjacent *Arctostaphylos*.

Figure 8.14. Diagram of the dynamic relations between *Calluna, Arctostaphylos,* and *Cladonia* ("British Soldier" lichens) in less-exposed places than those represented in 8.13.

Figure 8.15. The rugged terrain of the Bitterroot Range of the Rocky Mountains, along the Montana-Idaho border.

and in the higher valleys they give way to hemlock forests called *hemlock draws*. High ridges are occupied by either rhododendron thickets, open grassy areas called *balds*, or spruce-beech associations on the highest ridges. Various plants, like rhododendrons, occur as components in all the other places where they do not dominate. The whole mountain vegetation can therefore be considered a coherent, stable configuration (Plate 8.2)

A General Model to Answer Community Questions

Community ecologists deal with systems that operate in many different situations, with different suites of species and across a range of spatiotemporal scales; their challenge is to create a model that applies across such a wide range of situations. The general community model proposed here will indicate how other subdisciplines of ecology might

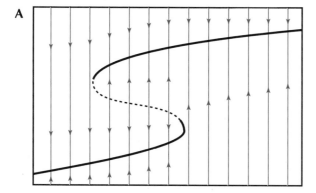

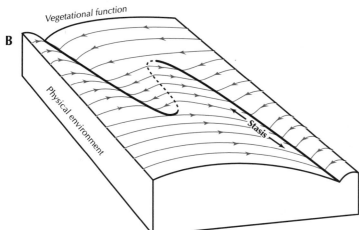

Figure 8.16. General model of a community using the metaphor of a folded surface. Upper diagram **A** illustrates a bird's-eye view of the surface contours on **B.** Imagine a ball placed on the landscape. The ball represents an association of species. For most starting places, the ball rolls (in the direction of the arrows) down to only one of the two solid black lines (which represent distinct communities). If the ball is placed on the dotted line, it can easily roll toward either solid line.

lend assistance in answering questions within the community conception. To capture community dynamics, the general model of community here uses the metaphor of a folded surface (Figure 8.16). Often the ultimate contribution of mathematical formalities is a justification for metaphor, so the algebra that gives the surface is superfluous.

The fold in the surface represents two communities over an environmental gradient (Figure 8.17A). The ecologically important physical environment at a site can be summarized by a representative single number, an index often relating to moisture. Although many species occupy a site, it is also possible to account for vegetation at a site by some summary number, say an index calculated as a ratio of species types. The folded surface occurs on a plot between the index for the physical environment and the summary index for the state of the dominant vegetation. Index values used to plot the surface will be specific to the two communities being compared, and the range of the physical environment across which they occur. For example, at the interface of grassland and forest, the general measure for vegetation could be the amount of woody

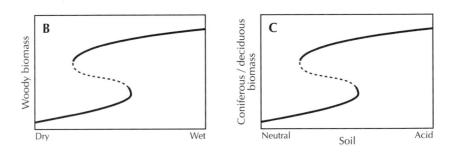

Figure 8.17. A. Surface fold representing two communities over an environmental gradient. **B** and **C.** Examples of index values comparing the two communities and the range of physical environment across which they occur.

biomass: low for the grass community, but high for the forest (Figure 8.17B). In another example at the edge of the cove forests and pine forest on ridges in Whittaker's study of the Great Smoky Mountains, the number that characterizes the vegetation might be a ratio of the biomass of coniferous species over the biomass of deciduous species—a high number for the pine community on the ridges, but a low number for the tulip-tree cove forests (Figure 8.17C).

Vegetation dynamics in the region where one community type gives way to another is regulated by the interaction of two competing strategies that govern the working of the two adjoining communities. The strategies could be fire adaptation on the part of grasses as opposed to shading strategies of trees, but any pair of competing strategies for any pair of associated vegetation types could be considered. The value of the number that characterizes the vegetation indicates which strategy has the upper hand in the tussle between the communities. The strategies are each self-reinforcing, which is the flip side of the fact that the two different vegetational strategies compete. For instance, in the case of deciduous forest and grassland, a high value on a woody biomass measure would indicate that the shading strategy is winning out over the fire strategy (Figure 8.17B). In the case of Whittaker's cove forests and pine-covered ridge communities, the strategies could turn on soil acidity and water use: soil-acidifying and drought-inducing pines, as opposed to soil-neutralizing tulip trees with little fuel to carry fire. A high value on an index for coniferous biomass would indicate that the soil-acidification and soil-drying strategy of the pines is working to the detriment of the tulip poplar strategy of keeping things moist and pH neutral (Figure 8.17C).

The folded surface tracks changes in community composition. For some changes in environment along the abscissa, the number that characterizes the vegetation on the ordinate changes continuously, such as small changes in tree species present. Here the community remains on one surface of the fold (Figure 8.18A). Other changes in environment produce a discontinuous change in vegetation, as when a forest is suddenly replaced by a grassland. The potential for sudden change occurs in the middle of the environmental range, in the zone of the fold itself and at its two edges. The fold indicates that more than one stable configuration of species is possible over the range of environment where the surface is actually folded. The upper part of the fold indicates the potential presence of one community type, and the lower part of the fold indicates the possibility of another stable community type (Figure 8.18B). The discontinuous change in vegetation occurs as the vegetation flips from the lower surface of the fold to the upper, or vice versa. There are two possible causes for the vegetation changing type. One is an environmental change that takes the system outside the region of the fold (Figure 8.18C). One strategy is thus taken out of its environmental region, and should the vegetation be in that phase when the environmental shift occurs, the other community type comes to assert itself after the environmental change. The second is some sort of disturbance (Figure 8.18D). Both community types remain an option in that general environment, but one has replaced the other by means of disturbance.

Away from the region of the graph where the fold occurs, environment is the dominant factor in explanation and prediction (Figure 8.19). An argument that turns on the environment determining vegetation is likely to help

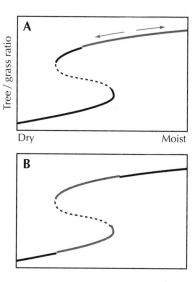

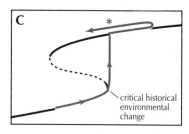

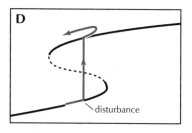

Figure 8.18. The folded surface model tracks changes in community composition. **A.** Continuous vegetation change inside a forest. **B.** Past disturbance and history determine which community option is present. **C.** Environmental change drives the flip to the other community. Note that a reverse change in the environment below point * cannot reverse the historical jump. **D.** Disturbance causes a flip to the other community.

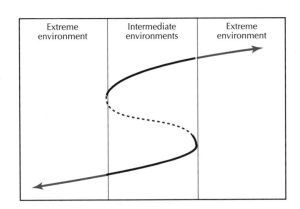

Figure 8.19. In intermediate environments, disturbance and history, not environment, explains vegetation. At extreme environments, the environment alone explains vegetation.

in extreme environments at the ends of the graph. In the case of fire adapted grasses against forest, at the dry end of the physical environment axis, grassland is the only possible vegetation. At the wet end of that axis, enduring drought is sufficiently uncommon that fire cannot keep vegetation open, so forest encroaches to the universal exclusion of grassland.

The region of the fold itself occurs in the middle of the physical environmental gradient. In the part of the graph where the fold occurs, the effect of physical environment becomes uncertain. Instead, details of history, disturbance, and feedbacks in species interaction explain the community found there (Figure 8.20). In the zone of the fold, these become the drivers of the direction of change, or alternatively the forces for stasis once the vegetation reaches the characteristic community type. In this way, history, disturbance and species interaction become the means of prediction. At intermediate environments, in the region of the fold in the middle of the graph, biotic factors are responsible for the vegetation clinging to the side of the fold on which it resides. This is an explanation that turns on biotic, as opposed to environmental, determinism. Vegetation stays forest or stays grassland because of biotic factors, more than environment,

Figure 8.20. Community situations where various other ecological discourses can assist community ecology. At extremes, physiology predicts communities. In intermediate environments landscape, population, historical or disturbance ecology all inform. Over the long term, paleoecology explains community origins and indicates prospective communities under future climate change.

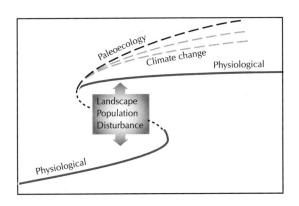

because the biota drive the environment toward a preferred state. The mix of all these causes was first captured in David Roberts dynamic synthesis theory (reported in Allen and Hochstra 1992).

The lines on the graph represent a stable configuration of species based on some shared ecological strategy across species, except for the dashed line connecting the upper and lower surfaces. The dashed line is the unstable middle ground between strategies. While one does find intermediate patches of vegetation between, say, forest and grassland, they are less common than other vegetational configurations. The reason is that intermediates are unstable, and local historical events will determine which stable strategy prevails. A jump from the upper to the lower surface in the folded region is often caused by local disturbance such as fire or drought, more than the prevailing context, like five-year average rainfall.

The unmistakable patterns of association upon which the community concept turns arise from shared strategies between species generating a positive feedback. The strategy alters the environment to suit further implementation of the strategy. For instance, trees generate shade and high humidity, and even increase local rainfall, encouraging an environment where tree seedlings thrive and fire is suppressed. Undisturbed vegetation remains in one type because of self-reinforcing strategies that keep the vegetation stable.

A **strategy** in this discussion refers to a way that a set of plants deals with its neighbors to secure advantage, such as trees shading out grasses that might otherwise carry fire into the forest. The strategies presented here characterize the major biomes of North America. **Biomes** are discussed in depth later in this chapter, but suffice it to say at this point that biomes are distinguished from communities by their defining plants being identified by life form (tree, shrub, herb etc.), not by species. A given biome has a certain look—desert, boreal forest, tropical rain forest, etc. In different geographic regions, these strategies butt up against each other to give the discontinuous behavior of a folded surface. Other more local strategies are responsible for heterogeneity inside the major types of vegetation. Some examples of these stable plant strategies that identify major community types are:

- *The strategy of shading with woody vegetation.* This strategy requires relatively high rainfall to support the biomass of the forest. Trees cannot grow and reproduce effectively in regions where the soil becomes functionally devoid of water for about a month (Figure 8.21A).
- *Low nitrogen and fire adaptation.* This is the strategy of the bunchgrasses that constitute the tall-grass prairies. Tilman and Wedin (1991 a, b) have shown that these grasses use several devices to keep nitrogen low (we'll return to this strategy later). First, the grasses in question are able to grow in areas where the soil is low in nitrogen, and left to their devices these species will draw down nitrogen to a level where competitors cannot survive. Second, they readily carry fire that leaves other mineral nutrients behind, but drives nitrogen into the air as an oxide

of nitrogen. Third, their low nutritional value discourages grazing beyond only a few weeks, so the nitrogenous waste of animals cannot reverse the direction of the positive feedback by adding nitrogen to the soil (Figure 8.21B).

- *High nitrogen and grazing.* Here the grasses again play a role, but a different set from those employing the low-nitrogen strategy. By adapting to grazing, these plants fuel animal mowing machines that keep out woody material. Relying on a nitrogen-rich soil, they grow faster than the low-nitrogen strategists, and crowd them out. They do not bring nitrogen levels down greatly, and by offering grazers nutritious food through the season, they keep high nitrogen input constant. The meadow is this community (Figure 8.21C).

- *Dry out the soil to reduce fuel.* In places where there is no danger of tree shading, shrub vegetation sucks all the water out of the soil. The effect of this is to reduce ground cover to isolated clumps, and minimize the risk of the spread of fire (Figure 8.21D).

- *Fire-adapted shrubs.* In the dry far west of the United States, seasonal rains grow shrub vegetation that burns every thirty years or so. The volatile oils of the vegetation ensure a good, hot burn which the shrubs via roots or seeds survive, but competitors do not (Figure 8.21E).

- *The desert strategy.* In very dry regions there are two stratagems. The first is water conservation in succulent perennials. The alternative is

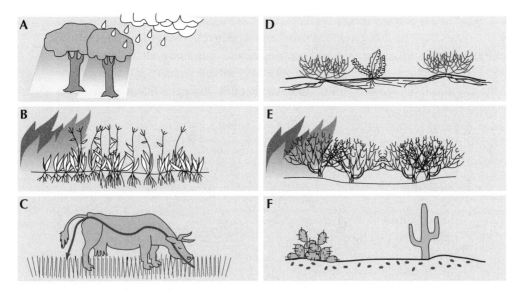

Figure 8.21. Examples of stable plant strategies that identify major community types. **A,** shading by woody vegetation; **B,** low nitrogen and fire adaptation; **C,** high nitrogen and grazing; **D,** dry out soil to reduce incidence of fire; **E,** adapt to fire; **F,** desert strategy.

to grow very fast and complete the entire life cycle on the water of a single hard rain, which causes the desert to bloom (Figure 8.21F).

Within the forest are substrategies that often relate to manipulating fire frequency one way or another. Around the globe there are other versions of being a tree, shrub or herb community. The above list is not intended to be complete, but is to illustrate how different species fall under one or another distinct life style that self-perpetuates at a site. The species composition is not climatically determined, but does have a relationship to physical environmental factors as ultimate constraints.

In its ideal climate a strategy will reign supreme, while others are prohibited. The physiological ecologist explains where the edges of tolerance occur, beyond which this or that strategy is excluded (see Chapter Five). In terms of the folded surface, the physiological ecologist can identify the edge of the zone of the fold, beyond which physical limits rule (Figure 8.20). Away from the fold, disturbance and history play a limited role. Away from the fold, disturbance takes the vegetation only temporarily away from a stable composition.

In environments in the zone of the fold itself, where alternative strategies are viable, physiological ecology is unlikely to be a powerful predictor. The overwhelming influence of biotic factors in the region of the fold invites population ecologists to contribute with their distinctive style of investigation. Population ecology, with its competitive hierarchies between species, can offer insights into the working of the feedbacks that sustain a given strategy. Separate from population biological reasons, history and disturbance also offer reliable explanations in the zone of the fold (Figure 8.20).

The exact form of the upper and lower surfaces of the fold is determined by the happenstance of which species are present as players in the local wars between strategies. The explanation here comes from **paleoecologists** who can say why there is a given mix in a given area. They, along with **climatologists** and specialists in long-term ecology, can indicate where the players will move next, and at what speed. While the community model presented here invites the contributions of all sorts of ecologists, the contributions of evolutionary biologists are conspicuously absent from the above discussion. Let us now turn to see where, and to what degree, evolution gives answers to community questions.

EVOLUTION IN COMMUNITY ECOLOGY
Macroevolution of the Parts

Macroevolution refers to the evolution of major distinctions, such as the emergence of a new order of animals, or the evolution of some general life form like a tree. There is relatively little evidence that communities are entities that arise significantly from evolution. Certainly all the species in a community are the products of evolution, but that need not imply that the community *itself* is more than a stable configuration of ready-made parts. The species in a community appear to be like pieces of a child's Mechano™

or Lego™ set, subunits that can be put together in any configuration that only needs to be stable. **Coevolution** is the evolution of biologically separate lines in which the species involved have continuously influenced each other to produce their special modern relationship. For example, noxious species that avoid predation are mimicked by other species that invest nothing in being unpleasant prey. Flowers with long spurs and their pollinators with tongues the same length are also products of coevolution. Communities may contain highly coevolved species pairs, but they are not usually the basis of the community.

Coevolved species pairs demand a different criterion that focuses more narrowly on those species, and the exquisite scaling that relates them over longer time periods than most temperate communities exist. Often coevolved pairs of species only interact at particular times, as when pollen needs distributing, or fruits need dispersing. There is no need to presume that a bee can tell a "stone" plant from an actual stone in a desert, unless the plant is flowering. To be important in communities at large, coevolution would have to affect coherently a wide range of happenings, more or less all the time. Let us look at community patterns to see the extent to which evolution exerts control over community structure.

Two reasons suggest a minimal role for coevolution in ordering communities as a whole. First, global disturbances break up communities too frequently for such large-scale coevolution. Modern plant communities have not yet recovered from the last ice age, in that various tree species are still moving away from the refugia to which they retreated through the last glacial period. For instance, hemlock has only been back in the Upper Peninsula of Michigan for the last 3,000 years, and the American beech tree even today reaches only as far as the western shore of Lake Michigan. Before the floristic influence of the last ice age has disappeared, another ice age should reset the clock. Evolution would require longer periods of time than it appears to have available for it to create community structure.

The second reason to doubt a large influence of coevolution is that the lines of connection upon which selection would have to work are too tenuous. Most clear coevolutionary relationships involve only two species. If communities are highly coevolved entities, we should regularly find clear mutualistic structures with three and four species, and generally we do not. Even in the narrowly specified world of mathematical equations, multiple-species coevolution is an invitation to very rich behavior, for reasons of chaos theory alone. Natural selection cannot work to generate structure in situations that are so complicated and given to frequent reversals in selection pressure.

The community, then, is not generally well considered as a coevolved entity. Even so, macroevolution generates the players and the strategy to which they subscribe. This macroevolution predates, and so is independent of, the communities in which the players occur today. The evolution that is pertinent to community strategies is convergent evolution, not coevolution. **Convergent evolution** arises when a general problem is solved by separate lines of evolution in a common fashion. The eyes of humans, mollusks and spiders are a common solution to the problem of seeing images. Insects have evolved a very different solution.

Addressing convergent evolution, Eric Knox has unraveled evolution of life forms and their attendant strategies for tropical mountain lobelias (Figure 8.22). The time window across which this evolution has occurred is twenty-five million years. That is far too long to apply to modern communities. In twenty-five million years, climates and landforms have come and gone; twenty-five million years ago Mount Kenya did not exist at all. Knox's research setting is so macroscopic, that the evolution of a life form has to wait for its mountain habitat to come into existence. By tracing the changes in the arrangement of genetic material in the organelles of plant cells, Knox has uncovered a consistent pattern of evolution between life forms.

In the lowlands of tropical Africa, the genus *Lobelia* exists as species of herbaceous plants (Figure 8.22C). In mountains scattered widely across four countries, there is a trend toward larger tree forms at higher altitudes (Figure 8.22A,B). So similar are species at particular altitudes, that a reasonable hypothesis is movement from one mountain to the next, to give a series of closely related sister species. Knox's genetic signal has unraveled the pattern, and it appears that convergent evolution, not long-distance dispersal, explains the similarity between species. The land literally rose to high altitudes, and species evolved to adapt to that condition each time it happened. Thus there is adaptive evolution in the emergence of given strategies that come with a given life form. The strategy of being a tree does appear to be evolved, even though the stable communities (forests) that turn on those strategies are not the result of coevolution. *Communities are stable configurations of ready made species, for evolution makes the pieces, not the whole.*

Figure 8.22. *Lobelias.* **A.** Shrub-like form. **B.** Tree-like form. **C.** Herbaceous rosette. Each mountain has its own version of these forms. Usually one form evolved separately on each mountain in response to the rising land surface. From photos by Martin Burd.

Microevolution Inside Communities

Microevolution is evolution as it is understood in common parlance. It is the result of natural selection applied to populations, such that local adaptations occur. Having asserted that communities are not principally ordered by coevolution, it is important to identify the lesser extent to which coevolution does lightly shade community form. The paleontological record of Britain indicates that the British flora and its associated communities differ between the interglacial periods in which they occur. **Interglacial periods** persist tens of thousands of years. In one interglacial scenario when the ice came off Britain early, much ice still persisted elsewhere, so the sea level was low, and the English Channel was dry land. In that case, hazel *(Corylus)* marched into the open habitat along with other species and took its place as a major vegetational component for that interglacial. In a later interglacial, the ice came off Britain late, the sea level was high and the English Channel full of water. Hazel has large nuts that do not disperse well, and so was left behind in what is now France. It did colonize later, but as a rare species.

The remarkable phenomenon is that, if it makes a late start, hazel then remains a minor component of the British flora for the many thousands of years of that entire interglacial. There must be some reason that hazel cannot recover from a marginal start. Evidently the initial conditions leave a permanent mark on community structure for the entire interglacial. This appears to be a matter of *founder effects* and microevolution, whereby later species stand in the shadow of the first arrivals. The original order of arrival leaves an inherited mark on community members through the entire interglacial. Hazel would indicate that coevolution and assembly rules are indeed relevant to forest communities covering thousands of square of miles for millennia. This invites the contributions of population biologists in studying the apparent resistance to deformation of the folded surface in the general community model laid out earlier.

In communities, the loser is the phenomenon of interest, not the winner. Why doesn't the loser lose it all? John Harper came close to expressing that position in his study of weed communities, although in the final analysis he casts his findings in population terms. Harper was able to show that a built-in negative feedback keeps the rare species in a community from extinction, and that this process would have a microevolutionary explanation.

Harper's study turns on the relative abundance of five poppy species as weeds in agricultural fields. At turn of the century, Impressionist painters of wheat fields were daubing golden-yellow canvases, adding splotches of bright red for the poppies. The poppy fields of Flanders saw some of the fiercest fighting of the First World War. But all that was before the general application of broadleaf weed killers, and Harper was lucky enough to catch the tail end of English wheat fields in their full natural beauty. He noted five species of British poppy, one common, *Papaver rhoeas,* and the others less so. Often all five species were present at a site. Counter to first expectations, the less-common species as a group were represented by at least one of their number as regularly as was the common species. Generally small populations were no predictor of occurrence in the weed community.

BOX 8.3. DEMONSTRATING COMPETITION WITH DE WIT EXPERIMENTS

Harper and McNaughton (1962) experimented on four British poppy species. They used the device of the De Wit replacement series experiment to find out if competition within species is the same as competition between species. The De Wit method has its critics, but it is still an intuitively appealing design that makes the point we need to establish here. Two species are sown at specified densities in monoculture. These are the two reference conditions, against which all others are set. In the replacement series, other conditions are intermediate between the two pure stands. In the middle of the series, some individuals or propagules of one species are substituted by individuals of the other in mixed stands. For example, the exactly intermediate condition is sown at exactly half the density of the respective monoculture sowing, but both species are sown and grown together.

The plants are allowed to grow mixed in the various relative sowing densities. The resulting productivity of all these pure and mixed stands are graphed according to the proportion of each species sown. The result of the experiment is the harvest of each species in each mixture. The yield of each pure monoculture is put at its respective end of the abscissa, representing the full density sowing of one species and zero sowing of the other. The growth in biomass or number of individuals resulting from the half-and-half density treatment is plotted exactly in the middle, while the other proportions are plotted on the abscissa according to their respective representation of each species (Figure 8.24). In the specific work on poppies, Harper and his colleague were particularly interested in the effect of common species on rare species, so they tested ratios that made one species common and the other rare in about an eight to one ratio: 87.5% A with 12.5% B, and vice versa.

On the graph of the results of a De Wit experiment, the yields of the monocultures are indicated, and a straight line is drawn from the productivity of the monoculture in

question to the zero point of the other monoculture at the other end of the abscissa. The zero point indicates the absence of that species from the monoculture of the other. These straight lines, one for each species, cross to represent the expected yield of the intermediate conditions. The assumption in the expectation is that substituting the other species makes no difference; it implies that competition within species is the same as competition between species, so the presence of the other species only switches one competitor for its equivalent.

Of course, often the intermediate condition in the replacement series deviates from straight line expectation, and that is where one looks for experimental signals. If competition within species is greater, then the plants in the intermediate condition benefit from the substitution of some of their fellows with members of the other species. If competition between species is more intense than that within species, then the substitution lowers yield below expectations. If one species is the superior competitor, its yield in the intermediate conditions is greater than expectation, at the expense of the other species, which are suppressed below expectation.

In the case of the poppies, the rare species in nature were thrust into the common condition by the experimenter sowing them at 87.5% of all seeds. In those experimentally contrived situations, the species that are rare in nature interfere with themselves more than with *P. rhoeas*, the common species in nature. The rare species in nature, finding themselves common, suppress their own species members with a vengeance (Box figure).

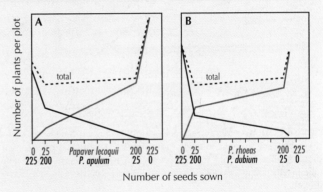

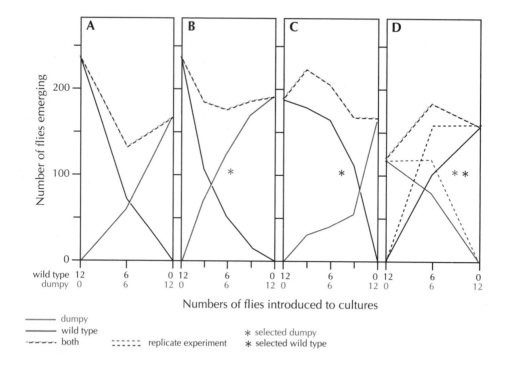

Figure 8.23. The results of a replacement experiment involving wild type and dumpy *Drosophila melanogaster* grown in pure cultures and mixtures. The two forms were prevented from mating with each other. Results show that fruit flies can be coevolved in milk bottles to avoid competition between strains. Redrawn from Seaton & Antonovics 1967. In **A** two naive populations compete strongly with each other. In **D** two coevolved populations avoid each other. **B** and **C** mix naive and coevolved populations, to the disadvantage of the respective naive populations.

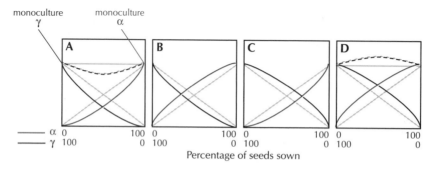

Figure 8.24. DeWit series replacement series showing expectations and results. The straight lines (gray) are expected results for each species and total production. In **A,** between-species competition is greater than that within species, as shown by the sagging line for total production, which is less in mixed culture. In **D,** the reverse pertains, as shown by the total production as a bowed-up line, with increased production in mixed culture. **B** and **C** show one species winning and the other losing.

From the results of competition experiments on poppies (see Box 8.3), Harper showed that success is self-contained in plant populations. The rare species can afford to evolve individuals with a greater capacity for interference with neighbors, because immediate neighbors are usually not members of their own species. Selection is therefore towards greater interference for the rare species, and toward less interference for the common species. Harper argues for a long-term seesaw, where the capacity for greater interference is alternately selected for and against: for, when the species is rare; against, when the species has some trick to make it common. Winning in competition between species is a double-edged sword. The means of becoming dominant make it harder for dominance to lead to extinction of the losing species. Competition increases within the winning species by virtue of a greater encounter rate for individuals of that species.

Seaton and Antonovics (1967) showed how flies can be coevolved in milk bottles to avoid competition between strains (resource partitioning see Chapter Seven). Like Harper and McNaughton (1962), they too used De Wit experimental protocols (Box 8.3, Figure 8.23) Therefore, Harper's suggestions could apply at least as a subsystem inside animal communities. However, the notion of encounter rate in animals works more often at the level of the individual as it encounters something in its travels. Patterns of encounter in plants are more driven by sheer numbers in populations, because of the large random component of where seeds land. Animals can adjust encounter rates with others of their own species by behaviors that either seek out or avoid conspecifics. The experiment with flies evolved strains that behaved differently—laying eggs in only part of the milk-bottle environment—to avoid competition with the other strain. The rules of engagement appear to be very different within the plant kingdom, as opposed to the animal kingdom. Animal communities appear to function with another layer of control that adjusts simple processes of interference. In a community, the link between individual animals and their populations is different from that which applies to plants. For an individual, the balance between membership in a population and membership in the community is different for plants and animals.

THE USE OF MODELS IN COMMUNITY ECOLOGY
Analog Approximation Versus Digital Representations

Manipulation of whole communities is often out of the question. Only the most ambitious experimental protocols can make observation of the whole community system the test. At Hubbard Brook, whole forested watersheds were cut and treated with herbicides. These experiments have been so important that we take their findings as common knowledge. Nonetheless, the forest has been treated more as an ecosystem with a nutrient budget rather than as a community with a species constellation. On July 4th, 1976, by the Flambeau Flowage in Northern Wisconsin, a massive, tornado-like downdraft blew down the "Big Block," the biggest and best stand of hemlock in the state. Before the downdraft, this stand was a fenced experimental reserve that was to probe the effect of deer browsing by exclosure of large herbivores. After the event, it became a very impressive jumble of fallen timber, left in place to study **regeneration**. After more than 20 years it is still a pile of tree trunks, each as thick as a man is tall, with little regeneration. Although presenting an opportunity to learn, such natural experiments, save for their catastrophic beginnings, unfold too slowly and too uniquely to be standard protocols to deliver understanding of forested communities at large.

Given these limitations, some sort of modeling is needed. Two fundamental classes of models can be distinguished in the same terms as can analog and digital technology. *Analog models* are small and manipulable, much as the model boats used in yacht design. The limitations of analog systems are defined by the sometimes problematic differences between model system behavior and the full-sized system of interest. *Digital models* are representations inside computers. Their limitations come from the lack of any sort of material connection to, and derivation from, the working of the material system itself.

Regular recording tape is analog, in that magnetic material on the tape exists in a certain state by virtue of a magnetic pulse that came from the recording head. The source of that pulse was an electrical signal that ultimately had its source in the vibrations of the musical instruments. Analog recording has a direct material link to the original material vibrations that were recorded. In digital recording, between the original performance and the music coming from the speakers is some sort of symbolic representation. The sound is represented by an arbitrary coding system that gives each vibration period a number. The sound is reproduced by decoding that number.

Experimental manipulation of a material ecological system employs analogy. In analog modeling, some physical material system is used as an analog for a large, inconvenient system. The physical model system might be a collection of real organisms in an experimental container (eg. Seaton's flies in a milk bottle), whereas the unwieldy system that is modeled might be a whole forest. Digital modeling uses some symbolic representation of the forest, with no material organisms or ecological processes in the model. Instead of a tree being modeled by some other more convenient organism such as a unicellular alga in a flask, in digital modeling, the tree exists only as a piece of computer code coming from a programmer's mind. Robert Rosen (1991) has pointed out that when two systems can be coded in to and out of some representation, then the two material systems become models of each other (Figure 8.25). Thus one can either manipulate the small material system substitute as an analog of the big system, or one can play with the representation that arbitrarily links the coded abstraction of the scientist with the big, inconvenient system of interest. Playing with the small material system is what goes on in regular experimentation. Playing with the representation in between is what goes on in computer simulation modeling.

Figure 8.25. When two systems can be coded in to and out of some representation, then the two material systems become models of each other. Redrawn from Ahl & Allen 1996.

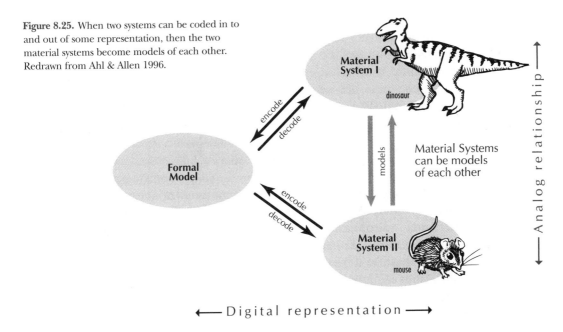

Material Manipulation

First consider small material systems as would be used in a laboratory or field experiment. This is equivalent to the model boat used to design the real thing. Communities of microscopic organisms are a possibility here: they fit into flasks, can be readily manipulated, and can be replicated under relatively controlled conditions. Jim Drake (1990) at the University of Tennessee has been doing some interesting work with small model communities. He builds plankton communities one species at a time to work out the rules for the assembly of communities. He has found that the order in which species arrive in his flasks (read communities) makes a difference to the outcome of the surviving species' mix and relative abundance. He can build communities as many times as he needs, and he can do it under highly controlled conditions that show what matters to give replicated results. Failure to achieve the same result across replicates indicates factors and patterns that are incidental.

This sort of experimental work on model systems can translate to more tangible communities. David Tilman started his career as an experimental algal community ecologist, but he has since moved into grassland community experimental studies. His aquatic origins show in his focus on mineral nutrition as the mediator of competition in prairie research sites at the University of Minnesota. Once again, the experimental systems are models small enough to be replicated as plots. As communities go, they are depauperate, with only a few species. Later in this chapter, we will describe Jonathan Roughgarden's island "communities" of lizards, where the islands had either one species—a population—or two species, making a depauperate community of sorts (Roughgarden and Pacala,

1989). Roughgarden's lizard systems are analog models for animal communities, much as Tilman's plots are analog models of whole prairie communities.

Tilman has been using his analog models to conduct experiments on competition in situations that are more complicated than those common in population work. He should therefore be seen as working the interface between populations and communities. Tilman and David Wedin (1991a,b) have experimentally addressed the success of a small number of bunchgrass species, which dominate grasslands from the Deep South's steamy climate to the wind-blasted cold of Canadian parklands. The dilemma is that, despite their abundant success across North America, bunchgrasses appear unable to return to a site after just a few years of agriculture. Tilman and Wedin have shown it is feedbacks reducing prairie soil nitrogen that are responsible. When the bunchgrasses achieve success, the feedback works in one direction to lower nitrogen (Figure 8.26). However, that same feedback works in the other direction, toward grazing and high nitrogen, after fertilizers increase nitrogen in a one time reversal of the feedback. We laid out the working of that feedback in the previous section on shared strategies of community members (Figure 8.21). Low-nitrogen prairies we labeled strategy 2, and high-nitrogen pasture as strategy 3. Those insights come from employing the results of Tilman and Wedin (1991a,b) as they observed their model prairie plots.

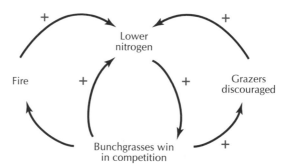

Figure 8.26. The positive feedback of bunchgrasses in tall-grass prairie

Digital Computer Models

Often the difficulties in running whole community experiments are insurmountable, but we still have questions about specific whole communities that need experimentation. Let us turn from material, analog modeling toward representational or simulation modeling, and the experimentation that occurs inside the computer.

Theoretical ecologist Buzz Holling once said he was not particularly pleased with his celebrated budworm model, because he had so much difficulty in making it fail! The point of models is not so much to get them to **work** (predict an outcome in nature), as to play out various assumptions to see their consequences. Simply giving an ecological entity a name makes the assumption that it is possible to freeze some dynamical situation and capture the whole thing. External processes are going to be included, and the boundary is going to leak molecules across it. Thus we cannot in fact freeze the process, and the assumption that we can is wrong. All models have assumptions, and all assumptions are wrong; nonetheless, the scientist is forced to make assumptions to encapsulate any understanding of experience. Therefore, science cannot learn which assumptions are wrong, but rather it exposes which assumptions one cannot afford to make, because they interfere with prediction. The point is to distinguish between significant and insignificant assumptions about how the material system works. The community ecologist does that by having computer-based models fail. *We find out about the material system by modeling to investigate the assumptions that underlie our understanding.*

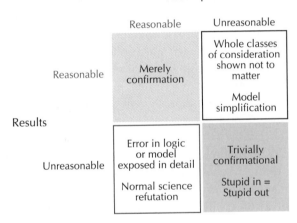

Figure 8.27. Probing assumptions with computer models. Only surprises instruct science about assumptions. Confirmation is not entirely trivial, but it only helps in the engineering phase to get the model to work well enough to set the stage for the surprise.

The two strategies for making assumptions

There are two strategies to probe assumptions with computer models. One is to make an assumption that is grossly and demonstrably false, and see if it matters. The surprise is if the model works. The other strategy is to make assumptions that seem entirely reasonable, and see if they work to give predictions. The surprise is if the model fails. In both cases it is the failure of our reasonable expectations that is instructive. Only surprises instruct science (Figure 8.27). Confirmation matters only in the preliminaries, as one calibrates a model.

- *Unreasonable assumptions.* A model of an ungulate community could assume that animals can move at the speed of light; that is, they can graze anywhere they like, taking successive mouthfuls miles apart. Of course real animals cannot do that, but there is nothing to stop us from modeling that situation anyway. If the model with such an assumption still gives reasonable grazing levels and species mixes, then the message is that movement of grazers does not matter in predicting outcomes, so you can ignore it. This first strategy is coarse grained in that it weeds out whole classes of reasonable assumptions as irrelevant. These assumptions can be eliminated from subsequent models not as wrong, for all assumptions are wrong, but as completely unimportant. The scientist has just found out what apparently does not matter, and simplifies models accordingly. In the simpler models that follow this strategy, the focus turns to assumptions that do matter. Tests become less ambiguous. This strategy allows gross simplification without creating models that are simplistic.

• *Reasonable assumptions.* The other modeling strategy for dealing with assumptions is to make reasonable assumptions. For example, assume that animals move at realistic speeds, and graze making choices of food in order of preference and availability. If that model gives reasonable results, then nothing is learned. If, however, such a reasonable model gives unreasonable results, then we have learned that either the subtleties of animal movement are more important than we thought, or the food choices are based on some other criterion than preference interacting with availability. We can then tinker with assumptions about how availability and preference of food work, eliminating finer reasonable assumptions for the meantime, until a very specific model emerges as our working model. Thus the second strategy for dealing with assumptions is fine grained. The trick in both strategies is to find a good reason for the observed results by elimination.

Population versus individual-based models

Another important distinction inside the class of representational simulations turns on whether or not there is feedback at the level of the individual in its functioning in the community. The distinction is between the orthodox **statistical population model** and the **individual-based model**. The former models populations as a whole, whereas the latter keep the identity of each individual in the computer for the entire simulation.

First consider the orthodox statistical *population models*. Progress is made in science when a counter-intuitive result pertains. Intuition would say that common species are less likely to go locally extinct than uncommon species. Somehow we are prepared to overlook examples like the passenger pigeon. Consider the following observation of Alexander Wilson, the father of American ornithology (about 1806):

"...They were traveling with great steadiness and rapidity, at a height beyond gunshot, several strata deep, very close together, and from right to left as far as the eye could reach, the breadth of this vast procession extended...from half-past one to four o'clock in the afternoon...the living torrent rolled overhead, seemingly as extensive as ever." Wilson estimated the flock that passed him to be two hundred and forty miles long and a mile wide—probably much wider—and to contain two billion, two hundred thirty million, two hundred and seventy-two thousand pigeons. On the supposition that each bird consumed only a half-pint of nuts and acorns daily, he reckoned that this column of birds would eat seventeen million, four hundred and twenty-four thousand bushels each day.

Passenger pigeons "obscured the light of noonday as by an eclipse" in the Midwest of the United States in the last century, only to have the last individual of the species drop off her perch in the Cincinnati Zoo early this century.

Over a decade ago in an appendix to one of his papers, Alan Hastings identified the critical importance of investing resources in colonization of new sites for populations in regions where habitat destruction is significant. In a recently published model in the population style, Tilman simulates populations using that idea of Hastings. The crucial assumption is that there is a tradeoff between competitive ability, once a plant has arrived, and the resources invested in getting to new sites. Given that tradeoff, Tilman's model shows that under conditions of habitat destruction, the better competitors appear to be more vulnerable. Counter-intuitively, the model thus suggests that common species may be more vulnerable to extinction. Tilman's result is not a quirk of some local condition prescribed in his simulation model, but appears to be robust under many conditions. The argument is that rare species persist by investing more in dispersal to new sites than in outcompeting others in any given site. This harkens back to Harper's findings about poppies. Under changing conditions, as habitat destruction ensues, those species investing resources differentially in invasion of new sites win out.

The implications for global climate change in the Hastings-Tilman model are significant. Conventional wisdom says that we should be more concerned about the survival of rarer members of communities under climate change. Tilman's model suggests the opposite; he indicates we worry more about the unexpected, that is how common species may disappear wholesale. The message is, common or rare, species which invest little effort in invasion are vulnerable. In communities we are interested in how the rare species do not lose it all, and that concern is rewarding.

Now let us ask what makes the style of modeling used by Alan Hastings and David Tilman statistical- and population-based, as opposed to individual-based. The conventional population-based models make the assumption that when you have modeled one individual, you have modeled them all. It is remarkable how often that assumption holds,

BOX 8.4. STATISTICAL MOMENTS

Statistical moments calibrate the properties of frequency distributions. A frequency distribution is just an economical means of recording data about a collection of individuals. It puts a sequence of values such as height or weight along the abscissa, and lumps the individual numbers into classes, such as 50 to 59 kilograms, 60 to 69 kilograms, and so on. The frequency distribution assigns each individual to one of the classes. It is called a frequency distribution because, as the individuals pile up in the classes, the height of the pile indicate the frequency with which one encountered individuals in that size or height class (Figure 8.28.) Normal frequency distributions follow a symmetrical bell-shaped curve.

The first statistical moment of a frequency distribution is the *mean,* which precisely ignores everything about individuals. The mean identifies the characteristics of the mythical average individual (Figure 8.29a). The second moment acknowledges differences in individuals by recording variation, but the variation of all individuals is always in the same terms: the average of the squared deviations from the mean. A sin-

gle number, the *variance,* records the total variation, with all individual variations subsumed in the one number (Figure 8.29b). In normal distributions, the value of the mean tells you nothing about the variance; they are independent.

The third moment is called the *skew.* It acknowledges that there may be more variation on one side of the mean than the other. The skew is again a single number in which the average differences between the individuals on either side of the mean is recorded. (Figure 8.29c). The normal distribution is symmetrical, and so it has a skew of zero. Usually the third moment will do it, but one can go further. There is a fourth moment called *kurtosis,* which refers to how flat the distribution is across the top, and how sharp are its shoulders (Figure 8.29d). The thing to notice is that in all these moments, one number tells all. Individuals are explicitly ignored in a summing operation—a different operation for each moment, but a sum nevertheless. Any time we add up numbers, there is an assumption that the units are the same, i.e. there are no critical differences hidden inside the grand total.

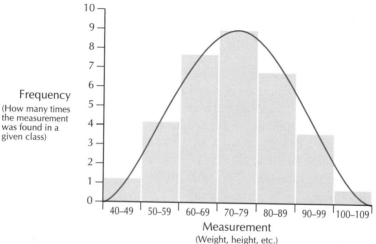

Figure 8.28. A frequency distribution.

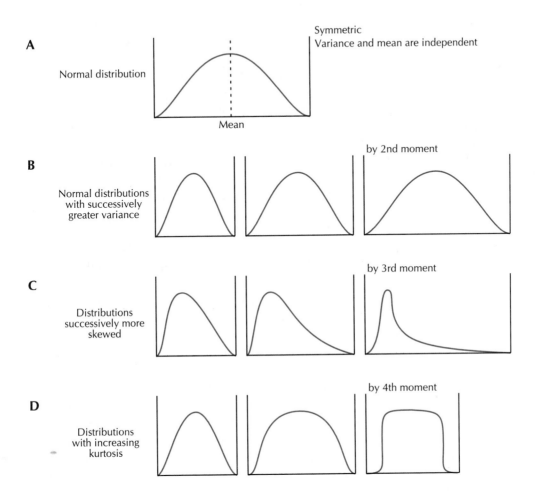

Figure 8.29. Moments of a frequency distributions. See Box 8.4 for discussion.

for one can often successfully model a material situation while ignoring differences between individuals altogether. The general assumption takes the form of various statistical moments of distributions (Box 8.4). The strategy is to find some single number that represents some property of the whole population. All the differences between individuals are subsumed under that one number.

How do *individual-based models* differ from population-based models? Throughout all the operations inside individual-based models, the individual retains its identity. One should use this sort of model when differences between individuals keep getting larger because of an inequivalence between individuals. Individual-based models are the model of choice when there is feedback in the differences between individuals. An example is size feeding back on itself. Here a large individual is likely to grow larger, faster, because it is large in the first place. In such situations, differences amplify over time. The differences between sizes of individuals are themselves different; the difference between small and medium individuals may have less significance with regard to growth in body length than does the difference between medium and large body length. In this sense, the original differences and final differences are not in the same units when the implications for growth of individuals are considered. Therefore one cannot add the differences together to achieve a meaningful result.

The third moment, the *skew*, can capture some of this sort of feedback, in that skew gets larger as the few large individuals at the top of the distribution pull away from the pack because they have more with which to grow (Figure 8.29C). However, skew assumes that, even if size increases across the entire range differ (bigger gets bigger faster), the differences between classes all have the same mathematical relationship. All size increases have the same effect, but with different magnitude. If, however, there is a threshold effect, and being larger has no effect below a certain size, then the only difference in body length that matters is the one that crosses the threshold. Skew treats all size differences the same, and cannot ascribe special significance to particular differences. Individual-based models keep track of all individuals, precisely so that important differences can be given a special status.

Mike Huston and his colleagues at Oak Ridge have nicely characterized individual-based models. Consider two fish species with roughly normal distributions of size amongst their individuals, but with one species' mean size being slightly larger. Over time, the fish will eat and get bigger, so the size of the mean individual of both populations will increase. Using a population-based model, the two species populations should move to the right on the graph (Figure 8.30). Growth in small to moderate sized fish is at first exponential. That is to say, larger fish of both species have more biomass on which to grow and so the variation in size between individuals, as measured by the second moment, will get larger over time. The same differential growth will take the roughly symmetrical fish distributions and skew them, so that a small number of fish in both species become disproportionately larger. Therefore a population-based model will predict that both species persist, both grow larger, and both become more skewed.

Figure 8.30. Population-based model predicts all species grow and survive.

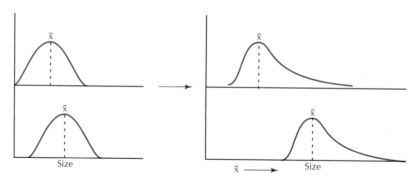

The result from an individual-based model might be different. Say that, at the outset, the larger individuals of the larger species are large enough to eat the smaller individuals of the smaller sized species, but other differences in body length are not sufficient to allow predation between size classes. The smallest fish at the bottom end of the population with small average body size are eliminated (Figure 8.31). Now here comes the feedback. Because the bigger fish have eaten heartily, they get bigger quickly, whereas all other fish in both populations are not big enough to be effective predators, and so they do not grow in body length. Because the biggest fish are now bigger, they can eat larger prey. This leads to the largest fish in the population with the larger mean size being able to eat the middle-sized fish of the other species, and the smallest fish of their own species. Round the feedback loop again, and the elimination of the two smaller classes of fish means that the biggest fish keep on growing (Figure 8.32A). At this point they finish off the other species altogether, and consume their own middle-sized brethren. The outcome is therefore very different from the population-based model, in that only one species survives, and we have a small population of large individuals (Figure 8.31).

Individual-based forest simulators

Forest stand simulators are individual-based models that are an important part of community ecology. The working of FORET, Hank Shugart's general forest model, captures the essence of a family of models. It tracks hundreds of trees individually in the simulation of approximately one-tenth of a hectare of forest. FORET's central assumption is that the most important resources for vegetation come from above. In forests, the resource is light above ground, and in the grassland derivatives of these models, it is water moving down through the soil. Being an individual-based model, FORET sees the forest as having critical feedbacks: the larger the tree, the greater its chance of growing even larger. Conversely, failure to grow allows other individuals to grow past the slow growing trees and suppress them further (Figure 8.32B).

No assumptions are ever correct, so one should not be distracted by the fact that certain assumptions in the stand simu-

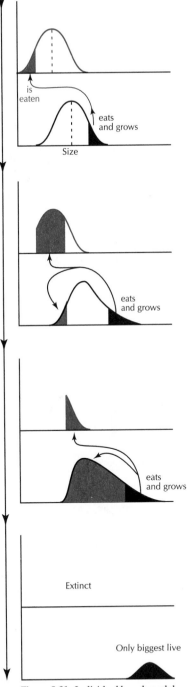

Figure 8.31. Individual-based model predicts only the largest individuals grow and only one species survives.

A

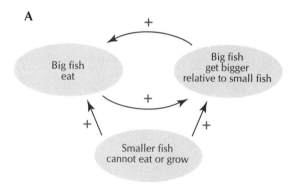

B

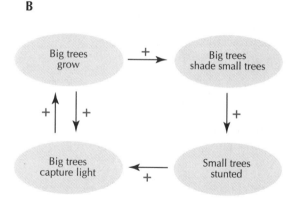

Figure 8.32. Positive feedback loops.

lators are patently false. All the trees are modeled as growing from one spot in the center of the one tenth hectare unit (Figure 8.33A). The area simulated is only implied by the amount of vegetation that is allowed to grow in the plot. Each tree in the model is subjected to the shade of the canopy of all taller trees in the plot. Shade tolerant species are assigned a greater probability of survival and growth than a shade intolerant species, for a given low light regime. Probabilities for death or growth are modified by moisture and temperature.

FORET has been some criticized for not giving a perfect simulation. Another complaint is that the parameters need to be finely tuned, independent of actual forest data, to give the desired results. If the simulation must be tweaked until it gives the desired result, then the result gives little insight into detailed processes that grow forests on the ground. These are valid criticisms at one level, but it is still impressive that a model with such strong simplifying assumptions can give such rich and reasonable results.

With the arrival of ever more powerful computers in the last decade, it has been possible to run the alternative strategy of testing of assumptions with a new class of model called SORTIE. SORTIE demands that the computer keeps track of the different placements of all plants, and relates each to its neighbors according to proximity (Figure 8.33B). Unlike FORET, SORTIE relies heavily on actual field data from forests resampled after many years. The payoff is that in SORTIE, it is possible to include extensive reasonable assumptions about spatial issues, such as distance of seed dispersal, or trees in the Northern Hemisphere casting shadows generally in a northerly direction. The model dissects and tests the detailed mechanisms of individuals relating to each other to give whole forests. This is something FORET cannot begin to achieve.

The downside of SORTIE is that its computational demands are so gigantic as to press even modern limits. Also, we do not have sufficient field data to apply the model in many settings. That means SORTIE cannot readily be used to calculate vegetational patches repeatedly to fill in a regional landscape. Meanwhile, limited by

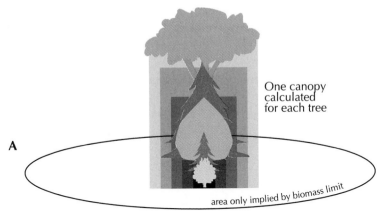

A

One canopy
calculated
for each tree

area only implied by biomass limit

Unreasiltic assumption about trees in space in FORET

Figure 8.33. Examples of
individual-based forest
simulator models.

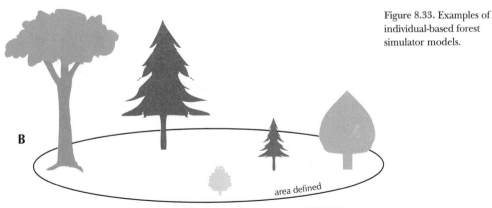

B

area defined

Realistic assumptions about trees in space in SORTIE

the computational technology of a decade ago, Smith and Urban (1988) took advantage of simplifying assumptions of FORET-type simulations to grow simulated forests at a regional scale. They simulated a landscape grid wherein simulated plots were able to influence their neighboring simulated plots. With the need to put a vegetational component into large-scale climate-change models, we must live with models that make gross simplifying assumptions. Each model has strengths and weaknesses, and each serves the community ecologist for its particular purpose. No model is good for everything.

Stand simulators and complex behavior

At least one form of the FORET model showed **hysteresis**, a character of non-linear models of complex systems. Hysteresis occurs when change in one direction follows a different track than does an equivalent reverse change. Hysteresis would occur in our folded-surface model for community, where, under increasing water, the drought-loving species persist until it is very wet. Conversely, established moisture-loving species persist under drying conditions until extreme drought (Figure 8.34).

While hysteresis may seem an esoteric point, it has implications for sustainability of forests. If the environment of the upland forest for East Tennessee is changed to favor American Beech, or is changed to favor it less, the change in the population of Beech is asymmetric. The lower altitude limit for Beech is different depending on whether Beech is encroaching or retreating. The implications for sustainability are that it is easier for a declining population to hold its own than for it to re-establish once it is gone. It therefore appears more important to expend resources holding on to the desired cover for desired sites than it is to expend those same resources attempting to mitigate loss of desired populations after they have disappeared. Note that this recommendation comes from an unreasonable minimal model, but one

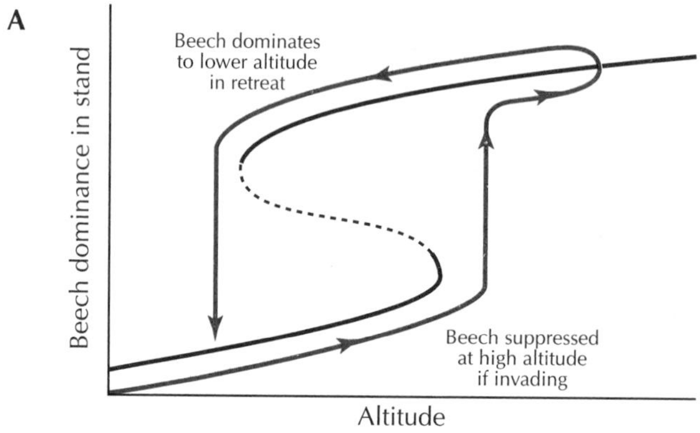

Figure 8.34. A. Hysteresis for Beech in the FORET simulation. B. Beech trees on a ridge are easily recognized by their smooth, bright gray bark. From a photo by T. Allen.

which inspires confidence by virtue of displaying convincing richness and surprising accuracy, relative to field data.

TECHNIQUES SPECIFIC TO COMMUNITY ECOLOGY
Data Collection

Experimental approaches to community ecology are not that distinctive, but data collection is a different matter. A particular feature of communities is the requirement that species, perhaps a large number of them, be identified. Since individuals of different species will vary in size and form, one of the challenges of community ecology is to collect data about different species in comparable and relevant terms.

In plant community ecology, the very identity of the community turns on Roscoe Pound's invention of the **quadrat** at the end of the last century (Pound went on from ecology to become a Justice of the United States Supreme Court). The quadrat started

A Square quadrat

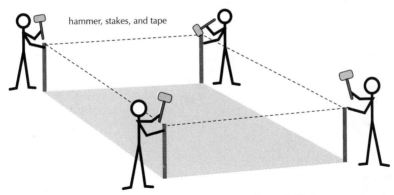

hammer, stakes, and tape

Figure 8.35. Square quadrats in forests are more work than circular quadrats or distance measures.

B Circular quadrat

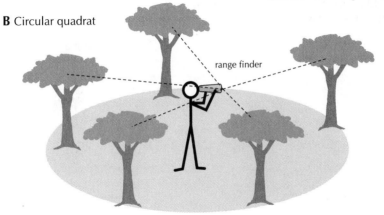

range finder

as a square sampling unit in which plants were counted. For herbaceous vegetation studies employing quadrats, the one-square-meter size quickly became the standard. In forests, however, large circular quadrats are faster to set up, and a range-finder indicates whether trees are close enough to the center to be included in the sample (Figure 8.35).

There are two standard procedures for collecting community field data, **density** and **frequency**. Each has its own advantages, as argued by John Curtis and his student R. P. McIntosh (Curtis and McIntosh, 1950). Density counts all individuals in the quadrat of a specific size, then the number of individuals per unit area are calculated. Collecting density data can be tiresome, particularly if it is difficult to assess what constitutes an individual, as in grasses (Figure 8.36A). Avoiding such dilemmas, a quicker procedure collects frequency data. Frequency requires only that the observer determine whether the species is present or not in any given quadrat. The frequency is the percentage of quadrats in which a given species occurs (Figure 8.36B). The disadvantage of frequency is that, unlike density, frequency changes with the quadrat size. Certainly more individuals are found in larger quadrats, but one divides by that larger area in the calculation of density, so the increase in numbers cancels out. For frequency, larger quadrats give a higher percent frequency, and there is no correction for area.

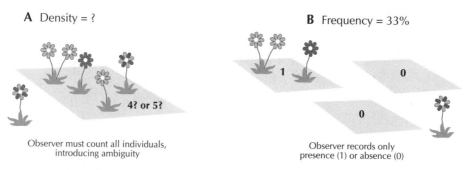

A Density = ?

4? or 5?

Observer must count all individuals, introducing ambiguity

B Frequency = 33%

1   0

0

Observer records only presence (1) or absence (0)

Figure 3.36. A. Collecting density data via quadrats. **B.** Collecting frequency data via quadrats. **C** and **D,** distribution of the study species will affect calculation of frequency, but not density. Density stays at 1 in **C** and **D**, but frequency increases as plants are distributed more evenly. Individuals outside of quadrats are ignored.

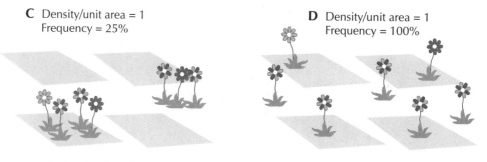

C   Density/unit area = 1
Frequency = 25%

Aggrigated or clumped placement

D   Density/unit area = 1
Frequency = 100%

Regular placement

Another disadvantage of frequency is that it increases if the distribution of plants on the ground tends to be even, that is regular. Conversely, clumped distributions will record a lower frequency for a given plant density (Figure 8.36C, D). Average density is unaffected by departures from random placement of individuals on the ground, but the variance of the density does increase when the quadrat is the same size as the aggregations of individuals on the ground. Peter Greig-Smith (1952) seized the bull by the horns, and instead of bemoaning how variance changed with quadrat size in nonrandom populations, he used those changes to work out the scale of the aggregations and regularities.

For forest data, density and frequency are available, but when measuring trees, a new measure called **dominance** becomes useful. To measure dominance, one wraps a tape at breast height around a tree in the quadrat, and from that circumference one can calculate the cross-sectional area of the tree trunk. This not only estimates the biomass of the species, but it also corresponds to the proportion of the ground which is shaded by each species. The area covered by leaf shade may also be measured directly with what is called **cover**. Cover can be measured in herbaceous vegetation by inserting long pins down through the vegetation, and recording which species touch the pins. Cover, like frequency, is expressed as a percentage. The percentage of pins touched is an indication of the percentage of ground shaded. If the ecologist records not just touch or no-touch, but the number of times the pin is touched by a given species, the measure is called **cover repetition**. Much as dominance, a biomass measure, approximates cover in trees, cover repetition in herbaceous vegetation is roughly a measure of biomass. For shrub data collection, a good method is called **line intercept,** a hybrid between cover and frequency. Here one lays out a line of perhaps 100 meters, marked into segments of perhaps one meter. The measure is of how many segments are intercepted by the species in question. The line intercept is the percentage of segments that are under the canopy of the shrubs.

A particularly fast measure of density in trees can be achieved through the **distance measures**, pioneered by University of Wisconsin's Grant Cottam with John Curtis. Distance measures depend on the intuitively obvious relationship between density and the proximity of individuals. Individuals of a high-density species will be close together. Therefore, measures of the proximity of trees also gives density.

Each species may respond to the various measures differently, but it is possible to put things on a more level footing by expressing density, frequency, dominance or cover not as absolute values, but as relative measures. Thus *relative density* expresses species abundance not as actual density, but as a proportion of the sum of all the densities for all the species. The same relative relationships can be applied to frequency, dominance or cover.

Subjective methods of estimating vegetation using classes of dominance, and describing only good examples of the type of vegetation, are not satisfactory. If one samples vegetation using only "typical" vegetation, then of course the vegetation will appear to fall into distinctive types, because the vegetation that would provide ambiguity has been purposefully excluded. Randomization of sample points, placement of sampling units, and an unbiased choice of site in the first place are all crucial. Plan the general type of vegetation to be sampled, but avoid bias inside the operationalization of the plan. For example, John Curtis and Robert McIntosh, while choosing sites in their study of Wisconsin southern upland forests, had species-neutral criteria for selection: 15 acres or more of homogeneous, undisturbed, natural forest on uplands. So long as a stand met those criteria, it could contain any species. This last point ensured that the vegetation was chosen in an unbiased fashion, even if it was chosen according to a certain set of explicit restrictions.

Randomize sample points, and beware that the observers easily impose bias unconsciously, either by including rare species or impressive individuals, or by capriciously excluding remarkable plants while fighting the first bias. Sometimes unconstrained randomization of sample points is too time consuming to get a fair representation. In that case, it is valid to use a stratified random sample. Here, the total area is divided up into segments, and random sampling occurs equally in those segments (Figure 8.37). That way, the total area is sampled to give a properly representative sample in a workable amount of time.

A Random sampling

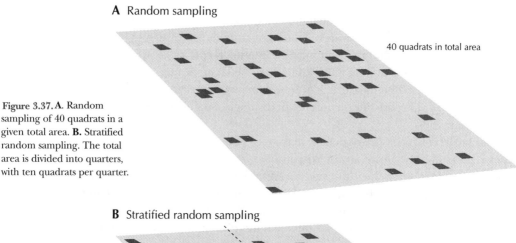

40 quadrats in total area

Figure 3.37. **A.** Random sampling of 40 quadrats in a given total area. **B.** Stratified random sampling. The total area is divided into quarters, with ten quadrats per quarter.

B Stratified random sampling

10 quadrats in each quarter

Zoologists use a different suite of methods, because animals move. Plants are ubiquitous, but animals represent a much smaller biomass, and spend relatively little time at any given spot on the landscape. Therefore, a straight animal count is usually not an option. **Mark and recapture** is one method in common use for animal data collection. Animals are caught and tagged or marked in some fashion, then released. As others are captured, it becomes possible to calculate the population size from the tagged animals recaptured. In other protocols for animal data collection, the observer walks a compass line and listens or observes individuals. Unlike counting plants in a quadrat, listening for birds hazards the distinct chance of unknowingly counting the same individual more than once (see Figure 8.3). Each encounter is scored as a hit, the logic being that, because animals move onto spots as much as they move away, the individual identity of each animal is beside the point.

Another sampling method involves pitfall traps—destructive of the animals sampled, but effective. On the assumption that animals such as insects, spiders, amphibians or small mammals move around and blunder into hazards in a random fashion, pits made of jam jars and the like are laid out. After a fixed length of time, the dead animals in the pit are counted.

Data Analysis

A particular problem of community ecology is the richness of the data. Often a community study will have a large number of sites occupied by a large number of different species. It is hard to see the pattern in the entire data set, so some sort of analysis or summary process is necessary.

Data analysis as interpolation

The summary of data is a matter of degree. All data analysis is some version or another of a process of **interpolation**. Connecting two points with a line is an interpolation, for it asserts a series of unmeasured values in between the points. Interpolation is also a matter of degree. We could make a plot that connects all data points in a tight interpolation. An equation to describe data can be so elaborate that it that goes through every data point. This is hardly a summary at all, in that all the data of species occurring in sites are preserved and presented (Figure 8.38A). Conversely, one could draw a straight line across a graph at the value of the mean. The line is influenced by all the data points, but it is a loose fit of all of them (Figure 8.38B). At one extreme of analysis, the complete interpolation misses no points, and at the other extreme the loose interpolation misses all points.

Variously tight or loose interpolations exist between the extremes. Very tight and very loose interpolations going through either one or none of the points are usually undesirable. An intermediate interpolation would be a waving line that makes its best

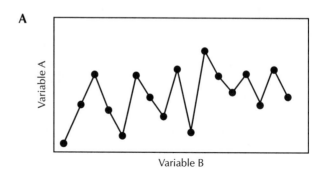

Figure 3.38. **A.** Tight interpolation: the line goes through every point. Summary equation will probably have 16 terms plus a constant, e.g., $a = c_0 + c_1b + c_2b^2 + c_3b^3 + \ldots + c_{16}b^{16}$ **B.** Loose interpolation. A straight line going only through the mean. Summary equation is the sum/N and a single term to express slope. **C.** An intermediate interpolation.

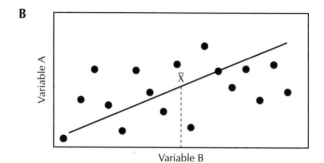

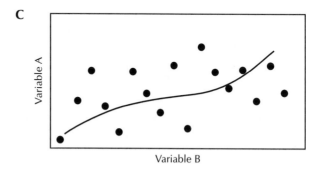

shot at getting close to all points but probably goes through none (Figure 3.38C). Even so, the use of single indices, as very loose interpolations to summarize complex data sets, is common. A popular summary index of this type in ecology is some measure of diversity, perhaps something as simple as S, the number of species at a site. Even more extreme is the average number of species at all sites.

Assumptions in data summary

The weakness of the complicated equation going through every point exactly is that it makes no assumptions; one might as well present the entire data set as a mind-numbing spreadsheet. Conversely, the weakness of any single summary index, including diversity indices, is that they all are heavily burdened with many important assumptions, so many that the list of significant assumptions is almost as long as the original data. The summary indices are more dangerous, in that the single index can be used as a summary without regard to the assumptions, which may remain hidden, and whose implications are often ignored. Since attention to, and testing of, assumptions is the name of the game in science, both extremes are of limited utility. Community ecology has so many summary methods, that some practitioners specialize in being methodologists or statisticians.

Powerful standard statistical methods of testing (described in Chapter One) are offered in statistical packages for microcomputers. They all work closer to the end of summary indices, giving a loose interpolation. For example, analysis of variance greatly compresses data to a small number of indices (within and between type effects), but the beauty of that approach is that the assumptions are laid out absolutely explicitly. More common in community ecology are classification schemes that group similar types of vegetation or animal assemblages together. A more explicitly geometric approach places particular sites on a small series of summary *gradients,* instead of a small number of classes. These too are powerful summary descriptions, but their assumptions are often not pressed to the fore.

Data summary through gradient analysis

At the outset of our discussion of community ecology, we suggested that adequate description is crucial. **Gradient analysis**, a member of the general class of interpolation methods, is a powerful descriptive method. One can place sites on gradients to compare vegetations or animal assemblages. For example, Roger Bray and John Curtis (1957) performed a gradient analysis that arranged the sites of Southern Wisconsin forests on three mathematically defined gradients, to compare the ecology of different sites. Alternatively, species of plants or animals may have their occurrences summarized on gradients so that we can compare their ecologies. Bob Gittins not only arranged British grassland vegetation on gradients, but he also performed an analysis that arranged the species on summary gradients as a complementary analysis (Figure 8.39A).

The summary gradients can be directly derived from environmental data, as in Whittaker's analysis of the Great Smoky Mountains. Alternatively the summary may come from continuous changes in species composition across different sites (Bray and Curtis 1957) (Figure 8.39B). In systems that exhibit rapid change in composition over time, it is possible to analyze the system as a best summary of community change over time (Bartell, Allen and Koonce, 1979) (Figure 8.39C).

Figure 8.39. A. Summary plot redrawn from Gittins (1965). Instead of plotting vegetation at sites, Gittins shows a scatter plot of species, where species close together occur at similar sites, and so have similar ecology. **B.** Left, redrawn from Whittaker's (1956) analysis of the Great Smoky Mountain species composition, derived from environmental data. Right, redrawn from Bray and Curtis's (1957) summary from continuous changes in species composition across different sites. Line color and dash indicate same species. The similarity is remarkable. **C.** Redrawn from the summary of Bartell et al. (1979) of the path of change of phytoplankton in Lake Wingra over a year.

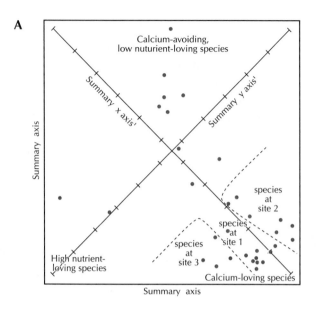

A.

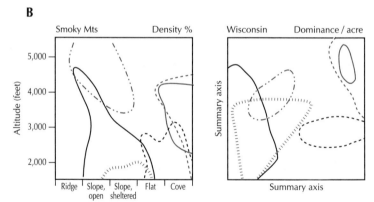

B.

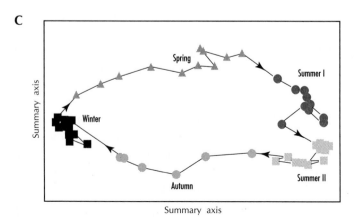

C.

All of these techniques amount to compressing data onto the **summary axes**. Various procedures are used for compression, but most of them can be intuited by a geometric projection. This is like projecting the shadow of a three-dimensional structure, such as your hand, onto a two-dimensional screen (Figure 8.40). The shadow is a two-dimensional approximation of a three-dimensional space. Note that the shadow cast from your hand changes as the light source comes from different directions, or as your hand is rotated relative to the light. Some gradient methods calculate the projection by explicitly rotating the cluster of data points inside the computer. Principal-components analysis does exactly that; it is akin to rotating a long cigar in a three-dimensional space such as a shoe box, so that instead of propped diagonally in a box, the cigar is rotated to lie flat on the bottom of the box, as it would in a cigar box. The point of all of these techniques is to present data of a point cluster in the most informative way.

If two species are set up as axes, all the sites can be plotted as points on that surface by their respective quantities of the two species. With the addition of a third species, the sites move into a three dimensional space (Figure 8.41). Although one cannot literally visualize it, the introduction of a fourth species moves the data points into a four-dimensional space. Once we have crossed the boundary into an abstract four-dimensional space, it is no more difficult to include all the other species as axes, and cast the data points in a space with as many dimensions as are necessary (n-dimensional space). The gradient summary techniques cast a two- or three-dimensional shadow of

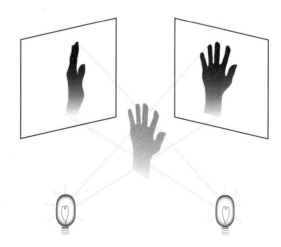

Figure 8.40. Some shadows are more informative than others.

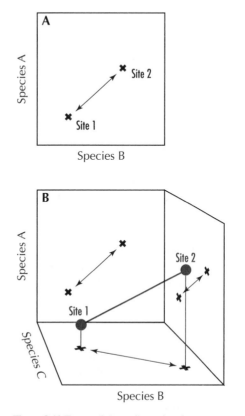

Figure 8.41. Two- and three-dimensional spaces.

Figure 8.42. Hand as cluster analysis.

this n-dimensional point cluster, sometimes in exactly the manner of your three-dimensional hand being cast as a shadow on a two-dimensional screen. The clustering or classification methods of analysis are like clustering the index and middle finger tips together, because they are closest. Then cluster the little finger with the third finger. At a higher level in the classification, the two pairs of fingers are all put in one class, called the fingers. Last of all, the thumb can be clustered in to make a yet larger entity— a whole hand (Figure 8.42).

Robert Whittaker pioneered direct gradient analysis with his summary of the vegetation of the Great Smokey Mountains (Figure 8.39B, left). When defining the community concept, we referred to this work as an example of a very large diverse community. His summary environmental axes are altitude and soil moisture. The four corners of the plane on which he summarizes his vegetation are low valleys, high valleys, low ridges and high ridges. Examples of vegetation occur all across the space, with many intermediate environments. In this way he is able to plot all his sites on a two-dimensional environmental plane. When he superimposes the major species on this environmental space, he can see the relationship between community variants, and between variants and the physical environment.

At about the same time as Whittaker performed his environmental gradient analyses, John Curtis and his students at Wisconsin were developing the so called indirect methods that cast the smaller dimensional summary space as a shadow from the n-dimensional species space. The concurrence of the two analyses is remarkable, as Bray and Curtis (1957) point out in their original paper. Although altitude differences in Wisconsin are nothing like those in the Smokies, the associations of the species are remarkably similar in the two analyses (Figure 8.39B, right). This indicates that altitude is summarizing a large number of important factors that appear to influence many species in some equivalent manner. In both direct and indirect analyses, the point is to achieve a powerful summary of unwieldy masses of data. Generally, the purpose of these methods is to achieve description so that hypotheses can be developed. Then the community ecologist knows what to test in subsequent experimentation.

Diversity indices as extremely loose interpolations

Community ecology deals with several species at the same time, but the exact number of species may vary greatly. The differences in the richness between communities is a matter of diversity of species. Diversity is an issue in itself in community ecology. The literature on diversity is largely concerned with how to measure it (see Box 8.5 on measuring diversity). Diversity is a coarse measurement, and the successful theory to predict it is limited to coarse explanations that depend on general issues of invasion and extinction rates in highly constrained situations, such as islands (MacArthur and Wilson, 1967) or lakes. The meaning of diversity in more open situations is often not at all clear, because diversity can change for a very large number of specific reasons. Diversity indices are blunt instruments in summarizing communities, and the assumptions underlying such heavy handed treatments are likely to be hidden. Nevertheless, the literature on diversity is extensive, and it is appropriate to lay out the assumptions that underlie this general approach.

Whittaker (1972) identified three sorts of diversity: alpha, beta and gamma diversity. The distinction between them is a matter of scale. Alpha diversity is the diversity that occurs in some natural community type. *Alpha diversity* applies to the exemplars of the abstract community, the concrete community, the community as a thing in a place. *Beta diversity* is concerned with which patches of vegetation are adjacent, and is the diversity that occurs between artificial sampling units laid out across space. *Gamma diversity* is the diversity that occurs over a wide range of community types in a general region (Figure 8.43).

One of the appeals of diversity is that it is an energetic simplification of very complicated situations. It works at a level where unaided perception can see things happen. Patterns of gamma diversity at a gross scale are fairly clear, even to the uninformed t-

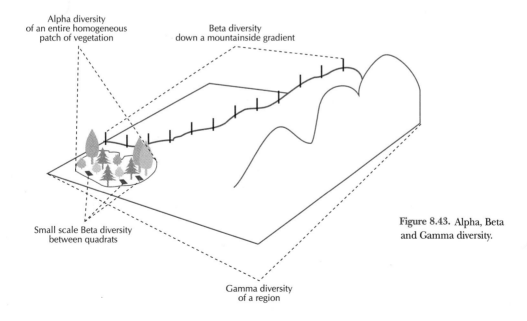

Alpha diversity
of an entire homogeneous
patch of vegetation

Beta diversity
down a mountainside gradient

Small scale Beta diversity
between quadrats

Gamma diversity
of a region

Figure 8.43. Alpha, Beta and Gamma diversity.

raveler. For instance, at high latitudes diversity appears low. The coniferous forests of the far north are dominated by only a few species. Diversity is even lower in the tundra. Conversely, diversity is very high in tropical rain forests and some of the grassland communities of Australia. The explanation is generally that the more challenging physical environments put tight constraints on vegetation, constraints that can only be tolerated by a few species. It is not a simple matter of latitude, because deserts too have lower diversity. Challenging environments press organisms up against ultimate limits, such as no water or short growing seasons. These limits allow less productivity and invite extinction. Thus the trends in global diversity are trivially explained.

In situations where there is no overriding stringent limit, evolution has some productivity with which to work, and things are less straightforward. The differences in diversity across moist sites at middle latitudes are not great. The problem is that the variety of mid-level latitude circumstances is great, and most middle latitude situations are not comparable is any simple way. The assumptions underlying the whole diversity approach have different implications in each place. Processing community data with gamma diversity indices across these middle latitudes reveals a lack of equivalence in the biological implications of the mathematical operations applied to different places. We are adding apples and oranges. As a result, little convincing or explicable pattern emerges. This leaves situations which are very local, considered by alpha and beta diversity. Here, at least, there is an equivalence between sites for which diversity is calculated. Problems arise when diversity becomes such a heavy-handed summary of the biota that local differences in biology are too small to be likely to survive the ferocity of the data summary (see Box 8.5).

APPLICATIONS OF COMMUNITY ECOLOGY

The history of the community concept goes back to F. E. Clements in Nebraska (Pound and Clements, 1901) as well as his contemporary in Chicago, H. C. Cowles (1899). Clements and Cowles introduced dynamic concepts that were missing in the biogeographical studies through the last century. Cowles' dynamics came from dynamical substrate in his sand dunes. Clements' dynamics focused on plant succession as different species with different adaptations entered the site. The definition of a plant community emerged as an attempt to put the adaptation (not necessarily Darwinian) of organisms belonging to species in the context of a biogeographical setting. The urgency for both Cowles and Clements was their prophetic vision of a coming century where humans appeared likely to change the natural world radically.

How right they were for their respective systems. Cowles' study sites are now in the lee of the steel mills of Gary, Indiana. There are still some dunes around Lake Michigan, but almost all of Cowles' original sites are long gone. As to Clements, he was born in Nebraska at the time Custer's defeat at the Little Bighorn, just one state farther west. He was entering the University of Nebraska at the time of the massacre at Wounded Knee,

---

BOX 8.5. CALCULATING DIVERSITY

Diversity indices turn on two issues. One of them is simply the number of species. Indeed, this is the diversity index S called "species richness." Here numbers of individuals in each species is assumed to be unimportant. Therefore an assumption exists about the equivalence between species that is probably—and importantly—unwarranted.

The second issue in diversity indices is the notion of evenness. Here the relative number of individuals is important. Assume site A has exactly the same number of species as site B. If site A's species are each represented by exactly the same number of individuals, that site is defined as even. On the other hand, if site B's individuals are nearly all members of one species, with all the rest of the species represented by a single or a few individuals, then that site scores low on evenness. Site A has a higher diversity. The logic here is that diversity should be a measure of the richness of experience of the individuals at a given site. Individuals wandering around at random at a site are more likely to encounter individuals of another species if the site is even.

One index that uses this general model of evenness is H'. The proportion of the whole community that comprises one species is mul-

tiplied by the logarithm of that same proportion. The outcome of this operation is summed across all n species,

$$H' = -\sum_{i=1}^{n} p_i \log p_i$$

where p_i is the proportion of an individual of species i in the whole community. Some authors use log base 10, others log base 2 or natural logarithms. The different logs only change the units in which diversity is measured. There is a link across to species number in that, if species at a site are completely even, $H' = \log S$.

Most diversity indices are used for *alpha diversity*, the type that aims to characterize the intrinsic diversity of homogeneous vegetation units. It is important to note that these standard diversity indices give a higher value as the size of the area to which they apply is increased. Therefore, alpha diversity, as measured by the standard indices, is as heavily influenced by the placement of arbitrary sampling units as is beta diversity. The measurements of alpha diversity found in the literature often do not indicate a fundamental diversity of a natural unit, but rather reflect decisions about sampling.

---

even closer across the border in South Dakota. As he began his work, the steel plow was breaking the plains for the first time (Plate 8.3C). Both Cowles and Clements viewed as their mission the recording of their respective natural places, which were about to be dismembered. Toward the end of his career, Clements watched the Dust Bowl consume the whole landscape that he knew as a boy and student.

Part of the mission of community ecology and its methods is description of natural places. The powerful descriptive methods of community ecology work just as well in places that are being compromised, so we can document the decline and perhaps attenuate it. Species extinction occurs not only in a physical environment, but also in a biotic setting, and a proper understanding of that context is crucial. While animal ecologists probably defer too much to plant community ecologists to define their habitats, plant community ecology allows us to address the context of animal species that might otherwise become extinct.

Community ecologists probably pay more attention to **rare species** than do other kinds of ecologists. Species are a natural point of focus, in that species associated one with another make up the community. We cannot help but notice that in a community dominated by certain species, other species occur as rarities. There is therefore a clear point of comparison, where we can answer the question, "Rare, as opposed to what?" While moderately uncommon species are probably in no greater danger of extinction than those that are common, clearly very rare species are vulnerable to some local catastrophic happening. Extinction has occurred throughout history, but we live in a special time where extinction is rampant. The implications for pharmaceutical considerations are clear. In a less pragmatic vein, extinction is forever, and there are moral implications that must also be our concern.

The modern community work of Tilman and his colleagues suggests the mechanisms that underlie processes of decline. While description is crucial, management actions should be predicated on an understanding of the overriding processes that are responsible for the state we address and hope to achieve. There are too many species to manage individually, and the community is the perspective from which management can be conducted. Tilman suggests that we should look at the processes of feedback in nutrient status that are responsible for the behavior of a whole suite of species, not just one. Community ecology deals with a large number of species in concert. It addresses the context of populations, and shows how elaborate are the workings of landscape units. While single-species plantations do not need much input from a community ecologist, most of the forests that are used for production of wood are natural regenerations, or at least mixed-species units—thus most forest management is a matter of sequestering resources from communities. The same applies to rangelands, so most renewable natural resources are extractions from communities, and community ecologists offer the scientific basis for the majority of the action of resource managers.

LINKS TO OTHER KINDS OF ECOLOGY
Communities and Biomes

Biomes differ from communities in that biomes are not conceived as being composed of species. Many other categories besides species can be used to group plants. For example, all succulent plants have a common form and often share special physiological adaptations for drought resistance. Not only do succulents belong to many species, but they also often come from different plant families. Cacti are but one family (Cactaceae), other succulents belong to the stoneworts (Crassulariaceae), the sunflower family (Asteraceae) and the Poinsettia family (Euphorbiaceae). The general term for such a group of similar organisms is **life form**. Another life form is the deciduous tree, as distinguished from a coniferous tree. Herbs, shrubs, and bulbs are three other life forms. Of course biomes contain species, but species are not the category used to group plants as they are found in biomes. *Biomes are identified by the dominant life form.* Some desert biomes are dominated by big succulents. The boreal forest biome is dominated by confers. Because each life form looks so distinctive, biomes are recognizable by how they look on first glance. The first impression of vegetation is called **vegetational physiognomy**.

Biomes also have characteristic animals that groom the vegetation. Whereas a community is usually seen as being either plant or animal, a description of a vegetational biome is incomplete without its dominant animals. For example, the boreal forest which is dominated by coniferous trees is affectionately known as the Spruce-Moose biome. The species mix is a separate issue involving community, not biome, conceptions. Large ungulates browse the saplings and understory, while epidemic leaf-eating insect infestations, like the spruce budworm, wreak havoc on great tracts of forest. The animals contribute to the vegetational physiognomy.

Physiology normally dictates the life form of plants, and that physiology is a reflection of environmental limits. For instance, succulents have a particular physiology. The size and degree of woodiness in shrubs gives them a physiological capacity to survive drought to which trees would succumb. Since the life form of the dominant plants in a biome is environmentally determined, biomes per se are environmentally determined, in a way that community composition is not.

Standing in stark contrast to biomes, populations are biotically determined. Most of the behavior or pattern seen in a population comes from the processes of competition, reproduction and death as moderated by predation and resource capture. Of course, the physical environment affects those biotic processes, much as biotic processes place the vegetation on the ground in environmentally determined biomes. Which is the overriding predictor, biotic processes or environmental limits? In the upshot, *populations are reliably biotically determined, whereas biomes are environmentally determined.*

In biomes, the requirement for identifying plants to species is relaxed. In populations, the identity of the species in question is usually a central issue. Communities occupy a realm in between populations and biomes. At one extreme, depauperate communities are merely complications of populations, a strongly species-focused situation. At the other extreme, species-rich communities can be identified by a small subset of the full suite of species that occur in that community. In species-rich communities, the required presence of many species is greatly relaxed, and community conceptions give way to system characterization as biomes. Communities are a conception where species identity is important, but where physical environment plays an important role. Thus, in some communities the system may be tied to a particular environmental regime. In other communities, species interactions in biotic terms appear the dominant consideration.

The community falls exactly between the conceptions of biome and population (Figure 8.44). Accordingly, most communities yield neither to environmental determinism nor biotic determinism alone. In communities, we must deal with the switches in causality between the physical environment and biotic processes. Communities are therefore very difficult to predict, because the class of explanation of their pattern changes willynilly between physical and biotic causes. Community prediction is well known to be dif-

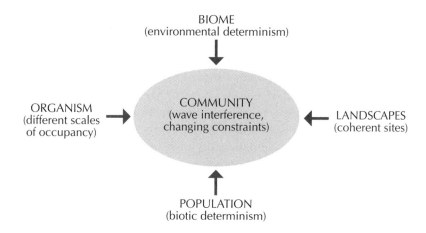

Figure 8.44. The community falls between conceptions of biome and population, and is an accommodation of different species to landscapes.

ficult, exactly because of changing constraints. The few overtly predictive community studies are unusually dominated by either environmental causes or biotic causes, but not both. A case in point is Paine's work on the intertidal, discussed in Box 8.1, along with Kimmerer's work on bryophyte communities. In intertidal animal communities, all the prime causes come from only one side, biotic determinism, as they do for Kimmerer's lowly plants. In both cases space is the constant limiting factor, so the different outcomes are a reliable result of biotic processes of competition for space. But in most communities, the ecologist is not so lucky—physical factors limit only occasionally, and the on-again off-again physical limits muddy the outcome of biotic processes.

Community Links to Behavioral and Landscape Ecology

Communities are an accommodation of organisms of different species to landscapes (Figure 8.44). Each organism belongs to a species, and that assignment predicts the organism's size, longevity, mobility, time to reproductive age, and reproductive effort. In other words, the complex scaling of an organism is shared by all members of its species. The German behavioral biologist Jacob Von Uexkull made much of this point, and identified each species as having its own "Umwelt," a word that translates into English as "self-world" (or "environment" or "milieu"). Each species defines its own environment in time and space, depending on how it reads and responds to its local environs (Figure 8.45).

The organizing principle for landscapes is spatial arrangement. That is to say, the landscape ecologist's concern is how far apart are the happenings at two points on the landscape. In that sense, landscapes are about how far is "a long way away" and how far is "close." But the biological significance of distance and the time taken to traverse a given distance depends on the scale at which organisms occupy time and space. Von Uexkull (1934) discusses at length the life journey of organisms, noting that birds have long life journeys, whereas a grain beetle spends almost all its life traversing a single grain from point of entry to the exit hole.

Communities are composed of many species, and seen from the point of view of the different species, the very meaning of distance is different for each species. In a Euclidean space, distance is measured in the same units everywhere, whereas the very units of space itself change as one traverses a non-euclidean space, so that a long distance will not be the sum of the shorter segments of distance of which it is composed. In geometric terms, the landscape for a community is non-euclidean; it depends on the behavioral characteristics of the organism about whom we ask, "How far is far." Each species uses its own metric for distance as determined by its Umwelt. For example, a long, thin landscape unit, perhaps a hedgerow, would operate as a corridor down which a mouse might move; but for another species, perhaps a deer, that same landscape feature is a barrier that limits rather than facilitates movement. All species see the local environs as different, because the local place defines a different space for each species. Therefore, communities do not fit easily onto landscapes.

If the landscape has non-euclidean characteristics for a community ecologist, what happens when a community is forced into a particular Euclidean area by a definition that sees the community as a thing in a place? The contrasting patterns of occupancy

Figure 8.45. Each species occupies the landscape according to a spatial and temporal scale that is defined by how the organism reads its context. Behavioral biologist J. Von Uexkull strolls through the world of grasshoppers, here showing how individuals respond to the sound of the recorded song, not to the sight of the individual singer under the "dome of silence." From Von Uexkull's original 1934 drawing.

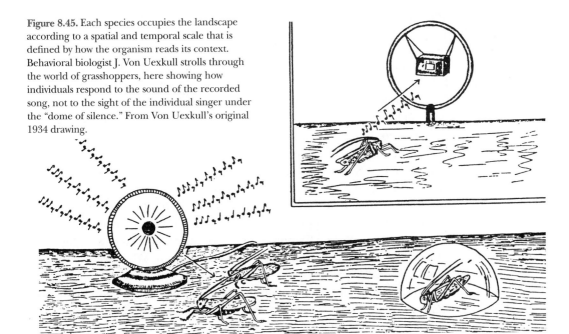

mean that the place in space is not occupied by all the species. Some species only occupy such a small part of the space that they are absent from most of the area. Other species in the general vicinity of the given area occupy a space so large, that no individual of that species happens to be in the defined spatial unit at the time of inspection (Figure 8.5). Once again, the example of the Serengeti grazer community pertains (Box 8.2). Forcing the community onto a place in the landscape means that both those species with a very large Umwelt and those with a very small Umwelt are functionally absent from most of the community area most of the time. This is a good reason for abandoning the concept of a community as a thing in a place.

Community and Population Ecology

As a stark contrast to population ecology, community ecology invokes entities (communities) with heterogeneous parts (many species). Community ecologists have been intuiting for some time that communities are not often readily reducible to sets of populations; understanding all the populations in the community in population terms would not solve many of the issues that the notion of community raises. However, it takes someone respected from the population side of the argument to press the point home for other population ecologists. Such a scientist is Jonathan Roughgarden.

Roughgarden's work was an investigation of community structure, using lizards in the Lesser Antilles—the chain of islands that circles southward from Puerto Rico to South America (Figure 8.46). While the original intention of the work was not a comparison of populations and communities, it emerges as the classic example that addresses the comparison between the two types of ecology. The plan at the outset was to look at the simplest community as it is forming, so that a generalization could be made toward the more complex and mature systems that are studied by most community ecologists. In the end, Roughgarden failed to meet his agenda; but as stated earlier, it is through failure that scientists learn, particularly when the failure is unexpected. What appeared to be simple communities at first glance, proved to be turnover in populations, a transient with no stable community forming at all.

On the face of it, the system would appear to span the threshold from populations to minimal communities. Some islands have one lizard species, while other islands have two species, apparently forming a minimal lizard community. When Roughgarden found one species alone on an island, the lizards were always close to 55 mm long. Fifty-five millimeters appears to be some sort of optimum length for the climate and the lizard resource base for the region. If he found two lizard species on an island, Roughgarden at first presumed he had the beginnings of a lizard community. Whenever there were two species on an island, one was larger than the climatic optimum—about 100 mm long—and the other species 45 mm, smaller than the presumed climatic optimum size. At first it seemed as if the two-lizard condition was indeed the beginnings of a community. In all cases, the larger lizard appeared to be the later arrival, and the decrease in

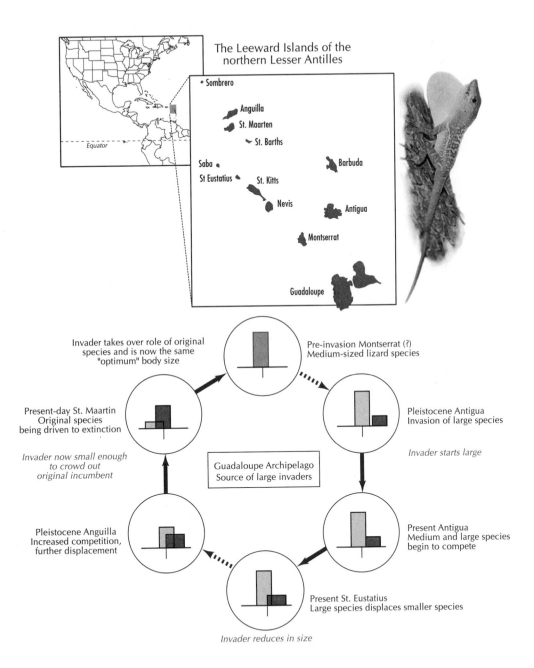

Figure 8.46. A taxon cycle involving *Anolis* lizard species in the northern Lesser Antilles (note the proximity of the islands in the study). One island species evolves an "optimum" body size. The island is invaded by a second species larger than the first. The invader evolves toward the size of the native species, competing with it and eventually displacing it. Redrawn from Roughgarden 1989.

size of the original species by 10 mm indicated natural selection for smaller lizards that could minimize competition with the new arrival. The technical name for this sort of accommodation is "**character displacement**," and it would be the basis of community structure, were communities to emerge from coevolution.

On closer examination, it appears that there is no community here, and none is developing. The character displacement is not part of the emergence of a community where species have evolved to coexist. Fossil lizard deposits, corresponding to 10,000 years of occupancy of favorite sites, manifest different sizes of teeth at different levels, indicating changes and reversals of changes in tooth size over time. These oscillations imply that the size of lizards never stabilizes into a community configuration. Furthermore, if an island possesses two lizard species, and one is restricted in its range, the restricted lizard is always the small species. Consistently, the smaller lizard species appears to be retreating in a losing battle up toward the peaks of the mountains. From these and other lines of evidence, we deduce that the two-lizard islands are not emergent communities, but are populations caught in the act of competitive exclusion. No community is in the process of emerging; rather, populations are failing to accommodate sufficiently.

Roughgarden's scenario is that isolated lizard species, one to an island, have evolved to the optimum size. In communities, physical limits are only a part of the full set of selection pressures; thus it often happens that no species is selected to be climatically optimal. On larger islands such as Hispaniola, competitive exclusion between lizard species cannot reach its conclusion, because the competitive regime is muddled by the greater area's wider range of environmental variation. The situation is further complicated by a large number of species as players, and the changing patterns of species interrelationships over time. If one of these larger-than-optimal species from a large island arrives on one of the small islands of the Lesser Antilles, it then out-competes the smaller resident, which responds by character displacement, but to no avail. The smaller species is driven to extinction. The victor is then selected to become smaller until it is the optimal size for the climate (Figure 8.46). Thus one is not dealing with a community, but with naked populations that have been excised from their communities. In the absence of community constraints and environmental heterogeneity, population properties have a rare opportunity to express themselves unmasked, in a natural setting.

SUMMARY

Community ecology is at the core of all of ecology. Indeed, the emergence of community studies at the turn of the century led ecology to become a self-conscious discourse, independent of biogeography and physiology. Communities are what one experiences on a walk through the woods; the community is the first approach to nature. However, that intuitive quality to communities is as much a liability as a benefit. The temptation is to conceive of communities as collections of organisms in a place. We are free to define anything the way we want, but nature does not obey definitions. Even so, a collection in a place is barely a working entity with processes that hold it together. So if a community needs to be more than a collection, what is it?

Since all types of ecology converge on communities, communities are not juxtaposed to entities that define other types of ecology, but a set of conflicting notions where different types of ecology pull at each other across the community concept (Figure 8.44). Since communities are composed of many species, the place in which a concrete community resides is variable, depending on which species we ask—some species function in large places, others work in small, but both can be in the same community. Thus the first tension is between landscape conceptions and different organismal scales of landscape occupancy.

The second tension in communities is between biotic and abiotic determinism. Clearly some happenings in the community come from, and are therefore determined by, biotic processes, such as breeding and competition. However, other aspects of the community are environmentally determined, as such factors constrain biotic processes. The population is the naked expression of biotic determinism, whereas the biome expresses environmental determinism. As those two sets of processes tussle for control, community composition and behavior emerge from switches in constraints. The community is therefore best not conceived in concrete terms through its exemplars on the ground; it is more powerfully considered as an abstract thing with patterns that have multiple causes. These causes "tag-team wrestle" the system, so that the emergent whole behaves according to an unpredictable sequence of causes.

The regularity with which we find community patterns appears to be only indirectly the result of evolution. Of course the constituent organisms and species are evolved, but that does not mean that the community pattern itself is a result of evolution. Communities are too ephemeral to be evolved entities—a few thousand years is not long enough for large constellations of species to coevolve. Rather, communities are stable configurations of ready-made evolved parts.

The stability that gives pattern to communities comes not from coevolution, but from positive feedbacks that reinforce the persistence of certain species mixes. Once in the domain of attraction, species composition moves by itself toward the more common species associations in that community type. The reinforcement is because different species in the community subscribe to certain aggressive strategies that favor species with like strategies, and work against species with different strategies. For example, the shared strategy of shade tolerance and an engineering of shaded environments keeps trees growing together in a forest community, and keeps grassland dominants excluded. Disturbance can take a plot and convert it to some other stable community type in environments where multiple communities are possible.

A change in environment can cause shifts from communities that are disallowed to those which can flourish in the new environment. The particular species composition of the abstract community in a given region is a reflection of the time since the last major climate change, and the distance from the refugia of the various species in the prior climate regime. Experimental work indicates that assembly rules for communities can change the differences between present and past communities by the sequence of founder effects. There may be some microevolutionary aspect to the persistence of founder effects.

The heterogeneity in communities, and their often-large size, make experimentation difficult. Microcosm experimentation is possible, as are experimental plots in herbaceous communities. Computer simulations are often a preferred option. The two styles of simulations are population models and individual-based models. Population models use statistical moments, such as the mean, to summarize behavior of the parts. Individual-based models keep track of all the individuals in a plot, and require fast computers to perform the simulation.

The methods of community study are dominated by the rich heterogeneity of the system. Data collection procedures need to be exquisitely conceived and almost ritually executed. The methods must be applied consistently, for the essential heterogeneity of the system does not allow the community ecologist to easily spot a mistake being made, or some inconsistency or unexpected lack of equivalence within the investigation. The methods of analysis of community data vary considerably, but they can be unified and put in perspective by considering all data analysis as a process of interpolation. In all modeling and data analysis, the whole point is an investigation of the workability of assumptions. This can easily be overlooked in the overt complexity of communities. Precisely because the community is such a riot of experience, precision of thought is essential in approaching them.

There is a joy in studying communities that the author of this chapter hopes has rubbed off on the reader. The common conception of ecology, and the intuitive appeal of it, both turn on the things that community ecologists consider. Community ecology is the science of a walk through the woods, of paddling in a tide pool on a seaside holiday. Come on in, the water is fine.

RECOMMENDED READING

T. F. H. ALLEN AND THOMAS W. HOEKSTRA. *Toward a Unified Ecology*. Columbia University Press, New York. 1992.

> *The organization and content of this book inspired the organization of this chapter.*

PAUL COLINVAUX. *Why Big Fierce Animals are Rare: An Ecologist's Perspective*. Princeton Paperbacks, Princeton University Press, Princeton, New Jersey. 1978.

> *A readable, accessible and provocative set of essays about ecology, especially about community ecology.*

ALDO LEOPOLD. *A Sand County Almanac, and Sketches Here and There*. Oxford University Press. New York, New York. 1949.

> *This classic of ecological and wildlife management literature presents some of the most beautiful writing on conservation, ecological ethics, and the observation and celebration of nature.*

JOHN T. CURTIS. *The Vegetation of Wisconsin: An Ordination of Plant Communities*. Wisconsin University Press. Madison, Wisconsin. 1959.

> *This book provides a wealth of data on the distribution and association of plants in the Badger State. It is a milestone in the community ecology of plants.*

PETER GREIG-SMITH *Quantitative Plant Ecology*. Third ed. University of California Press, Berkeley and California. 1983.

> *Very broad-based and technically rigerous account of methods of data collection and analysis.*

EVELYN C. PIELOU. *Population and Community Ecology*. Gordon and Breach. New York, New York. 1979.

> *This book is a rigorous account of the community-population interface, with a good account of diversity and statistical methods.*

References

Bold page numbers following references
indicate in-text citations of the first author.

Ahl, V. and T. F. H. Allen. 1996. *Hierarchy*. Columbia University Press, New York, NY. **349**

Alden, H. 1992. Watersheds, landscapes, biodiversity, fire and forest management. The S. J. Hall Lectureship in Industrial Forestry, University of California, Berkeley. **68, 385**

Allen, T. F. H. and T. W. Hoekstra 1992. *Toward a Unified Ecology*. Columbia University Press, New York, NY. **ix, xii, 315, 332**

Alliende, M. C. and J. L. Harper. 1989. Demographic studies of a dioecious tree. I. Colonization, sex and age structure of a population of *Salix cinera*. Journal of Ecology 77:1029–1047. **243–244**

Altmann, J. 1980. *Baboon Mothers and Infants*. Harvard University Press, Cambridge, England. **43, 385**

Ames, F E. 1910. "Addresses given at the 1910 joint supervisors' meeting for the Northwest region (Oregon and Washington)." Forest Service Research Compilation Files, National Archives, Region VI, Entry 115, Box 136. **69–70**

Ames, F. E. 1915. Letter to forest supervisors. Forest Service Research Compilation Files, National Archives, Region VI, Entry 115, Box 139. **69–70**

Andow, D. A., R. J. Baker and C. D. Lane. 1994. Karner blue butterfly—a symbol of a vanishing landscape. University of Minnesota Miscellaneous Publication 84. **293**

Arcese, P., L. F. Keller and J. F. Cary. 1997. Why hire a behaviorist into a conservation or management team? Pp. 48–71. In J. R. Clemmons and R. Buchholz (eds.). *Behavioral Approaches to Conservation in the Wild*. Cambridge University Press, Cambridge, England. **231**

Arp, A. J. and C. R. Fisher. 1995. Introduction to the symposium: life with sulfide. American Zoologist 35:81–82. **167**

Attenborough, D. 1995. *The Private Life of Plants: A Natural History of Plant Behaviour*. Princeton University Press, Princeton, NJ. **309**

Bahn, P. and J. R. Flenley. 1992. *Easter Island, Earth Island*. Thames and Hudson, London, England. **309**

Baker, W. L. 1989. Landscape ecology and nature reserve design in the Boundary Waters Canoe Area, Minnesota. Ecology 70:23–25. **309**

Baker, W. L. 1992. Effects of settlement and fire suppression on landscape structure. Ecology 73:1879–1887. **309**

Bartell, S. M., T. F. H. Allen and J. Koonce. 1979. An assessment of principal components analysis for description of phytoplankton periodicity in Lake Wingra. Phycologia 17:1–11. **309**

Bell, R. H. V. 1971. A grazing system in the Serengeti. Scientific American 225:86–94. **328**

Belovsky, G. E. 1978. Diet optimization in a generalist herbivore: the moose. Theoretical Population Biology 14:105–134. **328**

Belovsky, G. E. 1981. Food plant selection by a generalist herbivore: the moose. Ecology 62:1020–1030. **328**

Bergelson, J. 1996. Competition between two weeds. American Scientist 84:579–584. **328**

Bess, H. A., R. van den Bosch and F. A. Haramoto. 1961. Fruit fly parasites and their activities in Hawaii. Proceedings of the Hawaiian Entomological Society 17:367–378. **279**

Boomsma, J. J. and A. Grafen, A. 1990. Intraspecific variation in ant sex ratios and the Trivers-Hare hypothesis. Evolution 44:1026–1034. **229**

Botkin, D. B., J. F. Janak and J. R. Wallis. 1972. Some ecological consequences of a computer model of forest growth. Journal of Ecology 60:849–873. **56,**

Bowler, P. 1993. *The Norton History of the Environmental Sciences.* Norton, New York, NY. **40, 42, 46, 50, 66**

Bradbury, J. W., R. M. Gibson and I. M. Tsai. 1986. Hotspots and the evolution of leks. Animal Behavior 34:1694–1709. **223**

Bray, J. R., and J. T. Curtis. 1957. An ordination of the upland forest communities of Southern Wisconsin. Ecological Monographs 27:325–349. **367–368, 370**

Brockmann, H. J., A. Grafen and R. Dawkins. 1979. Evolutionarily stable nesting strategy in a digger wasp. Journal of Theoretical Biology 77:473–496. **xii, 217–218**

Budiansky, S. 1995. *Nature's Keepers: The New Science of Nature Management.* Free Press, New York, NY. **68**

Burke, I. C., E. T. Elliott and C. V. Cole. 1995. Influence of macroclimate, landscape position, and management on soil organic matter in agroecosystems. Ecological Applications 5:124–131. **vii, 102**

Burnett, T. 1958. Effect of host distribution on the reproduction of *Encarsia formosa* Gaham (Hymenoptera: Chalcidoidea). Canadian Entomologist 90:179–191. **272–273**

Burrough, P. A. 1982. *Principles of geographical information systems for land resources assessment.* Clarendon Press, Oxford, England. **90**

Bush, M. 1997. *Ecology of a Changing Planet.* Prentice Hall, New York, NY. **29**

Cade, W. 1979. The evolution of alternative male reproductive strategies in field crickets. Pp. 343–379. In M. S. Blum and N. A. Blum (eds.). *Sexual Selection and Reproductive Competition in Insects.* Academic Press, New York, NY. **217**

Campbell, D. G. 1996. Splendid isolation in Thingvallavatn. Natural History 105(6):48–55. **142**

Caro, T. M. and S. M. Durant. 1995. The importance of behavioral ecology for conservation biology: examples from Serengeti carnivores. Pp. 451–472. In A. R. E. Sinclair and P. Arcese (eds.). *Serengeti II: Dynamics, Management, and Conservation of an Ecosystem.* University of Chicago Press, Chicago, IL. **231**

Carpenter, S. R. and J. F. Kitchell. 1993. *The Trophic Cascade in Lakes.* Cambridge University Press, Cambridge, England. **viii, xii, 10, 133, 135, 170**

Carpenter, S. R., S. G. Fisher, N. B. Grimm and J. F. Kitchell. 1992. Global change and freshwater ecosystems. Annual Reviews in Ecology and Systematics 23:119–139. **viii, xii, 10, 133, 135, 170**

Caswell, H. 1989. Matrix population models. Sinauer Associates, Inc. Sunderland, MA. **266**

Caughley, G. 1970. Eruption of ungulate populations, with emphasis on Himalayan Thar in New Zealand. Ecology 51:53–72. **55**

Chase, A. 1995. *In a Dark Wood: The Fight over Forests and the Rising Tyranny of Ecology.* Houghton Mifflin Co., Boston, MA. **68**

Childress, J. J. 1995. Life in sulfidic environments: historical perspective and current research trends. American Zoologist 35:83–90. **167**

Christensen, N. L. 1989. Landscape history and ecological change. Journal of Forest History 33:116–124. **27, 50, 56, 66, 115, 157**

Christensen, N. L., A. M. Bartuska, J. H. Brown, S. R. Carpenter, C. D'Antonio, R. Francis, J. F. Franklin, J. A. MacMahon, R. F. Noss, D. J. Parsons, C. H. Peterson, M. G. Turner and R. G. Woodmansee. 1996. The report of the Ecological Society of America committee on the scientific basis for ecosystem management. Ecological Applications 6:665–691. **27, 50, 56, 66, 115, 157**

Clements, F. E. 1905. *Research methods in ecology.* University Publishing Co., Lincoln, NE. **50–52, 55–56, 320, 330, 372–373**

Clements, F. E. 1904. Developments and structure of vegetation. Botanical Survey of Nebraska 7. **50–52, 55–56, 320, 330, 372–373**

Clements, F E. 1916. *Plant Succession: An Analysis of the Development of Vegetation.* Publication/Carnegie Institution, Washington, D.C. Vol. 242. **50–52, 55–56, 320, 330, 372–373**

Clutton-Brock, T. H., A. W. Illius, K. Wilson, B. T. Grenfell, A. D. C. MacColl and S. D. Albon. 1997. Stability and instability in ungulate populations: an empirical analysis. American Naturalist 149:195–219. **247, 263**

Cohen, J. E. 1995. Population growth and Earth's human carrying capacity. Science 269:341–346. **151, 192**

Cook, R. E. 1977. Raymond Lindeman and the trophic-dynamic concept in ecology. Science 198:22–26. **16**

Cooper, W. S. 1926. The fundamentals of vegetational change. Ecology 7:391–414. **318–320**

Coulson, R. N., C. N. Lovelady, R. O. Flamm, S. L. Spradling and M. C. Saunders. 1991. Intelligent geographic information systems for natural resource management. Pp. 153–172. In M. G. Turner and R. H. Gardner (eds.). *Quantitative methods in landscape ecology.* Springer-Verlag, New York, NY. **90**

Cowie, R. J. 1977. Optimal foraging in great tits *(Parus major).* Nature 268:137–139. **213**

Cowles, H. C. 1899. The ecological relations of the vegetation of the sand dunes of Lake Michigan. Botanical Gazette 27:95–117, 167–202; 282–308; 361–391. **51, 372–373**

Cronon, W. 1983. *Changes in the Land: Indians, colonists, and the ecology of New England.* Hill and Wang, New York, NY. **32–33, 78**

Cullen, E. 1957. Adaptations in the kittiwake to cliff-nesting. Ibis 99:275–302. **207**

Curtis, J. T. and R. P. McIntosh (1950) The interrelation of certain analytic and synthetic phytosociological characters. Ecology 31:434–55. **78–79, 154–155, 331, 362–364, 367–368, 370**

Curtis, J. T. (1959) *The vegetation of Wisconsin.* University of Wisconsin Press, Madison, WI. **78–79, 154–155, 331, 362–364, 367–368, 370**

Curtis, J. T. 1956. The modification of mid-latitude grasslands and forests by man. Pp. 721–736. In W. L. Thomas (ed.). *Man's role in changing the face of the earth.* University of Chicago Press, Chicago, IL. **78–79, 154–155, 331, 362–364, 367–368, 370**

Curtis, J. T., and R. P. McIntosh. (1951) An upland forest continuum of the prairie-forest border region of Wisconsin. Ecology 32:476–496. **78–79, 154–155, 331, 362–364, 367–368, 370**

Cyr, H. and M. L. Pace. 1993. Magnitude and patterns of herbivory in aquatic and terrestrial ecosystems. Nature 361:148–150. **134–135**

Daily, G. (ed.). 1997. *Nature's Services.* Island Press, Washington, DC. **147**

Dale, V. H., R. V. O'Neill, M. Pedlowski and F. Southworth. 1993. Causes and effects of land-use change in central Rondonia, Brazil. Photogrammetric Engineering and Remote Sensing 59:997–1005. **111**

Darwin, C. 1964. *On the Origin of Species.* [Facsimile of 1st edition]. Harvard University Press, Cambridge, MA. **2, 21, 221, 225, 233, 236**

Davies, N. B. 1991. Studying behavioural adaptations. Pp. 18–30. In M. S. Dawkins, T. R. Halliday and R. Dawkins (eds.). *The Tinbergen Legacy.* Chapman and Hall, London, England. **12, 200, 202, 231**

Davies, N. B. and Houston, A. I. 1981. Owners and satellites: the economics of territory defense in the pied wagtail, *Motacilla alba.* Journal of Animal Ecology 50:157–180. **12, 200, 202, 231**

Davis, M. B. 1983. Quaternary history of deciduous forests of eastern North America and Europe. Annals of the Missouri Botanical Garden 70:550–563. **86**

DeAngelis, D. L. 1992. Dynamics of nutrient cycling and food webs. Chapman and Hall, New York, NY. **137, 187**

DeAngelis, D. L. and L. J. Gross (eds.). 1992. *Individual-Based Models and Approaches in Ecology.* Chapman and Hall, New York, NY. **137, 187**

Denevan, W. 1992. The pristine myth: the landscape of the Americas in 1492. Annals of the Association of American Geographers 82:369–385. **32**

Dennis, B., and R. F. Constantino. 1988. Analysis of steady-state populations with the gamma abundance model: application to *Tribolium.* Ecology 69:1200–1213. **xiii, 261**

Detenbeck, N. E., C. A. Johnston and G. J. Niemi. 1993. Wetland effects on lake water quality in the Minneapolis/St. Paul metropolitan area. Landscape Ecology 8:39–61. **100**

Drake, James A. (1990) Communities as assembled structures; Do rules govern pattern? Tree 5:159–164. **349, 387**

Dunlap, T. R. 1988. *Saving America's Wildlife: Ecology and the American Mind, 1850–1990.* Princeton University Press, Princeton, NJ. **55**

Egerton, F., ed. 1977. *History of American Ecology*. Arno Press, New York, NY. **46–47, 49**

Egerton, F. 1973. Changing concepts of the balance of nature. Quarterly Review of Biology 48:322–350. **46–47, 49**

Ehrlich, P. and A. Ehrlich. 1981. *Extinction*. Random House, New York, NY. **245, 297**

Ehrlich, P. R., A. H. Ehrlich and J. P. Holdren. 1973. *Human Ecology: Problems and Solutions*. W. H. Freeman and Company. San Francisco, CA. **13**

Elton, C. 1927. *Animal Ecology*. Sidgwick and Jackson, London, England. **10, 19, 52**

Elton, C. 1930. *Animal Ecology and Evolution*. Oxford University Press, New York, NY. **10, 19, 52**

Emlen, S. T. 1990. The white-fronted bee-eater: helping in a colonially nesting species. Pp. 487–526. In P. Stacey and W. Koenig (eds.). *Cooperative Breeding in Birds: Long-Term Studies of Ecology and Behavior*. Cambridge University Press, Cambridge, England. **226–227, 231**

Emlen, S. T. 1995. Can avian biology be useful to the social sciences? Journal of Avian Biology 26:273–276. **226–227, 231**

Emlen, S. T. and P. H. Wrege. 1989. A test of alternative hypotheses for helping behavior in white-fronted bee-eaters of Kenya. Behavorial Ecology and Sociobiology. 25:303–319. **226–227, 231**

Fenner, F. and K. Myers. 1978. Myxoma virus and myxomatosis in retrospect: the first quarter century of a new disease. Pages 539–570. In E. Kurstak and K. Maramorosch, editors. *Viruses and Environment*. Academic Press, New York, NY. **301**

Finegan, B. 1984. Forest succession. Nature 312:109–114. **50**

Fisher, R. A. 1930. *The Genetical Theory of Natural Selection*. Clarendon Press, Oxford, UK. **127, 164, 167, 223**

Flenley, J. R. and S. M. King, 1984. Late quaternary pollen records from Easter Island. Nature 307:47–50. **30–31**

Forbes, S. 1887. The lake as a microcosm. Peoria Scientific Association Bulletin 77–87. **52**

Forman, R. T. T. 1995. *Land mosaics*. Cambridge University Press, Cambridge, England. **81, 91**

Foster, D. R. 1992. Land-use history (1730–1990) and vegetation dynamics in central New England, USA. Journal of Ecology 80:753–772. **78–79**

Foucault, M. 1980. *Power/knowledge: selected interviews and other writings, 1972–1977*. ed. and trans. by C. Gordon. Pantheon Books, New York, NY. **70**

Franklin, J. F. 1993. Preserving biodiversity: species, ecosystems, or landscapes? Ecological Applications 3:202–205. **116**

Frost, T. M., S. R. Carpenter, A. R. Ives and T. K. Kratz. 1995. Species compensation and complementarity in ecosystem function. pp 224–239. In C. Jones and J. Lawton (eds.), *Linking Species and Ecosystems*. Chapman and Hall, London, England. **142**

Gittins, R. (1965) Multivariate approaches to a limestone grassland community II: a direct species ordination. Journal of Ecology 53:403–409. **367–368**

Gleason, H. A. 1926. The individualistic concept of the plant association. Bulletin of the Torrey Botanical Club 53(1):7–26. Reprinted in Edward J. Kormondy (ed.) *Readings in Ecology*, Prentice Hall, Englewood Cliffs, NJ. **56, 320, 330**

Goodall, J. 1986. *The Chimpanzees of Gombe: Patterns of Behavior*. Belknap Press of Harvard University Press, Cambridge, MA. **43**

Goudie, A. 1994. *The Human Impact on the Natural Environment*. M. I. T. Press, Cambridge, MA. **28, 67**

Gould, S. J. and R. C. Lewontin. 1979. The spandrels of San Marco and the Panglossian paradigm: a critique of the adaptationist programme. Proceedings of the Royal Society of London B, 205:581–598. **204**

Grant, P. R. 1986. *Ecology and evolution of Darwin's finches*. Princeton University Press, Princeton, NJ. **viii, xii–xiii, 280, 363**

Grayson, D. K. 1993. *The Desert's Past: A Natural Prehistory of the Great Basin*. Smithsonian Institution Press, Washington, DC. **72**

Greeley, W. B. 1925. The relation of geography to timber supply. Economic Geography 1:4–5. **33**

Greig-Smith, P. 1964. *Quantitative Plant Ecology*. Second ed. Butterworths, London, England. **316, 363**

Grumbine, R. E. 1992. *Ghost Bears: Exploring the Biodiversity Crisis*. Island Press, Washington, DC. **68**

Gunderson, L. H., C. S. Holling and S. S. Light. 1995. *Barriers and Bridges to the Renewal of Ecosystems and Institutions*. Columbia University Press, New York, NY. **142–143, 148, 156–158**

Gwynne, D. T. 1981. Sexual difference theory: mormon crickets show role reversal in mate choice. Science 213:779–790. **xii, 224**

Gwynne, D. T. 1984. Courtship feeding increases female reproductive success in bushcrickets. Nature 307:361–363. **xii, 224**

Gwynne, D. T. 1990. Experimental reversal of courtship roles in an insect. Nature 346:172–174. **xii, 224**

Hamilton, W. D. 1964. The genetical evolution of social behaviour. I, II. Journal of Theoretical Biology 7:1–52. **224–227**

Hamilton, W. D. and M. Zuk. 1982. Heritable true fitness and bright birds: a role for parasites. Science 218:384–387. **224–227**

Haraway, D. 1989. *Primate Visions: Gender, Race, and Nature in the World of Modern Science.* Routledge, New York, NY. **42**

Harper, J. L. and I. H. McNaughton. 1962. The comparative biology of closely related species living in the same area VII. Interference between individuals in pure and mixed population of *Papaver* species. New Phytologist 61:175–188. **xiii, 243–244, 344–345, 347, 353**

Hassell, M. P., J. H. Lawton and R. M. May. 1976. Patterns of dynamical behavior in single species populations. Journal of Animal Ecology 45:471–486. **257**

Hassett, R. P., B. Cardinale, L. B. Stabler and J. J. Elser. 1997. Ecological stoichiometry of N and P in pelagic ecosystems: Comparison of lakes and oceans with emphasis on the zooplankton-phytoplankton interaction. Limnology and Oceanography 42:648–662. **176**

Hastings, A. 1980. Disturbance, coexistence, history, and competition for space. Theoretical Population Biology 18:363–373. **353**

Hecht, S. and A. Cockburn. 1990. *The Fate of the Forest: Developers, Destroyers, and Defenders of the Amazon.* Harper Perennial, New York, NY. **28**

Henderson, M. T., G. Merriam and J. Wegner. 1985. Patchy environments and species survival: chipmunks in an agricultural mosaic. Biological Conservation 31:95–105. **93**

Hilborn, R. and C. Walters. 1992. *Quantitative Fisheries Stock Assessment: Choice, Dynamics and Uncertainty.* Chapman and Hall, New York, NY. **22, 145, 148, 306**

Hilborn, R., D. Ludwig and C. Walters. 1995. Sustainable exploitation of renewable resources. Annual Review of Ecology and Systematics 26:45–67. **22, 145, 148, 306**

Holland, H. D. 1995. Atmospheric oxygen and the biosphere. Pp. 127–136. In C. G. Jones and J. H. Lawton (eds.). *Linking Species and Ecosystems.* Chapman and Hall, New York, NY. **132, 139, 239, 248**

Holland, H. D. and U. Petersen. 1995. *Living Dangerously: The Earth, Its Resources, and the Environment.* Princeton University Press, Princeton, NJ. **132, 139, 239, 248**

Holling, C. S. 1992. Cross-scale morphology, geometry and dynamics of ecosystems. Ecological Monographs 62:447–502. **141, 350**

Houghton, J. T., L. G. Meira Filho, B. A. Callander, N. Harris, A. Kattenberg and K. Maskell. 1996. *Climate change 1995. The science of climate change.* Cambridge University Press, Cambridge, England. **119**

Houseal, G. A. and B. E. Olson. 1995. Cattle use of microclimates on a northern latitude winter range. Canadian Journal of Animal Science 75:501–507. **183**

Howard, R. D. 1978.The evolution of mating strategies in bullfrogs, *Rana catesbeiana.* Evolution 32:850–871. **217**

Huffaker, C. B. 1958. Experimental studies on predation: dispersion factors and predator-prey oscillations. Hilgardia 27:343–383. **291–293**

Hughes, D. 1993. *Pan's Travail: Environmental Problems of the Ancient Greeks and Romans.* Johns Hopkins University Press, Baltimore, MA. **34**

Hulbert, S. H. 1984. Pseudoreplication and the design of ecological field experiments. Ecological Monographs 54:187–211. **20**

Huston, M., D. DeAngelis and W. Post (1988) New computer models unify ecological theory. Bioscience 38:682–691. **356**

Intergovernmental Panel on Climate Change. 1995. *Impacts, Adaptations, and Mitigation of Climate Change: Scientific-Technical Analyses.* Cambridge University Press, Cambridge, England. **134**

Ives, A. R. 1988. Aggregation and the coexistence of competitors. Annales Zoologici Fennici 25:75–88. **ix, xiii, 262, 287**

Ives, A. R. 1991. Aggregation and coexistence in a carrion fly community. Ecological Monographs 61:75–94. **ix, xiii, 262, 287**

Ives, A. R. 1995. Measuring resilience in stochastic systems. Ecological Monographs 65:217–233. **ix, xiii, 262, 287**

Jones, C. G. and J. H. Lawton. 1995. Linking Species and Ecosystems. Chapman and Hall, New York, NY. **141, 159**

Kareiva, P. and M. Andersen. 1988. Spatial aspects of species interactions: the wedding of models and experiments. Pages 38–54 in H. A, editor. *Community Ecology*. Springer-Verlag, New York, NY. **xi–xii, 294**

Keyfitz, N. 1989. The growing human population. Scientific American 261:119–126. **245**

Kidd, K. A., D. W. Schindler, R. H. Hesslein and D. C. G. Muir. 1995. Correlation between stable nitrogen isotope ratios and concentrations of organochlorines in biota from a freshwater food web. Science for the Total Environment 160/161:381–390. **189**

Kitchell, J. F. 1983. Energetics. Pp. 312–338 In P. Webb and D. Weihs (eds.). *Fish Biomechanics*. Praeger Publishers, New York, NY. **viii, xii, 11, 133, 135, 172**

Krebs, C. J., S. Boutin, R. Boonstra, A. R. E. Sinclair, J. N. M. Smith, M. R. T. Dale, K. Martin and R. Turkington. 1995. Impact of food and predation on the snowshoe hare cycle. Science 269:1112–1115. **12, 200, 202, 274**

Krebs, J. R. and N. B. Davies. 1993. *An Introduction to Behavioral Ecology*. Blackwell Scientific, London, England. **12, 200, 202, 274**

Lamberson, R. H., B. R. Noon, C. Voss and K. McKelvey. 1994. Reserve design for terrestrial species: the effects of patch size and spacing on the viability of the Northern Spotted Owl. Conservation Biology 8:185–195. **268**

Lampert, W. and U. Sommer. 1997. *Limnoecology: The Ecology of Lakes and Streams*. Oxford University Press, New York, NY. **168–169**

Langston, N. 1995. *Forest Dreams, Forest Nightmares*. University of Washington Press, Seattle, WA. **vii, xii, 11, 49, 51, 61, 69**

Lank, D. B., Smith, D. M., Hanotte, O., Burke, T., Cooke, F. 1995. Genetic polymorphism for alternative mating behaviour in lekking male ruff *Philomachus pugnax*. Nature 378:59–62. **215**

Lefkovitch, L. P. and L. Fahrig. 1985. Spatial characteristics of habitat patches and population survival. Ecological Modeling 30:297–308. **93–94**

Leopold, A. 1933. *Game Management*. Charles Scribner's Sons, New York, NY. **2, 26, 55, 124, 155**

Leopold, A. 1953. *Round River*. Oxford University Press, Oxford, England. **2, 26, 55, 124, 155**

Levin, S. A. 1992. The problem of pattern and scale in ecology. Ecology 73:1943–1967. **56, 126–127**

Levin, S. A. and R. T. Paine. 1974. Disturbance, patch formation, and community structure. Proceedings of the National Academy of Sciences, USA. 71:2744–2747. **56, 126–127**

Likens, G. E. 1992. *The Ecosystem Approach: Its Use and Abuse*. Ecology Institute, Oldendorf/Luhe, Germany. **124, 140, 147**

Likens, G. E., C. T. Driscoll and D. C. Buso. 1996. Long-term effects of acid rain: response and recovery of a forest ecosystem. Science 272:244–246. **124, 140, 147**

Likens, G. E., F. H. Bormann, N. M. Johnson, D. W. Fisher and R. S. Pierce. 1970. Effects of forest cutting and herbicide treatment in nutrient budgets on the Hubbard Brook watershed. Ecological Monographs 40:23–47. **124, 140, 147**

Loucks, O. L., M. L. Plumb-Mentjes and D. Rogers. 1985. Gap processes and large-scale disturbances in sand prairies. Pp. 71–83. In S. T. A. Pickett and P. S. Whites (eds). *The Ecology of Natural Disturbance and Patch Dynamics*. Academic Press, Orlando. **56, 58**

Ludwig, D., B. Walker and C. S. Holling. 1996. Sustainability, stability and resilience. Conservation Ecology 1 (http://www.consecol.org). **148**

Luecke, C., C. C. Lunte, R. A. Wright, D. Robertson and A. S. McLain. 1992. Impacts of variation in planktivorous fish on abundance of daphnids: A simulation model of the Lake Mendota food web. Pp. 407–426. In J. F. Kitchell (ed.). *Food Web Management. A case study of Lake Mendota*. Springer-Verlag. New York, NY. **19, 21**

MacArthur, R. H. 1955. Fluctuations of animal populations and a measure of community stability. Ecology 36:533–536. **16, 117, 299–300, 371**

MacArthur, R. H. and E. O. Wilson. 1967. The theory of island biogeography. Monographs in Population Biology 1:1–215. **16, 117, 299–300, 371**

MacArthur, R. H. and E. O. Wilson. 1967. *Island biogeography*. Princeton University Press, Princeton, NJ. **16,**

117, 299–300, 371

MacInnes, P. F., R. J. Naiman, J. Pastor and Y. Cohen. 1992. Effects of moose browsing on vegetation and litter of the boreal forest, Isle Royale, Michigan, USA. Ecology 73:2059–2075. **103**

MacLulich, D. A. 1937. Fluctuations in numbers of the varying hare *(Lepus americanus).* University of Toronto Studies, Biology Series 43:1–136. **13, 273**

May, R. M. 1973. *Stability and Complexity in Model Ecosystems.* (2nd ed.). Princeton University Press, Princeton, NJ. **287, 299–300, 310**

May, R. M. 1981. Models for two interacting populations. Pp. 78–104. In R. M. May (ed.). *Theoretical Ecology.* Sinauer Associates, Sunderland, MA. **287, 299–300, 310**

May, R. M. and G. F. Oster. 1976. Bifurcations and dynamic complexity in simply ecological models. American Naturalist 110:573–599. **287, 299–300, 310**

Maynard-Smith, J. 1982. *Evolution and the Theory of Games.* Cambridge University Press, Cambridge, England. **216, 219**

Mayr, E. 1982. *The Growth of Biological Thought. Diversity, Evolution, and Inheritance.* Harvard University Press, Cambridge, MA. **201**

McCleery, R. H. and C. M. Perrins. 1985. Territory size, reproductive success and population dynamics in the great tit, *Parus major.* Pages 353–374 in R. M. Sibly and R. H. Smith (eds.). *Behavioural ecology. Ecological consequences of behaviour.* Blackwell Scientific, Oxford, England. **247, 263**

McCune, B. and T. F. H. Allen (1985) Will similar forests develop on similar sites? Canadian Journal of Botany. 63:367–376. **331–333**

McGarigal, K. and B. J. Marks. 1995. *FRAGSTATS: Spatial pattern analysis program for quantifying landscape structure.* USDA Forest Service General Technical Report PNW-GTR-351. **114**

McIntosh, R. P. 1985. *The Background of Ecology: Concept and Theory.* Cambridge University Press, Cambridge, England. **46, 362, 364**

McKibben, B. 1996. *Hope Human and Wild: True Stories of Living Lightly on the Earth.* Hungry Mind Press, St. Paul, MN. **33–34**

McMahon, R. F. 1996. The physiological ecology of the zebra mussel, *Dreissena polymorpha,* in North America and Europe. American Zoologist 35:339–363. **196**

McNaughton, S. J. 1977. Diversity and stability of ecological communities: a comment on the role of empiricism in ecology. American Naturalist 111:515–525. **300, 345, 347**

McNeill, J. R. 1992. *The Mountains of the Mediterranean World: An Environmental History.* Cambridge University Press, Cambridge, England. **xiii, 34–37**

Mech, L. D. 1966. *The wolves of Isle Royale.* United States Government Printing Office, Washington DC. **269**

Meffe, G. K. and C. R. Carroll. 1994. *Principles of conservation biology.* Sinauer Associates, Inc., Sunderland, MA. **95, 116**

Merchant, C. 1993. *Major Problems in American Environmental History: Documents and Essays.* Heath and Co., Lexington, MA. **33**

Mitman, G. 1992. *The State of Nature: Ecology, Community, and American Social Thought, 1900–1950.* University of Chicago Press, Chicago, IL. **41**

Mitsch, W. J. and J. G. Gosselink. 1986. *Wetlands.* Van Nostrand Reinhold, New York, NY. **11, 178, 180**

Mladenoff, D. J., T. A. Sickley, R. G. Haight and A. P. Wydeven. 1995. A regional landscape analysis and prediction of favorable gray wolf habitat in the northern Great Lakes Region. Conservation Biology 9:279–294. **xii, 95**

Moehlman, P. 1979. Jackel helpers and pup survival. Nature 277:382–383. **12**

Mooney, H. A., J. H. Cushman, E. Meidna, O. E. Sala and E.-D. Schulze. 1996. *Functional Roles of Biodiversity: A Global Perspective.* Wiley, New York, NY. **143**

Mumme, R. L. 1992. Do helpers increase reproductive success? An experimental analysis in the Florida scrub jay. Behavioral Ecology and Sociobiology 31:319–328. **227**

Naiman, R. J., C. A. Johnson and J. Kelley. 1988. Alteration of North American streams by beaver. Bioscience 38:753–761. **39–40**

Naiman, R. J., J. M. Melillo and J. E. Hobbie. 1986. Ecosystem alteration of boreal forest streams by beaver *(Castor canadensis).* Ecology 67:1254–1269. **39–40**

National Research Council. 1992. *Restoration of Aquatic Ecosystems.* National Academy Press, Washington, DC. **133, 154, 156**

Nichols, G. E. (1923) A working basis for the ecological classification of plant communities. Ecology 4:11–23. **317–319**

Nicholson, A. J. 1933. The balance of animal populations. Journal of Animal Ecology 2:132–178. **270–273**

Nonacs, P. 1986. Ant reproductive strategies. Quarterly Review of Biology 61:1–21. **229**

Noon, B. R. and K. S. McKelvey. 1996. Management of the spotted owl: a case history in conservation biology. Annual Review of Ecology and Systematics 27:135–162. **xiii, 265, 389**

Odum, H. T. 1957. Trophic structure and productivity of Silver Springs, Florida. Ecological Monographs 27:55–112. **10, 67**

Ogden, J. G. 1966. Forest history of Ohio: radiocarbon dates and pollen stratigraphy of Silver Lake, Ohio. Ohio Journal of Science 66:387–400. **75**

Pacala, S., C. Canham, and J. Silander Jr. 1993. Forest models defined by field measurements I: the design of a northeast forest simulator. Canadian Journal of Forest Research 23:1980–1988. **349**

Paine, R. T. 1966. Food web complexity and species diversity. American Naturalist 100:65–75. **xi, 56, 324, 326, 376**

Paine, R. T. and S. A. Levin. 1981. Intertidal landscapes: disturbance and the dynamics of pattern. Ecological Monographs 51:145–178. **xi, 56, 324, 326, 376**

Parton, W. J., D. S. Schimel, C. V. Cole and D. S. Ojima. 1987. Analysis of factors controlling soil organic matter levels in Great Plains grasslands. Soil Science Society of America Journal 51:1173–1179. **102**

Pauly, D. and V. Christensen. 1995. Primary production required to sustain global fisheries. Nature 374:255–257. **149**

Pearson, S. M. 1993. The spatial extent and relative influence of landscape-level factors on wintering bird populations. Landscape Ecology 8:3–18. **97–98**

Pearson, S. M., M. G. Turner, L. L. Wallace and W. H. Romme. 1995. Winter habitat use by large ungulates following fires in northern Yellowstone National Park. Ecological Applications 5:744–755. **97–98**

Perrins, C. M. 1965. Population fluctuations and clutch size in the great tit. Animal Ecology 34:43–48. **205, 247, 263**

Perrins, C. M. 1979. *British Tits*. Collins, London, England. **205, 247, 263**

Perrins, C. M. and D. Moss. 1975. Reproductive rates in the great tit. Journal of Animal Ecology 44:695–706. **205, 247, 263**

Perry, D. 1988. Landscape pattern and forest pests. Northwest Environmental Journal 4:213–228. **xii, 45, 50, 57–58, 60, 73**

Perry, D. 1994. *Forest Ecosystems*. Johns Hopkins University Press, Baltimore, MA. **xii, 45, 50, 57–58, 60, 73**

Peterjohn, W. T. and D. L. Correll. 1984. Nutrient dynamics in an agricultural watershed: observations on the role of a riparian forest. Ecology 65:1466–1475. **99–100**

Pickett, S. T. A. and M. L. Cadenasso. 1995. Landscape ecology: spatial heterogeneity in ecological systems. Science 269:331–334. **22, 56, 81**

Pickett, S. T. A. and P. S. White (eds). 1985. *Ecology of Natural Disturbance and Patch Dynamics*. Academic Press, Orlando, FL. **22, 56, 81**

Pinet, P. R. 1998. *Invitation to Oceanography*. Jones and Bartlett, London, England. **167**

Platt, W. J. and I. M. Weis. 1985. An experimental study of competition among fugitive prairie plants. Ecology 66:708–720. **283**

Pollan, M. 1992. *Second Nature: A Gardener's Education*. Laurel: New York, NY. **45–46**

Ponting, C. 1992. *A Green History of the World: The Environment and the Collapse of Great Civilizations*. St. Martin's Press, New York, NY. **29**

Postel, S. L, G. C. Daily and P. R. Ehrlich. 1996. Human appropriation of renewable fresh water. Science 271:785–788. **150**

Pound, R., and F. E. Clements. (1901) *The phytogeography of Nebraska*. Second ed. The Seminar, Lincoln, NE. **361, 372**

Pulliam, H. R. 1988. Sources, sinks and population regulation. American Naturalist 132:652–661. **94, 109–110**

Pulliam, H. R., J. B. Dunning and J. Liu. 1992. Population dynamics in complex landscapes: a case study. Ecological Applications 2:165–177. **94, 109–110**

Pyne, S. 1982. *Fire in America: A Cultural History of Wildland and Rural Fire*. Princeton University Press, Princeton, NJ. **67**

Queller, D. C., J. E. Strassmann and C. R. Hughes. 1993. Microsatellites and kinship. Trends in Ecology and Evolution 8:285–288. **xii, 229–230**

Queller, D. C., J. E. Strassmann, C. R. Sol's, C. R. Hughes and D. M. DeLoach. 1993. A selfish strategy of social insect workers that promotes social cohesion. Nature 365:639–641. **xii, 229–230**

Queller, D. C., J. M. Peters. C. R. Solis and J. E. Strassmann. 1997. Control of reproduction in social insect colonies: individual and collective relatedness preferences in the paper wasp, *Polistes annularis*. Behavioral Ecology and Sociobiology. 40:3–16. **xii, 229–230**

Rackham, O. 1990. Ancient landscapes. Pp. 85–111. In O. Murray and S. Price (eds.). *The Greek City: From Homer to Alexander.* Clarendon Press, Oxford, England. **36, 38**

Ramcharan, C. W., D. K. Padilla and S. I. Dodson. 1992. Models to predict potential occurrence and density of the zebra mussel *(Dreissena polymorpha)*. Canadian Journal of Fisheries and Aquatic Sciences 49:2611–2620. **195**

Rasmussen, D. I. 1941. Biotic communities of the Kaibab Plateau, Arizona. Ecological Monographs 11:229–275. **53–55**

Ratnieks, F. L. W. 1988. Reproductive harmony via mutual policing by workers in eusocial Hymenoptera. American Naturalist 132:217–236. **229**

Ratnieks, F. L. W. and P. K. Visscher. 1989. Worker policing in the honey bee. Nature 342:796–797. **229**

Riitters, K. H., R. V. O'Neill, C. T. Hunsaker, J. D. Wickham, D. H. Yankee, S. P. Timmons, K. B. Jones and B. L. Jackson. 1995. A factor analysis of landscape pattern and structure metrics. Landscape Ecology 10:23–40. **114**

Romme, W. H. 1982. Fire and landscape diversity in subalpine forests of Yellowstone National Park. Ecological Monographs 52:199–221. **xii, 82, 106**

Romme, W. H. and D. G. Despain. 1989. Historical perspective on the Yellowstone fires of 1988. BioScience 39:695–699. **xii, 82, 106**

Rosen, R. 1991. *Life itself.* Columbia University Press, New York, NY. **348**

Roughgarden, J. and S. Pacala (1989) Taxon cycling among Anolis lizards populations: review of the evidence. In D. Otte and J. Endler (eds.). *Speciation and its consequences.* Sinauer, Sunderland, MA. **349–350, 378–380**

Sachse, N. D. 1965. *A Thousand Ages.* Regents of the University of Wisconsin, Madison, WI. **155**

Sauer, C. 1969. *Seeds, Spades, Hearths, and Herds.* MIT Press, Cambridge, MA. **67**

Scheffer, M. and J. Beets. 1994. Ecological models and the pitfalls of causality. Hydrobiologia 275/276:115–124. **145**

Schindler, D. E., J. F. Kitchell, X. He, S. R. Carpenter, J. R. Hodgson and K. L. Cottingham. 1993. Food web structure and phosphorus cycling in lakes. Transactions of the American Fisheries Society 122:756–772.

Schindler, D. W. 1977. The evolution of phosphorus limitation in lakes. Science 195:260–262. **133, 177**

Schmid-Hempel, P., A. Kacelnik and A. I. Houston. 1985. Honeybees maximize efficiency by not filling their crop. Behavorial Ecology and Sociobiology 17:61–66. **214**

Schneider, D. W. 1992. A bioenergetics model of zebra mussel, *Dreissena polymorpha,* growth in the Great Lakes. Canadian Journal of Fisheries and Aquatic Sciences 49:1406–1416. **196**

Schoener, T. W. 1983. Field experiments on interspecific competition. American Naturalist 122:240–285. **278**

Scott, J. M., C. B. Kepler, C. van Riper III and S. Fefer. 1988. Conservation of Hawaii's vanishing avifauna. Bioscience 38:238–253. **305**

Seaton, A. P. C. and J. Antonovics 1967. Population inter-relationships. I. Evolution in mixtures of *Drosophila mutans.* Heredity 22:19–33. **346–348**

Shelford, V. 1913. Animal communities in temperate America as illustrated in the Chicago region. Bulletin of the Geographical Society of Chicago. 5. Reprinted (1977) Arno Press, New York, NY. **52–53**

Sherman, P. W. 1988. The levels of analysis. Animal Behavior 36:616–619. **xii, 201**

Shugart, H. H. 1984. *A Theory of Forest Dynamics: The Ecological Implications of Forest Succession Models.* Springer-Verlag, New York, NY. **50, 56, 333, 357**

Shugart, H. H. Jr. and D. West. 1977. Development of an Appalachian deciduous forest succession model and its application to assessment of the impact of the Chestnut blight. Journal of Environmental Management 5:161–179. **50, 56, 333, 357**

Shugart, H. H. Jr., W. R. Emanuel, D. C. West and D. L. DeAngelis. 1980. Environmental gradients in a

Beech-Yellow Poplar stand simulation. Math. Bioscience. 50:163–170. **50, 56, 333, 357**

Simmons, I. G. 1996. *Changing the Face of the Earth: Culture, Environment, History.* Blackwell, Oxford, England. **28, 36**

Sinclair, A. R. E. and M. Norton-Griffiths. 1982. Does competition or facilitation regulate migrant ungulate populations in the Serengeti? A test of hypotheses. Oecologia 53:364–369. **243–244**

Smith, J. N. M., P. Arcese, and W. M. Hochachka. 1991. Social behaviour and population regulation in insular bird populations: implications for conservation. Pp. 148–167. In C. M. Perrins, J. -D. Lebreton and G. J. M. Hirons (eds.), *Bird population studies: relevance to conservation and management.* Oxford University Press, New York, NY. **231**

Smith, T. M. and Dean Urban. 1988. Scale resolution if forest structural pattern. Vegetatio 74:143–150. **359**

Stauffer, D. 1985. *Introduction to percolation theory.* Taylor and Francis Ltd., London, England. **96**

Sterner, R. W., J. J. Elser and D. O. Hessen. 1992. Stoichiometric relationships among producers, consumers, and nutrient cycling in pelagic ecosystems. Biogeochemistry 17:49–67. **138**

Stow, C. A., S. R. Carpenter, C. P. Madenjian, L. A. Eby and L. J. Jackson. 1995. Fisheries management to reduce contaminant consumption. BioScience 45:752–758. **153**

Strier, K. B. 1997. Behavioral ecology and conservation biology of primates and other animals. Advances in the Study of Behavior, 26:101–158. **231**

Summers, K. 1992. Mating strategies in two species of dart-poison frogs: a comparative study. Animal Behavior 43:907–919. **224**

Takacs, D. 1996. *The Idea of Biodiversity: Philosophies of Paradise.* Johns Hopkins University Press, Baltimore, MA. **xi–xii, 46**

Tansley, A. G. 1935. The use and abuse of vegetational concepts and terms. Ecology 16:284–307. **56, 124–125**

Temple, S. A. 1986. Predicting impacts of habitat fragmentation on forest birds: a comparison of two models. Pp. 301–304. In J. Verner, M. L. Morrison and C. J. Ralph (eds.). *Wildlife 2000: Modeling habitat relationships of terrestrial vertebrates.* University of Wisconsin Press, Madison, WI. **92**

Thoreau, H. 1975. Natural history of Massachusetts and Succession of forest trees. In Walter Harding (ed.). *Selected Works.* Houghton Mifflin, Boston, MA. **51**

Thornhill, R 1981. *Panorpa* (Mecoptera: Panorpidae) scorpionflies: systems for understanding resource-defense polygyny and alternative male reproductive efforts. Annual Review of Ecology and Systematics 12:355–386. **216, 223**

Thornhill, R. 1976. Sexual selection and nuptial feeding behavior in *Bittacus apicalis* (Insecta: Mecoptera). American Naturalist 110:529–548. **216, 223**

Tilman, D. 1977. Resource competition between planktonic algae: an experimental and theoretical approach. Ecology 58:338–348. **281–282, 300, 339, 349–350, 353, 374**

Tilman, D. 1996. Biodiversity: population versus ecosystem stability. Ecology 77:350–363. **281–282, 300, 339, 349–350, 353, 374**

Tilman, D. and D. Wedin. 1991a. Plant traits and resource reduction for five grasses growing on a nitrogen gradient. Ecology 72:685–700. **281–282, 300, 339, 349–350, 353, 374**

Tilman, D. and D. Wedin. 1991b. Dynamics of nitrogen competition between successional grasses. Ecology 72:1038–1049. **281–282, 300, 339, 349–350, 353, 374**

Tilman, D., C. L. Lehman and C. Yin.1997. Habitat destruction, dispersal, and deterministic extinction in competitive communities. American Naturalist 149:407–435. **281–282, 300, 339, 349–350, 353, 374**

Tinbergen, J. M., J. H. van Balen and H. M. van Eck. 1985. Density dependent survival in an isolated great tit population: Kluyvers data reanalyzed. Ardea 73:38–48. **200–202, 206–207, 248**

Tinbergen, N. 1958. *Curious Naturalists.* Basic Books, New York, NY. **200–202, 206–207, 248**

Tinbergen, N., G. J. Broekhuysen, F. Feekes, J. C. W. Houghton, H. Kruuk and E. Szulc. 1963. Egg shell removal by the black-headed gull, *Larus ridibundus* L. ; a behaviour component of camouflage. Behaviour 19:74–117. **200–202, 206–207, 248**

Tobey, R. 1981. *Saving the Prairies: The Life Cycle of the Founding School of American Plant Ecology, 1895–1955.* University of California Press, Berkeley, CA. **46, 50**

Trivers, R. L. 1971. The evolution of reciprocal altruism. Quarterly Review of Biology 46:35–57. **220, 228, 231**

Trivers, R. L. 1972. Parental investment and sexual selection. Pp. 136–179. In B. Campbell (ed.). *Sexual Selection and the Descent of Man.* Aldine, Chicago, IL. **220, 228, 231**

Trivers, R. L. and H. Hare. 1976. Haplodiploidy and the evolution of the social insects. Science 191:249–263.

220, 228, 231

Turchin, P. 1990. Rarity of density dependence or population regulation with lags? Nature 344:660–663. **275**

Turner, B. L. , W. C. Clark, R. W. Kates, J. T. Mathews and W. B. Meyer. 1990. *The Earth as Transformed by Human Action: Global and regional changes in the Biosphere over the Past 300 Years*. Cambridge University Press, London, England.

Turner, M. G. 1989. Landscape ecology: the effect of pattern on process. Annual Review of Ecology and Systematics 20:171–197. **viii, xii, 80–83, 106**

Turner, M. G., R. V. O'Neill, R. H. Gardner and B. T. Milne. 1989. Effects of changing spatial scale on the analysis of landscape pattern. Landscape Ecology 3:153–162. **viii, xii, 80–83, 106**

Turner, M. G., S. R. Carpenter, E. J. Gustafson, R. J. Naiman and S. M. Pearson. 1997. Land use. In M. J. Mac, P. A. Opler, P. Doran, C. Haecker and L. Huckaby Stroh (eds.). *Status and trends of our nation's biological resources. Volume1*. National Biological Service, Washington, DC. **viii, xii, 80–83, 106**

Turner, M. G., W. H. Hargrove, R. H. Gardner and W. H. Romme. 1994. Effects of fire on landscape heterogeneity in Yellowstone National Park, Wyoming. Journal of Vegetation Science 5:731–742. **viii, xii, 80–83, 106**

Turner, M. G., W. H. Romme, R. H. Gardner and W. W. Hargrove. 1997. Effects of patch size and fire pattern on early succession in Yellowstone National Park. Ecological Monographs 67:411–433. **viii, xii, 80–83, 106**

Turner, M. G., W. H. Romme, R. H. Gardner, R. V. O'Neill and T. K. Kratz. 1993. A revised concept of landscape equilibrium: disturbance and stability on scaled landscapes. Landscape Ecology 8:213–227. **viii, xii, 80–83, 106**

Urban, D. L., R. V. O'Neill and H. H. Shugart. 1987. Landscape ecology. BioScience 37:119–27. **4, 89, 359**

Varley, G. C. 1949. Population changes in German forest pests. Journal of Animal Ecology 18:117–122. **247, 262, 273**

Vitousek, P. M. and R. W. Howarth. 1991. Nitrogen limitation on land and in the sea: How can it occur? Biogeochemistry 13:87–115. **139–140, 149, 198**

Vitousek, P. M., H. A. Mooney, J. Lubchenco and J. M. Melillo. 1997. Human domination of Earth's ecosystems. Science 277:494–499. **139–140, 149, 198**

Vitousek, P. M., J. D. Aber, R. W. Howarth, G. E. Likens, P. A. Matson, D. W. Schindler, W. H. Schlesinger and D. Tilman. 1997. Human alteration of the global nitrogen cycle: Sources and consequences. Ecological Applications. 1998. **139–140, 149, 198**

Vitousek, P. M., P. R. Ehrlich, A. H. Ehrlich and P. A. Matson. 1986. Human appropriation of the products of photosynthesis. BioScience 36:368–373. **139–140, 149, 198**

Vollenweider, R. A. 1976. Advances in defining critical loading levels for phosphorus and nitrogen. Memorie dell'Istituto Italiano di Idrobiologia 33:53–83. **134**

Von Uexkull, J. 1955. A stroll through the worlds of animals and men. In C. H. Schiller (ed.) *Instinctive behavior: The development of a modern concept*. International Universities Press Inc, New York, NY. **376–377**

Waage, J. K. 1973. Reproductive behavior and its relation to territoriality in *Calopteryx maculata* (Beauvois) (Odonata: Calopterygidae). Behaviour 47:240–256. **222, 306**

Waage, J. K. and D. J. Greathead. 1988. Biological control: challenges and opportunities. Philosophical Transactions of the Royal Society of London, B 318:111–128. **222, 306**

Walker, J. C. G. 1984. How life affects the atmosphere. BioScience 34:486–491. **137**

Wallin, D. O., F. J. Swanson and B. Marks. 1994. Landscape pattern response to changes in pattern generation rules: land-use legacies in forestry. Ecological Applications 4:569–580. **xi–xii, 111–112**

Walters, C. J. 1986. *Adaptive Management of Renewable Resources*. MacMillan Publishers, New York, NY. **145, 171, 306, 389**

Watt, A. S. 1947. Pattern and process in the plant community. Journal of Ecology 35:1–22. **14, 56, 331–333**

Webb, D. A. 1954. Is the classification of plant communities either possible or desirable? Bot. Tidddkr 51:362–370. **329**

Wegner, J. and G. Merriam. 1979. Movements by birds and small mammals between a wood and adjoining farm habitats. Journal of Applied Ecology 16:349–357. **93, 389**

White, G. 1789. *The Natural History of Selborne*. E. P. Dutton and Co., New York, NY. **49–50, 56, 85**

Whitney, G. 1994. *From Coastal Wilderness to Fruited Plain: A History of Environmental Change in Temperate North America 1500 to Present*. Cambridge University Press, New York, NY. **33, 71**

Whittaker, R. H. (1956) Vegetation of the Great Smoky Mountains. Ecological Monographs 26:1–80. **56, 331, 333, 336–337, 367–368, 370–371**

Whittaker, R. H. (1972) Evolution and measurement of species diversity. In *"Origin and measurement of diversity."* Summer Institute in Systematics. V. Smithsonian Institute, Washington, DC. **56, 331, 333, 336–337, 367–368, 370–371**

Whittaker, R. H. 1953. A consideration of climax theory: the climax as a population and pattern. Ecological Monographs 23:41–78. **56, 331, 333, 336–337, 367–368, 370–371**

Wiens, J. A. and B. T. Milne. 1989. Scaling of landscapes in landscape ecology, or, landscape ecology from a beetle's perspective. Landscape Ecology 3:97–96. **xi, 81, 84, 115**

Wiens, J. A., N. C. Stenseth, B. Van Horne and R. A. Ims. 1993. Ecological mechanisms and landscape ecology. Oikos 66:369–380. **xi, 81, 84, 115**

Wiley, R. H. 1973. Territoriality and non-random mating in sage grouse, Centrocercus urophasianus. Animal Behavior Monographs 6:85–169. **221**

Wilkinson, G. S. 1984. Reciprocal food sharing in the vampire bat. Nature 308:181–184. **231**

Wilson, E. O. 1975. *Sociobiology.* Harvard University Press, Cambridge, MA. **viii, 117, 352–353, 371**

Wood, T. G. and W. A. Sands. 1978. The role of termites in ecosystems. Pp. 245–292. In M. V. Brian (ed.). *Production ecology of ants and termites.* Cambridge University Press, Cambridge, England. **233, 247**

Wootton, J. T. 1994. Predicting direct and indirect effects: an integrated approach using experiments and path analysis. Ecology 75:151–165. **xi, xiii, 294–297**

Worster, D. 1979. *Dust Bowl: The Southern Plains in the 1930s.* Oxford University Press, New York, NY. **xi–xii, 28, 34, 46, 49–50, 67**

Worster, D. 1993. The ecology of order and chaos. Pp. 156–170. In D. Worster (ed.). *The Wealth of Nature: Environmental History and the Ecological Imagination.* Oxford University Press, New York, NY.

Worster, D. 1994. *Nature's Economy: A History of Ecological Ideas.* Second Edition. Cambridge University Press, New York, NY. **xi–xii, 28, 34, 46, 49–50, 67**

Worster, D. 1996. The two cultures revisited: environmental history and the environmental sciences. Environment and History 2:3–14. **xi–xii, 28, 34, 46, 49–50, 67**

Wu, J. and O. L. Loucks. 1995. From balance of nature to hierarchical patch dynamics: a paradigm shift in ecology. Quarterly Review of Ecology 70:439–466. **56, 58**

Wu, J. and S. A. Levin. 1994. A spatial patch dynamic modeling approach to pattern and process in an annual grassland. Ecological Monographs 64:447–464. **56, 58**

Young, C. 1998. Defining the range: the development of carrying capacity in management practices. Journal of the History of Biology 31. 1998. **53, 55, 62, 64**

Zahavi, A. 1975. Mate selection—a selection for a handicap. Journal of Theoretical Biology 53:205–214. **224**

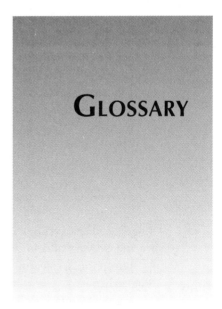

GLOSSARY

Abiotic. Physical-chemical conditions such as temperature of the habitat, matter available as nutritional resources, or light energy that can be used for photosynthesis.

Adaptive. (Re: genetic trait), The inherited trait favored by natural selection, manifested in an individual because that specific trait enabled past generations to produce more offspring compared to others that had a slightly different trait.

Adaptive characteristic. Inherited characteristic that enables an adapted organism to fit into its environment, in the sense of being able to survive and produce offspring.

Adaptive ecosystem management. Use of the findings of ecology to manage natural resources, not for maximum commodity production (a traditional industrial forest), or for preservation of current conditions (a traditional reserve), but for the perpetuation of patterns and processes that allow the ecosystem to persist. This management style stresses experimentation, collaboration and re-evaluation.

Adaptive management. Responding to changing ecological conditions so as to not exceed productivity limits of a specific place. For example, when crop growth slows, a good farmer learns to recognize ecological signs that tell either to add more manure or to allow a field to lie fallow. Adaptive management becomes impossible when managers are forced to meet the demands of outsiders who are not under local ecological constraints, as happened in Mediterranean Spain.

Aerial photography. Photographs taken of the Earth's surface from airplanes.

Aerobic. Conditions where oxygen is present.

Age class. Group of individuals of the same age in a population. To study species in which per capita reproduction and mortality rates depend on age, ecologists often divide a population into different age classes.

Age-structured model. Model that divides populations into different age classes.

Aggregation. The clustering of individuals in particular locations in space.

Ammonification. The process by which bacteria and fungi convert urea and amino acids to ammonia.

Autotroph. Organism that uses light energy to convert carbon dioxide to carbohydrate (photosynthesis).

Balance of nature. The concept that natural systems maintain themselves unless disturbed; i.e., disturbance "...was a transitory state, a foreign intrusion into an otherwise balanced and unchanging community that in the healthy ecosystem was quickly 'disposed of' in the same sense a healthy organism disposes of disease" (Perry 1995).

Biodiversity. The abundance, variety and genetic constitution of animals and plants in nature.

Biogeochemical cycling. The continuous flow and transformation through an ecosystem's living and nonliving components of essential elements such as nitrogen, phosphorus, calcium, potassium, and others.

Biogeochemistry. The study of the movement of elements or compounds through living organisms and the nonliving environment, usually at a large scale.

Biomagnification. The process whereby the concentration of a fat-soluble contaminant increases as it passes up the food chain. With each consumption step, the concentration of the contaminant in the consumer biomass is magnified.

Biomass. Weight per unit area of plants, animals or microbes.

Biome. Major vegetation association, distinguished from a community by its defining plants being identified by life form (tree, shrub, herb, etc.), rather than species. Examples include rain forest, boreal forest, desert, grassland, tundra, etc. Occasionally animals are included in the definition; e.g., "spruce-moose."

Biotic community. A local assemblage of species.

Carrying capacity. The size of the population that can be sustained by a given environment.

Change. (Re: landscape ecology), the structure and function of landscapes are not fixed through time, but rather are very dynamic.

Chaos. A mathematical term referring to population dynamics in which patterns never repeat themselves because there are no stable equilibrium points and no stable cycles. A system of equations that links too many long lags will appear chaotic, in both the technical and commonplace meaning of the word.

Character displacement. The change in heritable characters that occurs when two species coexist. These changes are believed to occur to prevent competition.

Climate. The composite, long-term weather across a region.

Climatologist. Specialist who studies long-term changes in climate.

Climax community. A stable community in which the vegetation is in equilibrium with the climate, a concept (developed by Howard Clements) now regarded as simplistic.

Climax. (Re: community), when individuals of species present at a site replace themselves with their progeny, and the vegetation is more or less in compositional stasis.

Coevolution. The evolution of biologically separate lines in which the species involved have continuously influenced each other to produce their special modern relationship. For example, noxious species that avoid predation are mimicked by other species that invest nothing in being unpleasant prey. Flowers with long spurs and their pollinators with tongues the same length are also products of coevolution.

Community. The community has been variously defined as a group of species, the concrete versus abstract community, a collection of populations, and interactions through periodic occupancy of space.

Competition. A struggle between two or more groups (often species) for a limited resource such as light, food, or space.

Competitive exclusion principle. Two or more species cannot coexist if they use the same limiting resources in the same way.

Conditional strategy. Strategy "chosen" or performed by an individual that is capable of following any of the potential strategies shown by its species, according to its abilities relative to those of others.

Constancy. A lack of change, or minimal change, in the numbers, densities or relative proportions of the elements of interest.

Constraint. Condition that defines whether the individual lives or dies, with its relative success expressed as growth and reproduction.

Convergent evolution. The situation when a general problem is solved by separate lines of evolution in a common fashion. The eyes of humans, mollusks and spiders are a common solution to the problem of sensing light, via organs that have evolved in different ways.

Corridor. A narrow connection between two patches of suitable habitat.

Cover. The area covered by leaf shade, expressed as a percentage. Cover can be measured in herbaceous vegetation by inserting long pins down through the vegetation, and recording which species touch the pins. The percentage of pins touched is an indication of the percentage of ground shaded.

Cover repetition. (Re: plant biomass measurement), the ecologist records not just presence or absence, but the number of times leaves of a given species overlap. Much as dominance, a biomass measure, approximates cover in trees, cover repetition in herbaceous vegetation is roughly a measure of biomass.

Critical threshold. (Re: landscape ecology), connectivity in random or structured maps developed from percolation theory.

Cultural eutrophication. The overfertilization of fresh water systems due in large part to excess nutrients (especially nitrate and phosphate) that unintentionally end up in lakes and streams.

Cultural lens. Mindset defined by cultural beliefs, perceptions, ideas, desires, dreams, ethics, religious ideals and scientific paradigms, which profoundly shapes the practice of science and the ways science is used to transform the Earth.

Decomposition. The breakdown of biotic matter into its organic components.

Demographics. Rates of birth, death, immigration and emigration.

Density-dependent population growth. A type of population growth where the per capita reproduction and mortality rates depend on population density.

Density-independent population growth. A type of population growth where the per capita reproduction and mortality rates do not depend on population density.

Density. Calculation of the number of individuals per unit area by counting all individuals in a quadrat or volume of a specific size.

Deterministic model. A model that holds that the direction of change is determined by laws that people can understand, and that any change will occur in a predictable manner and will lead to a constant endpoint. In ecology, deterministic models ignore unpredictable factors. These models are called deterministic because the population density in year t+1 is predicted using only the density in year t.

Detritus. Mass of dead organic matter.

Distance measure. A particularly fast measure of density in trees, pioneered by the University of Wisconsin's Grant Cottam and John Curtis. Distance measures depend on the intuitively obvious relationship between density and the proximity of individuals. Individuals of a high-density species will be close together; therefore measures of the proximity of trees also gives density.

Disturbance. A relatively discrete event that disrupts the structure of an ecosystem, community or population, and changes resource availability or the physical environment. Disturbances can also be defined as biological and physical factors that shape plant and animal communities. This includes processes such as fires, floods, insect epidemics, soil processes, nutrient cycles, erosion and the movement of water.

Disturbance regime. The spatial and temporal dynamics of disturbances over a long time period. Includes such characteristics as spatial distribution of the disturbances; disturbance frequency (i.e., number of disturbance events in a time interval, or the probability of a disturbance occurring); return interval (mean time between disturbances); rotation period (how long would it be until an area equivalent to the size of the study area was disturbed); disturbance size; and the magnitude, or force, of the disturbance.

Diversity. The variety or richness in the species composition of a community. Diversity has two components: species richness and species evenness.

Dominance. A measure used with forest data when measuring trees. One wraps a tape at breast height around a tree in the quadrat, and from that circumference one can calculate the cross-sectional area of the tree trunk. This estimates the biomass of the species, and also corresponds to the proportion of the ground which is shaded by each species.

Ecological model. The simplified representation of a system. The purpose is to understand the relationships, distribution or abundance of organisms or groups of organisms, or their interactions with the physical or chemical environment.

Ecology. The study of the relationships, distribution and abundance of organisms or groups of organisms, in an environment.

Economy of nature. An idea stemming from Linnaeus. He believed it critical to understand relations between species, relations he believed had been designed by the Creator specifically to implement order and stability.

Ecosystem. A complex system consisting of "the whole complex of physical factors forming what we call the environment" (Arthur Tansley). A spatially explicit unit of the Earth that includes all of the organisms, along with all components of the abiotic environment, within its boundaries.

Ecosystem management. 1. A new paradigm for land and resource management that emphasizes ecological systems as functional units, and stresses the long-term sustainability of those systems. 2. The application of ecosystem principles on a large scale that concerns society, but these scales may not correspond to useful physical, chemical or biological boundaries.

Ecotoxicology. The study of the effects of man-made chemicals on organisms. This includes lab bioassays to set standards for chemical compounds and field studies to determine the fate and effects of the chemicals in the environment.

Ectotherm. Plant or animal with internal temperatures approximately the same as those of the external environment. An ectotherm's metabolic rate follows the basic temperature dependence of rates of reaction for its enzyme systems.

Empirical testing. The testing of a model by collecting data and comparing those observations with the model's prediction. Such tests suggest general insights into the relationships between landscape patterns and ecological processes.

Endotherm. Animal that uses the heat generated by metabolic processes to establish and maintain a relatively constant body temperature, which is usually in the range of 35–38°C.

Enhanced greenhouse effect. The warming of the Earth's biosphere resulting from human-induced increases in greenhouse gases including carbon dioxide, methane and nitrous oxide, in the atmosphere.

Environmental history. The history of human interactions with the nonhuman world.

Equilibrium. 1. A condition of stability. 2. The situation where colonization from source to sink patches maintains the overall population on the landscape at a constant size.

Equilibrium population density (K). The population density in a deterministic model at which if the population density were started at K, it would remain at K indefinitely.

Estuary. Tidal ecosystem where a river enters the sea.

Eutrophication. The state of increased production from fertilization of streams and lakes, often caused by increased inputs of phosphorus to aquatic ecosystems.

Evenness. The relative number of individuals among the different species in a community. Several different mathematical formulae account for evenness in diversity calculations.

Evolution. Change over time. In biology, changes in characteristics such as form, physiology or behavior that can be passed from one generation to another. Evolutionary change is the natural result of 1) characteristics being inherited, 2) variation in characteristics among offspring, and 3) production of too many offspring, so that some live to reproduce and some do not.

Evolutionarily stable strategy (ESS). The point where the mix of two strategies in the population has equal payoffs in terms of fitness.

Exploitation or **scramble competition.** Competition occurring through the depletion of a shared resource such as food for animals, or water and nutrients for plants.

Exponential population growth. Population growth which occurs indefinitely at an ever-increasing rate.

Extent. (Re: ecological considerations), the size of the study area being considered; for example, whether a landscape is small or large.

Extirpate. The local extinction of a population (within a small geographic area).

Extrinsic factor. Factor external to a population that nonetheless influences changes in the population's density through time. Extrinsic factors typically include environmental factors such as temperature, rainfall and soil type.

Fitness. The number of viable and fertile offspring produced by an individual of a specific genotype, relative to other individuals in the population. More-fit organisms have more viable and fertile offspring than less-fit organisms of the same species.

Food web. The relationships of who eats whom; also a map of how energy flows through an ecosystem.

Forest patch. Small forested area separated by some distance from other forested areas.

Fossil fuel. Fuel such as coal, oil and natural gas, which is preserved in prehistoric times and is currently being mined for human use.

Frequency. Percentage of quadrats in which a given species occurs. The ecologist determines whether the species is present or not in any given quadrat.

Fugitive species. A plant that disperses rapidly into disturbed areas. Fugitive species use the same resources as the plants that eventually outcompete them, but are able to complete their life cycle quickly before eventual extirpation.

Function. The interactions among the spatial elements of the landscape; for example, movements of organisms, materials, or energy.

Fundamental niche. All of the requirements of a species. The fundamental niche is bounded by the upper and lower tolerance levels of all factors—for example, temperature and food abundance—that the species requires to persist.

Game theory. The quantitative analysis of two or more alternative behavioral strategies; for example, when the fitness of each depends on its frequency in the population.

Genetic engineering. The process of altering an organism's genome in the laboratory to produce plants or animals for domestic use, or bacteria that yield important products.

Geographic information system (GIS). A powerful set of computer tools for collecting, storing, retrieving, transforming and displaying spatial data from the real world for a particular set of purposes.

Gradient analysis. A member of the general class of interpolation methods. Gradient analysis is a powerful descriptive method that places sites on gradients to compare vegetations or animal assemblages.

Grain. (Re: ecological measurement), the spatial resolution of a study or data set; for example, data or samples obtained at a resolution of 1 m^2, 100 m^2, 1 ha, or 1 km^2.

Greenhouse effect. A well-established natural phenomenon in which the Earth's atmosphere retains heat by partially absorbing and re-emitting long-wave radiation, trapping energy and warming the surface of the Earth and its atmosphere.

Greenhouse gases. Include CO_2, methane and nitrous oxide; which absorb infrared radiation.

Habitat fragmentation. Reduction of the total amount of a habitat type in a landscape, and the apportionment of the remaining habitat into smaller more isolated patches.

Heritability. The degree to which a character can be inherited.

Heterotroph. Organism that consumes nutrients and various organic compounds for conversion into its own biomass.

Homeostasis. The maintenance of time, matter and energy budgets that allow for survival, growth and reproduction by the individual.

Homogeneity. Sameness or similarity.

Hypothesis. An "educated guess" about a relationship.

Hysteresis. A characteristic of non-linear models of complex systems. Hysteresis occurs when change in one direction follows a different track than does an equivalent reverse change.

Inclusive fitness. The fitness gained via assisting relatives, whether one's own descendants or not.

Individual-based model. A model which keeps track of the identity of each individual in the computer for the entire simulation. These models allow for feedback at the level of an individual's functioning in the community.

Instantaneous attack rate. The chance that an individual is attacked during a short period of time (say an hour) by a particular predator.

Interference or **contest competition.** Competition that takes the form of direct confrontations between individuals.

Interglacial period. Warm period between the glacial eras that persist tens of thousands of years. During an interglacial, ice sheets retreat, and plants and animals advance into the newly exposed habitat.

Interpolation. Insertion of values between observed values using a mathematical procedure.

Intersexual selection. The situation in which selection favors traits in one sex that attract the opposite sex.

Interspecific aggregation. The clustering of individuals of different species in particular locations in space.

Interspecific competition. The competition between two species when an increase in the population density of either species causes a decrease in the population density of the other species.

Intrasexual selection. The situation where males compete directly with one another for inseminations.

Intraspecific aggregation. The clustering of individuals of the same species in particular locations in space.

Intrinsic factor. Factor that directly involves the individuals within a population. Intrinsic factors include such things as competition among individuals for food, cannibalism of juveniles by adults and the shading out of seedlings by mature plants.

Intrinsic rate of increase. The population parameter r that defines exponential growth.

Island biogeography. A general theory to predict the number of species found on oceanic islands, based on the island's size and distance from the mainland.

Keystone species. A predator that increases species diversity by preying preferentially on the most successful competitor in a community.

Kin selection. A strategy to gain genetic fitness by helping a relative raise offspring, because relatives other than sons and daughters share one's genes.

Land cover. The vegetation or habitat type.

Landscape context. The effect of the surrounding area on the plants and animals in an area, when the habitat patch itself is suitable.

Liebig's Law of the Minimum. The rate of reaction proceeds in proportion to the availability of the reactant in shortest supply.

Life form. A common physical form common among species with similar physiological adaptations. Well-known life forms include succulent plants, deciduous trees, coniferous trees, herbs, shrubs and bulbs.

Life history strategy. The balances and tradeoffs an animal makes between traits affecting survival and traits relating to reproduction.

Lignin. A structural compound in wood that is difficult for most microbes to decompose; plant tissues with higher lignin contents tend to decompose more slowly.

Limiting resource. The resource that, if increased, will increase the population growth rate of a species.

Limnology. The study of lakes.

Line intercept. A hybrid measurement between cover and frequency often used for shrubs. Here one lays out a line of perhaps 100 meters, marked into segments of perhaps one meter. The measure is of how many segments are intercepted by the species in question. The line intercept is the percentage of segments that are under the canopy of the shrubs.

Litter. The leaves and other plant material that drop to the ground each fall.

Macroevolution. The study of evolution on a large scale. Refers to the evolution of major distinctions, such as the emergence of a new order of animals, or the evolution of some general life form such as a tree.

Mark and recapture. A method in common use for animal data collection. Animals are caught and tagged or marked in some fashion, then released. After a number of captures, it becomes possible to calculate the population size from the tagged animals recaptured.

Master factor. Factor that varies in time and space in ways that play a major role in the survival, growth and reproduction of organisms in the environment.

Metapopulation. A collection of subpopulations of a species, each occupying a patch of habitat in a landscape of otherwise unsuitable habitat.

Metapopulation dynamics. Situation in which populations are subdivided into smaller groups (subpopulations). Being partially isolated, each subpopulation's dynamics are partially independent of the dynamics in other subpopulations.

Metrics. Measurements used for the analysis of spatial pattern in landscape ecology.

Microevolution. Evolution as it is understood in common parlance. It is the result of natural selection applied to populations, such that local adaptations occur.

Model. Simplified representation of reality.

Model system. System in which spatial patterns can be manipulated experimentally at manageable scales and replicated.

Mosaic. The arrangement of habitats or vegetation, so called because, when mapped or viewed from above, they resemble the patterns in mosaic art created by small pieces of variously colored materials.

Mutualism. The situation in which an increase in the density of either member of a species pair increases the per capita population growth rate of the other species.

Natural selection. The process whereby animals produce more young than are necessary so that only some offspring populate the next generation. Occasionally defined as the non-random process of mortality.

Nonpoint source pollution. The pollution that does not come from a single source such as an outflow pipe.

Null hypothesis. Any apparent relationship is due only to chance resulting from the sampling technique and sample size.

Optimal foraging theory. Organisms behave in a way that maximizes food gathering for the minimum amount of time or effort.

Optimality theory. Natural selection will favor the variant of a behavior with the greatest net benefit; i.e., for which the value of the benefit of the behavior minus its cost is maximized.

Osmoregulation. Maintenance of water and ionic balance by an organism.

Paleoecologist. Specialist in long-term ecology who can indicate where the species will move next, and at what speed.

Paleoecology. The study of individuals, populations and communities of the plants and animals that lived in the past, and their interactions with and responses to changing environments.

Patch. A group of contiguous cells of the same habitat type. Also, a relatively homogenous area that differs from its surroundings.

Patch dynamics. The view of a community not as a stable, fixed assemblage of species, but as a mosaic of patches differing in successional stages.

Pattern. A series of ordered or repeated distinctions. Pattern emphasizes quality rather than quantity, structure more than dynamics.

Per capita mortality rate. The number of individuals that die over a specific time period, divided by the number of individuals in the population at the start of the time period.

Per capita population growth rate. The change in population size divided by the initial number of individuals in the population.

Per capita reproduction rate. The number of individuals that are born over a specific time period divided by the number of individuals in the population at the start of the time period.

Percolation theory. The theory of how the spatial arrangement of patches of suitable and unsuitable habitat affects the movement of organisms across a landscape.

Persistence. The continued presence of key elements of the landscape, which might include species, habitats or successional stages, or age classes of the vegetation.

Phenotype. All the characteristics of an individual.

Phytoplankton. Microscopic algae suspended in lake or ocean water.

Pioneer species. The first species to colonize a site after a major disturbance such as a fire that burns down all the trees. The first species to colonize are those that best exploit the conditions of the disturbed site.

Population. Collection of individuals from the same species that occupies some specified area.

Population density. The number of individuals in the population divided by the area covered by the population.

Population dynamics. Changes through time that occur in the number of individuals in a population.

Population growth rate. The change over a specified time period in the number of individuals in a population.

Population regulation. Processes that affect the patterns of population dynamics. The term is used by ecologists in a neutral way without the implication that something is doing the control or that the actions of the controller have some purpose.

Positive feedback cycle. Situation where an event creates conditions that favor the recurrence of that event, thus leading to rapidly accelerating changes.

Predation. Any interaction between two species in which one is benefited and the other is harmed. Predation includes classic predator-prey interactions and also parasite-host interactions.

Predator-prey cycle. Cyclic oscillations of predator and prey populations produced by the time lag between changes in prey density and predator density.

Primary production. The total amount of plant material produced within an ecosystem in some period of time, often a year. Primary production forms the basis for the food webs in ecosystems.

Primary productivity. The rate of production of new organic matter by the plant part of an ecosystem.

Probability density function. The theoretical distribution of probabilities that underlie a random variable.

Productivity. The amount of new biomass produced by an ecosystem's plants, animals and microbes.

Quadrat. Sampling unit of a site under investigation.

Radioactive isotope. Isotope of an element that decays to form a different element. Such isotopes provide a means for measuring rates of carbon fixation by primary producers (^{14}C) or rates of nutrient cycling (^{32}P) between trophic levels.

Random variable. A variable that is not described by one particular value, but instead is described by a probability distribution of values.

Rare species. Species that occur infrequently.

Raster format. The landscape expressed as a fixed grid of square or hexagonal cells of equal size. Irregularly shaped landscapes can be represented within a rectangular perimeter larger than the landscape itself.

Realized niche. The realized niche is that subset of tolerable conditions actually occupied by the organism. The conditions where the organism is observed to exist.

Regeneration. The return of a site to its natural state after some catastrophic event.

Remote sensing. Images of the Earth's surface obtained from satellites or airborne sensors.

Resilience. 1. The capacity to maintain ecosystem processes under stress. 2. The rapidity with which a stable equilibrium point or stable cycle is approached through time. Resilience in this sense has meaning only in terms of the rate of return to something that is stable. 3. The variability in the population density for a given level of extrinsic variability in stochastic systems.

Resilience to disturbance. The ability to undergo change and then return to a similar, but not exact, system configuration.

Resilient. Having resilience.

Resistance. Sensitivity of the per capita population growth rate to extrinsic variability.

Resolution. The smallest meaningful measurement that can be made of space or time.

Resource partitioning. General term describing how coexisting species use resources.

Either some of the resources they use can be different, or they can use the same resources in a somewhat different way.

Restoration. The process of returning a site to something like its natural state. Restoration may be expensive and time consuming and may require continual intervention to suppress invading species.

Return interval. (Re: forestry), the time between successive harvests.

Riparian zone. Areas of natural vegetation along the stream.

Savanna. Grassland with sparsely distributed trees.

Scope for growth. The possible growth based on a balanced energy budget. The energy left for growth and reproduction after maintenance needs have been met.

Serotiny. An adaptation in pine cones whereby the cones remain closed and retain their seeds until exposed to fire.

Sexual selection. Situation in which members of one sex are a *de facto* resource for which the other sex will compete.

Shifting mosaic. Situation in which the size of individual disturbance events is small relative to the size of the landscape, and disturbed areas usually recover before they are again disturbed.

Simulate. To resemble.

Sink patch. Area where the local mortality exceeds local reproductive success. The population in such patches will be extirpated unless it is regularly colonized by new individuals.

Size. The number of units measured. Size or extent refers to an entire length, area or time period.

Smog. A mixture of by-product gases from burning fossil fuels. The word is formed by combining "smoke" and "fog."

Source patch. Area where the local reproductive rate exceeds the local mortality rate, leading to a surplus of individuals that are potentially available to colonize other patches of suitable habitat.

Spatial heterogeneity. Differences or dissimilarities observed across an area of land or water.

Spatial pattern. Patches of different habitats within the landscape.

Spatial variability. The total variability that occurs from one geographical location to another within the spatial range of a population.

Spatially explicit. Representing the spatial locations of the quantities being modeled.

Species richness. The number of species in an area. Species richness counts each species equally, regardless of how abundant it is relative to the other species.

Stability. The way population dynamics evolve through time. An equilibrium point is stable if population densities move towards that point through time.

Stabilizing selection. Selection which acts most strongly against extreme individuals, favoring intermediate values of a trait.

Stable. Unchanging, at equilibrium, or a condition toward which systems move.

Stable isotope. Isotope of an element that does not decay. Stable isotope abundances, especially those of oxygen, carbon, nitrogen and sulfur, can be used to trace ecological processes.

Stable stationary population distribution. Equivalent of the equilibrium point in a deterministic model, but because the population dynamics are stochastic, population density must be described with a probability density function, rather than as a single point.

Stable two-point cycle. Cycle in which regardless of the initial population density, an alternating cycle between the two population values will develop.

Stationary population distribution. The probability density function for the population density. It is called stationary because it does not change through time.

Statistical population model. Traditional type of population model which models populations or communities as a whole.

Steady state. Situation in which the interactions of producers and consumers vary little in time and space.

Steady-state mosaic. Situation in which the vegetation at any given location changes through disturbance and succession, but the proportion of the landscape in various stages of succession remains relatively constant.

Stochastic model. Model in ecology that includes components of population dynamics that are unpredictable.

Stoichiometry. The ratio of abundance of different atoms in a compound (chemistry). In an ecological sense, stoichiometry applies to measures of the relative availability of different nutrients.

Strategy. A behavior pattern or morphology used in acquiring a scarce resource. The way that a set of plants deals with its neighbors to secure advantage, such as trees shading out grasses that might otherwise carry fire into the forest.

Structure. The spatial patterns of what is present and how it is arranged.

Succession. A hypothetically orderly sequence of changes in plant communities leading to a stable climax community.

Suitable habitat. An area having the conditions required for a given species to meet its needs for resources, survival and reproduction.

Summary axis. Graph axis onto which various procedures compress data. This can be intuited by a geometric projection, such as the shadow of a three-dimensional hand onto a two-dimensional screen. The shadow is a two-dimensional approximation of a three-dimensional space; the new axes are the summary axes.

Sustainability. The ability of ecosystem processes to maintain themselves into the future.

Temporal heterogeneity. Differences or dissimilarities through time.

Threshold. The point at which the habitat suddenly becomes either connected or disconnected, depending on point of view. Or, the minimum population size that can increase.

Toeslope. The base of a slope where materials tend to accumulate.

Transhumance. The practice by Spanish shepherds of moving sheep up into the mountains for the summer, where grass and water were more abundant, instead of keeping the sheep in the same low-elevation pastures year around.

Trophic cascade. Shift in top predators causing alternating waves of increase and decrease that pass through food webs.

Tundra. Alpine or polar ecosystems with low-growing plants, no trees and perennially frozen subsoils.

Two-point cycle. The stable alternating cycle in which the population density alternates between the same two values regardless of the initial population density.

Vector format. The landscape expressed as sets of coordinates to define the boundaries of polygons. The polygons are of variable size and shape, but a minimum mapping unit (i.e., level of resolution) is specified.

Vegetational physiognomy. The first impression of the landscape.

Vegetational succession. The regular patterns in the successional sequence of species change in patches of vegetation.

Watershed. A naturally bounded land area that is drained by a stream or river. Also, the boundary of the land area from which water flows into the lake.

Wave interference. A state of periodic interaction in physics where very small changes in the period or direction of one of the waves in a physical system can produce a very different wave interference pattern.

Wetland. Ecosystem in which the soil is saturated with water part of the year, such as a marsh or bog.

Work. (Re: modeling), a model works when it predicts an outcome seen in nature.

Zooplankton. Microscopic animals suspended or swimming in waters of lakes or oceans.

411

Kandis Elliot and James Jaeger prepared this book on Apple Macintosh computers at the Institute of Implied Science studio. All artwork was prepared electronically with Adobe Illustrator and Adobe Photoshop, with QuarkXPress used for the typesetting and compositing. Text font is New Baskerville. Illustration fonts are Optima and Futura Condensed. Spot color is Pantone 1807. Duotones, spot illustrations, and the color plates were all painted electronically using a Wacom digitizing tablet and Photoshop. Source images used as references or base images for the paintings and other illustrations are credited; if no credits are given, the image is a blank-slate creation.